Arthropod Phylogeny

WITH SPECIAL REFERENCE TO INSECTS

Arthropod Phylogeny
WITH SPECIAL REFERENCE TO INSECTS

H. BRUCE BOUDREAUX

Professor of Entomology
Louisiana State University
Baton Rouge, Louisiana

ROBERT E. KRIEGER PUBLISHING COMPANY
MALABAR, FLORIDA
1987

Original Edition   1979
Reprint   1987 w/Corrections

Printed and published by
ROBERT E. KRIEGER PUBLISHING COMPANY, INC.
KRIEGER DRIVE
MALABAR, FL 32950

Library of Congress Cataloging in Publication Data

Boudreaux, H. Bruce, 1914-
  Arthropod phylogeny with special reference to insects.

  Reprint. Originally published: New York : Wiley,
  © 1979.
  Bibliography: p.
  Includes index.
  1. Arthropoda--Evolution. 2. Insects--Evolution.
  3. Phylogeny.   I. Title.
  QL434.35.B68   1987   595.2'0438   84-3960
  ISBN 0-89874-746-5

10 9 8 7 6 5 4 3 2

Preface

This book is based on a course in entomology that I have taught at Louisiana State University for the past 20 years. The course has evolved over the years as an attempt to understand the nature of insects, with emphasis on searching out the possible evolutionary pathways that have led to modern forms. There have been numerous schemes of insect evolution proposed since serious study began with Darwin. No scheme yet proposed has been entirely satisfactory. In the present work, an attempt is made to understand the major plans of insect phylogeny that have been proposed, and to determine if there may be other better ones. Arthropods other than insects are treated in greater detail than a student of entomology usually encounters. I have done this because I believe that a fair understanding of other arthropod classes is necessary in order that insect phylogeny may be discussed intelligently. Yet, any conclusions one may reach are always subject to question. There seems to be no hope of ever solving the riddle of insect evolution, since the clues are so few. However, the desire to know is so strong that we will forever be trying.

The Voice of Authority is so compelling that many students are satisfied with it. It has been very disturbing to some of my students to learn that long cherished beliefs about insects are subject to question. Particularly disturbing has been the impossibility of reaching satisfactory conclusions about insect phylogeny into which all facts could fit. If one arrives at any conclusion, it must be one with which there is the least conflict. If this book espouses any particular conclusions, they merely happen to be the ones that in my opinion seem to offer the best compromise. There is no doubt that better conclusions will appear with more knowledge, and with better understanding of present knowledge. I hope that this book will stimulate further studies of the problem. If I have given the impression that I have merely substituted my own voice of authority for that of others, it is only because I have questioned some older conclusions and suggested better ones. I offer my conclusions with the expectation that they will be closely examined.

The principal thesis of the book is that the phylum Arthropoda represents a group that has arisen only once, all the members of which have descended from an unknown common ancestor that has given rise to no other creatures. No originality is claimed for this thesis. The treatment is based on the principles of phylogenetic analysis elaborated but not originated by Willi Hennig.

The phylogenetic relationships of taxa lower in rank than the class level, except for myriapods and insects, are beyond the scope of this book. The myriapod groups ordinarily considered to represent classes have been reduced in rank to below the class level in order to demonstrate the evolutionary splits that may have occurred, and because the myriapods and the insects seem to represent sister groups. This procedure is the reverse of the common tendency of taxonomists to elevate lower rank taxa to higher rank.

This book was made possible not only through the participation of many students in bringing up penetrating questions for discussion, but also through communication with colleagues. I am especially grateful for valuable suggestions offered by H. H. Ross, Willi Hennig, N. P. Kristensen, J. Kukalova-Peck, and Larry Rolston.

H. Bruce Boudreaux

Baton Rouge, Louisiana
December, 1978

Contents

CHAPTER 1

The Basis of Phylogenetic Studies

CLASSIFICATION AND PHYLOGENY

The biological discipline of classification includes attempts to arrange systematically concepts of the nature of living things in such a way that a name given to a group of organisms invokes a mental image of the group involved. The discipline of Nomenclature involves the construction of names for such groups. Such a group is called a taxon. When the activity of a biologist includes both nomenclature and classification, his actions are generally called Taxonomy. However, the word taxonomy is often made synonymous with classification, even though it may not imply the creation of names.

Beginning with the adoption of the Linnaean system of classification (Linnaeus, 1758), zoologists have developed a fairly standard classification scheme. Organisms are arranged in various taxa in a hierarchial array of ranked categories. The phylum category consists of classes, which in turn consist of orders divided into families, the divisions of which are genera. Individual species fit into the various genera. Originally the recognition of taxa relied mostly on the study of characteristics recognizable in dead preserved specimens. Ultimately taxonomists acquired additional information from living organisms that could be characterized, and now we have a branch of biology called Systematics, which is the study of the diversity of organisms (Mayr, 1969) as it bears on their taxonomy, evolution, and biology.

Animal classifications proposed up to the midnineteenth century had fairly well assembled the genera, families, and orders into classes. With the advent of the doctrine of organic evolution, remarkably few changes were made in the content of the various taxonomic categories, as the emphasis in classification shifted from mostly classifying together things which resembled each other most, to arranging the categories additionally to represent a presumed evolutionary sequence. At first it was the fashion to seek the ancestral type of one taxon from a living member of another taxon. However, living organisms must be regarded as the result of the evolutionary process up to the present. While one must admit that ancestral species probably were in many ways simpler than present descendants, it is hardly likely that any living species has remained living since its inception without having undergone any evolutionary change, unless it is a very recent species. It is unlikely that any genus or higher category today includes any member species which could be regarded as a type ancestral to another species in any other genus or higher category. Since the true ancestor of any taxon probably cannot be known beyond doubt, we must rely on clues which could suggest the nature of ancestral forms. The process of rearranging the known groups is continuing as new information is revealed and reinterpreted in an evolutionary sense. Phylogeny is literally the study of the evolution of a group. A classification that displays the evolution of a group may be characterized as a phylogenetic classification. The original meaning of the word has changed with usage, and a phylogenetic classification is often referred to as a phylogeny.

In order to preserve the familiar classical categories of classification, evolutionary taxonomists have found it necessary to increase the numbers of categories by employing prefixes ahead of established categories. For instance, classes may be asembled into a superclass. A class may be separated into subclasses, which in turn may be divided into infraclasses. The latter may even include divisions that may be called subterclasses, which finally may be divided into similar categories of the order rank, and so forth. Other authors have made use of other types of categorical names, such as branch, cohort, division, legion, phalanx, race, section, series, tribe, variety, and so on. The increase in categories came about with the need to express more clearly the evolutionary splits that are assumed to have taken place. Most taxonomists are reluctant to recognize evolutionary splits with taxonomic names because when this is done, an evolutionary classification becomes complicated and burdensome. One way out of this dilemma is to express evolutionary relationships using family trees as outlines of classification. But a family tree cannot provide all the information on which the tree is based. Further, it is not possible ordinarily to draw a family tree representing evolutionary relationships of all member species of a higher category, particularly of large groups, such as arthropods. Family trees therefore nearly always are used to represent selected portions of the classification. A phylum can be displayed easily if the twigs of the tree are limited to the class categories, while another tree is needed to express relations within a class only to its orders, and so on.

The family tree is merely an outline that indicates primarily the main direction and sequences of changes, generally called cladistic branching. Some family tree builders attempt to indicate the relative amount of change that has taken place by variously slanting the lines away from the vertical, or by the length of the lines (patristic distance). Other trees attempt to indicate the probable geological time of occurrence of each branching, by drawing the tree over a geological time scale.

The type of family tree used in this book is known as a cladogram. Such a tree is not designed to represent degree of change or time of origin. A cladogram should be restricted to indicating the most probable splits that occurred during the evolution of an organism. The splits can best be determined by attempting to recognize successive changes in homologous character transformation series (Hennig, 1953). While this method requires some speculation, it comes closest to objectivity because the presumption of cladistic splits is testable. The Hennig method assumes a split in an evolutionary line when it can be demonstrated that each line has acquired an advanced character state not found in the other line.

Recently, with the availability of electronic computers, there has arisen a movement generally called numerical taxonomy. Similarities and differences are assessed mathematically. Numerical taxonomists are looking for taxonomic methods that are as objective as possible, the results of which are reproducible. They advocate the use of large numbers of characters, all having equal significance. Classifications produced by numerical methods are commonly illustrated by a treelike diagram best called a phenogram, but often called a dendrogram. Such diagrams are not expected to reveal phylogenetic information, although the diagrams usually roughly approximate already existing evolutionary trees not constructed mathematically. The procedures are being modified with use, and it ultimately may be possible to use mathematical techniques for constructing evolutionary trees.

METHODS OF STUDY OF PHYLOGENY

A documented historical record of the changes that have occurred in the evolution of any organism is totally unavailable. The evidence that suggests that all organisms have evolved from more primitive types, however, is extensive. The theory of organic evolution has become universally entrenched in the study of living things. The science of genetics has produced useful concepts for explaining evolutionary change. It is generally assumed that all inheritable evolutionary changes result from changes in the genetic codes (mutations) resting in the reproductive cells. No evidence exists to support the notion that mutations occur in response to the need for a change. As far as is known, all genetic mutations occur unpredictably and at random.

The concept of homology originated before an understanding of genetics came about. But the concept was quite useful in early attempts at evolutionary classifications. Very simply, organs are considered homologous if it can be reasonably inferred that the nature of such organs is determined in part by a genotype that has a unique origin in some ancestral form. Since the exact original genotype must remain unknown, phylogenists employ criteria that are assumed to be valid, but are at best circumstantial evicence, in defining homology. Two organs are considered to be homologous if they (1) share a large number of similarities in structure, (2) originate in a similar position and manner ontogenetically, and (3) occur in organisms that possess many other additional homologous organs.

Because they are segmented animals, arthropods exhibit a special type of homology—serial homology—in which homologous organs are serially repeated on one individual. The paired locomotor appendages seem to have been genetically coded as such in the first arthropods. They may have a different final appearance or function on different segments of one animal (mouthparts contrasted to cerci), or they may differ when in the same position in different animals (grasping cerci in Japygidae and Dermaptera, sensory in many other insects). The mutations, which occurred controlling the final form of an original locomotor appendage causing it to develop into a mouthpart or a cercus, appear to be in addition to the fundamental code for "locomotor appendage," because locomotor appendages (legs) are to be found on insects that also bear mouthparts and cerci at places where there are no legs.

An important problem for a student of phylogeny is the recognition of convergence. Evolutionary convergence refers to the attainment of a great similarity between parts (or actions) of two organisms that occurred through presumed independent evolutionary processes. Convergent characters are evident when such are found in two organisms, each of which has been previously determined to be cladistically closer to widely related taxa. The raptorial forelegs of mantids and of mantispids are remarkably similar. But mantids are undoubtedly close relatives of cockroaches, whereas mantispids are classified in a specialized family of the order Neuroptera. In each case, the evolution of raptorial forelegs must have been completely separate events. In each case, the organs in question are homologous as forelegs, but their raptorial nature is convergent.

A phylogenetic classification ideally is one that includes in each taxon all the descendants and only the descendants of a presumed ancestral species of that taxon. Such a taxon can best be recognized when at least one, but preferably more than one, character state can be shown to be a unique derived state (synapomorphy), and is restricted to the members of the taxon. A derived character (apomorphy) may be defined as one which is a change from an ancestral state (plesiomorphy), but the term may be used also for a totally new character. The one or more derived character states shared by the members of the group frequently may give rise to further derived states within the group (autapomorphy), including reverse changes that may resemble the ancestral state. The insect taxon Pterygota includes all the descendants of a presumed original ancestral insect species that originally achieved the ability to fly by flapping two pairs of thoracic paranotal processes. Insect wings have variously evolved to the point that a single wing frequently is all that is needed to identify some species. The suppression of wings has occurred repeatedly in insect evolution, as can be noted in the wingless state of some species in almost every typically winged insect order. Some orders (fleas, biting and sucking lice, grylloblattids) consist entirely of wingless species, but these orders are retained in the Pterygota because their members seem to be derived from winged ancestors, as is indicated by the large number of characters they bear that are correlated with typically winged insects. An additional set of characters indicative of the monophyletic nature of the Pterygota is the fundamental similarity of the pterygote pleuron throughout the taxon. Other such characters include the tentorium formed by fusion of four cuticular invaginations and the temporary enclosure of the embryo by an "amnion."

Early classifications employed key characters to define taxa, and often such single key characters were almost the only definition of the taxa. For example, the order Aptera of Linnaeus included all the wingless arthropods known to him. The taxon Aptera, based mostly on primitive characters, can be characterized as polyphyletic. It included phylogenetically distant relatives, such as Arachnida, Crustacea, Myriapoda, and wingless insects. Another such taxon was Tracheata of Haeckel, which included groups that had acquired tracheae used in aerial respiration, apparently independently of each other: Arachnida, Onychophora, Insecta, and Myriapoda. Phylogenists insist that such polyphyletic taxa based on convergent characters must be avoided whenever their convergent nature is recognized.

Another type of taxon that phylogenetic taxonomists prefer to avoid is the paraphyletic taxon. A paraphyletic taxon typically includes other subordinate taxa all of which agree in possessing one or more primitive features. It is an unusual taxon whose set of characters does not include some persistent primitive ones along with derived characters. The insect taxon Apterygota is paraphyletic, and includes the primitively wingless insects, some of which are cladistically closer to the winged insects than they are to other apterygotes. The use of a paraphyletic taxon has little to recommend it for purposes of phylogeny. Many such taxa are still in use because they have become firmly entrenched in the literature. As these still-used taxa become better understood phylogenetically, their use will be abandoned as phylogenetic units, but they may still be useful in other ways. For instance, when it is desirable to discuss a grade of evolution characterized mostly by features that have not evolved much beyond the assumed primitive state, in comparison with an evolutionary line that has evolved rapidly much beyond its presumed ancestral state, then a paraphyletic group including the taxa with primitive features under discussion may be useful to define the grade of slowly evolving forms in contrast to the rapidly evolved group. While the Apterygota might include those insects that have not evolved wings, and so are primitive for that character, their winglessness represents a grade of evolution. But, as discussed later, these primitively wingless insects are highly evolved, each group in its own fashion. Their evolutionary relationships are hidden when they are simply lumped into the taxon Apterygota. The winged insects, Pterygota, appear to have arisen from an ancestor among the apterygote insects that had made several evolutionary steps not achieved by other apterygotes. The acquisition of flight seems to have been followed by a veritable evolutionary explosion in this line.

Other examples of paraphyletic taxa include Invertebrata (vs. Vertebrata), Pisces (vs. Quadrupeds), Reptilia (vs. Aves), Paleoptera (vs. Neoptera), Exopterygota (vs. Endoterygota), and so forth. Figure 1 illustrates the differences between monophyletic taxa based on shared unique derived characters, a polyphyletic taxon based on triple convergence, and paraphyly based on ancestral characters. Taxon 1 includes subordinate taxa A, B, and C, all and only descendants of ancestor X. Taxon 2 likewise includes D, E, F, and G, all and only descendants of ancestor Z. Both taxa 1 and 2 are monophyletic taxa, and at the same time are sister groups. Each sister group is defined by the derived features that all the members share and that were not developed in the other sister group. Taxon 3 includes subordinate taxa A, D, and F, and is polyphyletic, based on convergence in one or a few derived characters (e.g., Tracheata). Taxon 4 includes all except G. If the diagram expresses true cladistic relations, F and G share derived features originating in their immediate common ancestor W, and not anywhere else. But since F retains certain primitive characters also found in A, B, C, D, and E, a taxon including these but excluding G is a paraphyletic taxon (e.g., Apterygota).

PHYLOGENETICALLY USEFUL INFORMATION

The recognition and naming of taxonomic categories has had a long history, and relatively few groups have recently been discovered, particularly at the category level of order and above. The most recently recognized order of living insects is the Grylloblattaria, named in 1932. The main framework of insect classification has developed since the late nineteenth century through regrouping and separation of groups so as to embody evolutionary concepts. The process continues as new information is gathered and assimilated. There has been little radical change in groupings, but rather the process has been gradual.

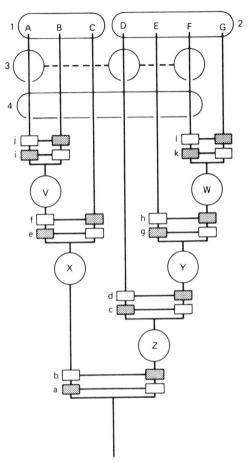

Figure 1. Cladogram illustrating types of taxa. 1, 2, monophyletic taxa; 3, polyphyletic taxon; 4, paraphyletic taxon; A-G, terminal taxonomic units; a-1, character states determining sister group splits; V-Z, common ancestors of descendant taxa. Shaded rectangles are apomorphic states, white rectangles are plesiomorphic states. Apomorphic character state "a" is restricted to ancestor "X" and its descendants. Apomorphic state "e" is restricted to ancestor "V" and its descendants. Apomorphic character state "i" is restricted to taxon "A," and so on.

The phylogenetic concepts that have developed are not universally acceptable to all. That there is disagreement among students of phylogeny illustrates that the evidence available for basing phylogenetic conclusions is insufficient, and that what evidence there is may be interpreted differently by different people. For about 100 years there have been proponents of the possibility of independent multiple origins of the Arthropoda from wormlike stocks, opposed to proponents of the idea that all arthropods are descended from a single evolutionary line beginning with an ancestral proarthropod. The polyphylists place greater emphasis on the large differences discernible between the end products of evolution, the present arthropod classes. Many of the similarities among them is envisioned as the result of convergent evolution, through the selection of changes favoring already established "functional needs." On the other hand, the monophylist viewpoint pays more attention to what appear to be basic similarities and presumed homologies. It envisions evolutionary divergence to be the cause of dissimilarities, while recognizing that evolutionary convergences are not uncommon.

The taxonomic catalogue of organisms has been so thoroughly worked over by past workers that it is nearly always possible to fit any newly discovered species into an existing order. New families and new genera similarly usually find their places in existing orders. Phylogeny, therefore, is usually devoted to the study of the evolution of already known taxa. In attempting to detect possible evolutionary sequences, the first task of the phylogenist is to determine homologies. Then he attempts to determine the sequence of changes that developed from the primitive original state. This assumes that it is possible to determine what was the primitive condition of a set of character states.

Morphology as a Source of Phylogenetic Clues

Morphology is the study of form. The word often is used synonymously with anatomy, but it should not be. The study of morphology is comparative, and while it may include anatomical studies, its main purpose is to discover homologies through comparison. After the establishment of homology between characters, the most probable primitive character state is searched for.

The primitive state of a particular character is generally assessed by procedures that many people use but are rarely spelled out. One notes the various character states of a particular organ that is assumed to be homologous in the taxa to be compared. The character state that consistently appears in many members of most of the taxa being studied is deemed most likely to be primitive for the group of taxa. The derived character states are then ranked in the order of presumed progression as elaboration or simplifications. The derived states are then assessed in terms of possible convergence, when compared with other characters of the group that have been similarly analyzed. In order to make a phylogenetic assessment, the sharing of derived states bears more weight than does the sharing of the primitive states. This is because when a particular character has not undergone much change among several taxa, the character has no value in determining advancing sequences. An example of such a primitive character is the presumed original winglessness of the insects united in the paraphyletic taxon Apterygota. As discussed later, other characters in this group show definite evidence of evolutionary change so as to suggest that the group contains taxa that have evolved in various directions, including one line bearing derived characters in common with the Pterygota. The Apterygota therefore consists of distant evolutionary relatives. The emphasis on derived characters in phylogenetic study has become known as the Hennig method because Hennig (1969) has essentially formalized the principles and demonstrated their application in suggesting evolutionary relationships among the insects. Nelson (1969) used the Hennig method for a classification of vertebrates, and Cracraft (1974) classified ratite birds by the Hennig method. Griffiths (1972) provided a phylogenetic classification of the Diptera, and Penny (1975) gave a phylogenetic concept of the Mecoptera. Others who have followed the Hennig method are Kinzelbach (1971) on Strepsiptera and Kristensen (1975) on insects in general.

Embryology and Ontogeny as Sources of Phylogenetic Clues

The developmental potential of an individual animal resides in its genome. The genome includes all the genes, but the influence of different genes on development appears at different times and places during ontogeny. It must be assumed that in mitotic growth, each daughter cell most probably acquires a duplicate of the original genome of the egg. Yet, in the end, different tissues display the influence of different sets of genes. There is no reason to assume that genes coding for "chitin" in an insect are absent in muscle tissue, or that genes coding for "muscle" are absent in epidermis. The genes for cuticle development are probably never activated in muscle fibers. However, chitin production is not restricted to cells of "ectodermal" origin. The peritrophic membrane of the mesenteron is produced in some insects by the mesenteron cells. There appears to be nothing sacred in the distinctness of the so-called ectoderm, mesoderm, and endoderm. The differences between these so-called germ layers are nothing more than the result of differential activation of different sets of genes in the various tissues. The processes that selectively activate different sets of genes at different times and places in the lifetime of an individual are unknown. But the study of comparative sequences of development are essentially morphological, and can serve to establish homologies to some extent.

With the establishment of the theory of biological evolution in the last century, the study of embryology was quickly employed for studying phylogeny. The Law of Recapitulation, popularized by Haeckel, assumed that each step in the development of an individual was representative of a once-existing ancestor of that individual. The idea of recapitulation was soon attacked for various reasons, but as a knowledge of genetics developed, it became apparent that what was formerly interpreted as recapitulation must be thought of as the expression of gene codes inherited from ancient common ancestors. The development of small paired outgrowths on the abdominal and cephalic parts of insect embryos is exactly similar to such outgrowths that become legs on the insect thorax, and on the trunk of other arthropods. On the insect head and abdomen, these rudiments may be called limb rudiments. The assumption is that the same genetic system that initiates the growth of limbs in the thorax also initiates the growth of limb rudiments in the head and abdomen. Those on the head of insects correspond to the limb rudiments of at least four metameres, the first pair of which are quickly suppressed. Remnants of the suppressed first metamere appear as the tritocerebrum of the brain, a few head muscles and the subesophageal body. The following three pairs continue to develop influenced by the expression of other genes controlling the formation of mouthparts. The insect maxillae develop with the greatest similarity to limbs, but the mandible remains simple and unjointed in insects, while labial limb rudiments fuse into a labium. There would appear to be no question of homology of insect mouthparts with thoracic limbs. The gnathal rudiments of the head are accompanied by segmental internal structures also appearing in body segments bearing legs. These are the embryonic postoral paired hollow coelomeres and the neuromeres. These strengthen the supposition that the insect head is a composite of at least four metameres which in the far past may have been separate body segments on an unknown ancestral form. Through observation of similar events in the embryonic head of myriapods, it appears that there is a similar suppression of the first head metamere (tritocerebral or intercalary segment), which strongly suggests that the mutation affecting this suppression occurred in an ancestor common to insects and myriapods. It follows that this common ancestor bore legs on its trunk segments and that the abdominal limbs became suppressed or became nonlocomotor structures in the first insects. However, these observations do not reveal the actual ancestor of these groups in any detail.

As in the case of morphological study of adults, ontogenetic study can be useful phylogenetically when one can detect modifications of a primitive pattern common to groups being studied. The derived features that are shared must also be analyzed in terms of probable convergence, and the apparent primitive states analyzed in terms of possible reversion from a derived state. Much has been made of the brief appearance of paired hollow "mesodermal" rudiments in the embryos of some spiders, insects, myriapods, and crustaceans. These appear in front of the tritocerebral segment, supposedly indicating the former occurrence of additional metameres in the heads of arthropods. However, the brief appearance of extra paired hollow rudiments in embryonic heads of such modern forms does not make these homologous with adult metameres of nearly similar origin. The gene code for "hollowed out mesoderm" (schizocoel) probably originated long ago in the ancestor of the coelomate protostomes, and it is a tenaciously persistent genetic code. This code must be in all nuclei of one individual, and such presumed coelomic sac rudiments can be interpreted as the result of temporary expression of the code in nonmetameric areas. That these are vestiges of former body segments that have become cephalized, but appear only as mesodermal coelomic sac rudiments, would be much more credible if embryonic events could be trusted as real indicators of ancestral states (DeBeer, 1971). All that we can reasonably trust about embryonic coelomeres is that their cells will probably differentiate into muscle, gland, or fat tissue, and that they seem to be involved in the induction of differentiation of neuromeres and appendage rudiments.

The term "coelomere" is introduced in this book to refer to the paired hollow internal masses of embryonic cells that are almost always found in arthropod and annelid embryos. These paired cell masses are the first indication of the future development of many segmentally repeated (serially homologous) organs such as limbs, muscles, glands, ganglia, and so on. The term is coined from the Greek *koilos* (hollow) and *meros* (part). It is not intended to have any morphological connotation, but is introduced as a substitute for commonly used terms for these structures, terms that have other meanings and have an unwarranted morphological implication. The word "somite" also is used by others synonymously with segment, somatome, merosome, and metamere, to mean an adult body segment with all its parts. As used in embryology, somite suggests that these are body segments, which they are not. Another term used for these is coelomic sac. Although the structures are saclike, their cavity is not yet a coelom. The space may become the definitive coelom in annelids. But no coelom ever develops in arthropods, if the classic definition of a coelom remains as "the body cavity of animals enclosed by a peritoneum, and arising in the embryo from enlargement of a closed embryonic space surrounded by 'mesoderm.' "

Aside from being the source of cells for muscle, gland, gonad, and other adult tissues, coelomeres seem to be the initial source of unknown events or substances that in articulate embryos result in the differentiation of neuromeres (future ganglia), and of the beginning of outpocketings of blastoderm which later differentiate into postoral metameric locomotor appendages or their homologues. In the embryonic acron, coelomeres seem to induce the differentiation of some neuromeres (antennal ganglia) and blastodermal outpocketings. The latter usually differentiate into sensory structures (antennules of crustaceans, antennae of insects and myriapods, palps of onychophorans and oligochaetes) whose homology with locomotor postoral limbs is in dispute. Since coelomeres contain only partially differentiated mesoderm cells, coelomeres are definitely not somites or metameres at their inception. They form through the pinching off of groups of mesoderm cells generated teloblastically near the hind end, pushed forward under the blastoderm in annelids, onychophorans, and crustaceans with yolk-poor eggs. In other arthropods the mesoderm bands develop from cells detached from the ventral blastoderm along the length of the embryo, and the bands segment into coelomeres later. Neuromeres and appendicular buds develop *after* the coelomeres assume their final position, suggesting that the first function of a coelomere is to provide a signal for initial differentiation of neighboring cells into nerve cells (future ganglia) or appendages. Whether the ganglia and appendages later differentiate into sensory, locomotor, or other types of organs depends on which of the alternate genetic programs become expressed. As many as three pairs of coelomeres push forward into some annelid prostomiums (Anderson, 1973), but the presomatic nonmetameric nature of the prostomium (acron) has not been seriously questioned. The arthropod embryonic acronal coelomeres designated as preantennular and antennular are not to be taken for cephalized segments. These coelomeres are merely the means of providing a signal for the initial induction process of acronal muscles, sensory organs (antennae), and their ganglia.

The significance of variations in embryonic development among members of various taxa must be assessed with caution. Evolutionarily important mutations may affect any stage of development, either in delaying, advancing, suppression, or modification of expression, or derepression of the old gene codes, or in producing totally new structures. The kinds of genetic change involved have not been sufficiently studied, and it is a task of the future to interpret much of the embryological data already known in terms of genetic homology. At any rate, the old law of recapitulation seems to be invalid. At best, the study of arthropod embryos may briefly allow the glimpse of the execution of gene systems that in the past controlled the primitive state of ancestral characters, and may be of limited usefulness in establishing homologies.

Paleontology as a Source of Phylogenetic Clues

The fossil record provides important clues on the nature of organisms of the past that are now extinct. Geology has established approximate ages of the various fossiliferous beds, and provides a rough chronology not only of possible evolutionary steps in the history of organisms, but also of local changes in the earth's surface and past climate. Fossil study may suggest places of origin of some groups through zoogeographic studies.

Unfortunately, fossilized remains consist mostly of hard skeletal parts. There is little left of fossil soft tissues. Further, older insect fossils consist mostly of impressions of wings. The principle of correlation of parts, popularized by Cuvier in the early nineteenth century, suggested that the parts of organisms are all correlated to the particular way of life of the organisms. Therefore, if one knew intimately the correlated adaptations of organisms, one could deduce from a selected part what the rest of the organism was like. Application of this principle made it possible to offer reconstructions of the entire animal with a fair degree of confidence, especially for vertebrate fossils. To a limited extent, fossil insect wings alone are sufficient to provide taxonomic treatment of the specimen, but since there may be extreme diversity in wing characters within many recent orders, some aberrant fossil wings cannot properly be related without more knowledge of the rest of the body. The characters of insect wings are not always closely correlated with other body features as are vertebrate bone. However, the fossil record of arthropods is in some ways a valuable aid in recognizing evolutionary trends. The oldest trilobites from the early Cambrian had free segments all along the trunk, and later ones tended to develop several of their posterior trunk segments into a tagma called the pygidium, consisting of segments that failed to separate from each other. The early xiphosurids of the Cambrian developed a full complement of 11 articulated segments behind the prosoma, as is found also in eurypterids and scorpions. Only seven free segments developed in others, and modern xiphosurids retain six that fail to separate from each other. The first of these has partly joined the prosoma, and the hinge between the prosoma and opisthosoma seems to be within the first opisthosomal segment between intersegmental lines.

The fossil record of insects, however, provides very little evidence of the nature of the first insects or their immediate ancestors. Other than a Devonian collembolan, the insect fossil record begins with insects bearing fully developed wings in the Upper Carboniferous. Since insects are mostly terrestrial animals, including the earliest known fossil insects, it would appear that the probablity of any individual being fossilized is very low. The number of known living species of insects far exceeds the total number of species of all other organisms combined. The fossil record of insects includes only a small number of the kinds that probably once existed. Therefore, caution must be exercised in trying to identify a fossil specimen as an ancestor of later species.

Paleontological data must be interpreted according to the same principles outlined for neontology. Homologies must be estimated and sequences of change must be recognized. A phylogeny based only on arthropod fossils is almost as difficult as reconstructing a tree from a few fragments of twigs and pieces of bark.

Comparative Biology as a Source of Phylogenetic Clues

Since the life processes as well as structure of organisms are restricted to the limits imposed by the genome of any organism, genetic changes resulting in evolution should be recognizable in terms of all living activites. When it is possible to recognize homology in the life processes, it is possible to deduce the probable sequence of change. Information from genetics, cytology, physiology, biochemistry, life cycles, host relations, ethology, functional morphology, and the like is dependable in suggesting phylogenetic relationships to the same extent as is information from morphology, embryology, or paleontology. The primitive state must be recognized and the derived states must serve to recognize the evolved groups.

From the discussion in this chapter, it is evident that all the evidence used in the study of phylogeny is circumstantial. The reliability of circumstantial evidence becomes more credible when all the sources together suggest the same conclusions. The many authors who have speculated on the phylogeny of arthropods have too frequently restricted their studies to one or a few sets of characters, with the result that the conclusions of the various authors are too often in disagreement. Even if there were universal agreement on evolutionary relationships, there would be no guarantee that such conclusions are truly correct, because they must be based on circumstantial evidence.

CHAPTER 2

The Origin of
the Arthropods

The study of evolutionary relationship among animal phyla has occupied the attention of zoologists for more than a century, and to date there is a variety of major theories upheld by various zoologists concerning the major lines of evolution in the animal kingdom. Various viewpoints are possible because all the evidence for evolutionary relationships is circumstantial. The fossil record indicates that most of the animal phyla were already in existence in the earliest Paleozoic period at least 600 million years ago. The presumed soft-bodied animals that existed before then in the Archeozoic (Precambrian) period have left few fossils, which are of no value for indicating possible primitive states. The only record that could indicate possible primitive states is difficult to interpret. This record is coded in the form of genes, some of which must be presumed to have originated in these ancestral animals. The manifestation of the execution of genetic codes is seen in the final product, the living animal. Since today's animals are all the product of much more than half a billion years of evolution, it is difficult to recognize primitive genetic codes as they are expressed in phenotypes and to distinguish such codes from more recent additions and changes. Zoologists have properly sought to guess at primitive states by studying the simpler animals of the various taxa that are assumed to have retained many primitive characters with little alteration. It must be assumed that the process of evolution does not occur at a constant rate, and that some lineages of organisms probably have evolved much less than others since their origin. Simplicity does not necessarily indicate a primitive state, because of the possibility of a derived state resulting from the repression of the genetic code for an elaborate primitive state. An example of this is the almost total lack of veins in the wings of certain small hymenopterous insects. However simple or primitive a present-day species might appear, it most probably has a great many new and altered genes that

15

were not present in an archeozoic ancestor. It is doubtful that the first meta-zoans could be classified in any known taxon, other than to be included perhaps as a unique class among the lower Metazoa, if such an animal could be known.

Although the true nature of the first Metazoa is still uncertain, and we cannot determine whether there were several lineages of original Metazoa or only one, there seems to be good evidence that annelids and arthropods possessed a common ancestor that departed from an unsegmented coelomate wormlike stock by becoming segmented in its own special way, and independently of the segmented condition acquired by the deuterostome line leading to the vertebrates.

THE COELOM AND SEGMENTATION

The origin of the coelom has been the subject of a great deal of speculation. Defined in its simplest form, the coelom is basically a closed pair of pouches of tissue surrounded by a continuous epithelium. In the group of coelomate phyla included among the Protostomia, the coelom characteristically forms in the embryo through the opening up of spaces within solid masses of embryonic cells, usually termed mesoderm. This mesoderm appears to form by mitosis from its first recognizable cell progenitor, cell 4d of the blastula of molluscs and annelids. This schizocoelous embryonic origin of the coelom is typical of Mollusca, Annelida, Onychophora, and Arthropoda. Therefore, since this feature in all its modifications is shared by these phyla, it is probably true that their common ancestor had acquired the genetic mutations coding for a schizocoelous coelom. The original function of the first coelom is unknown. One guess of such a function highly favored by many zoologists is that the coelom was the site of production of the gametes, and the supposedly original coelom is then termed gonocoel. Another possible original function may have been as a fluid skeletal mechanism, providing for easy local elongation and narrowing of the body, or local shortening and thickening as a means for creeping bottom locomotion and burrowing in soft mud. Still another possible early function was as a simple circulatory system, and another may have been excretory (nephrocoel). The functional implications of the coelom of early forms was discussed in detail by Clark (1964).

It appears that the acquisition of a schizocoelous coelom would not require much more than the establishment of simple genetic codes modifying a genetic system already present in noncoelomate metazoans. This is the genetic system controlling the formation of the simplest body cavity, the blastocoel of most embryos. The simple blastula with its blastocoel occurs in all animal phyla. The blastocoel remains as the functional body cavity in the pseudocoelomate phyla. The opening up of spaces between cells is a comparable process in later embryos and gives rise to the schizocoel.

Among the deuterostome phyla, the coelom arises through the closing up of spaces in an already formed cavity, the archenteron. The enteric pouches that become the coelom in these groups, exemplified by the echinoderms, hemichordates, and others, seem to be controlled by a different genetic system having an independent origin.

The origin of the metamerism, or body segmentation, typical of annelids, onychophorans, and arthropods is best sought in the immediate common ancestor of these phyla. It is very difficult to imagine that segmentation such as is found in the vertebrates and other deuterostomes originated from the same mutation that produced the segmented protostomes. Zoologists have promoted several theories on the origin of metamerism, each attempting to find a common basis for all the segmentation types found among animals. The *gonocoel theory* assumes that the original body cavities were multiple paired hollow gonads whose adjacent front and rear walls became the septa marking off the body segments, followed by local elaboration of the metameric organs such as excretory organs, nerve ganglia, muscles, and so on, as is indicated by the condition in many annelid embryos. The *enterocoel theory* assumes that the original body cavities were multiple (originally three) pairs of enteric lateral pouches that became closed off, forming in the beginning a short animal with three pairs of body cavities. This condition is exemplified in the oligomerous phyla, particularly the Brachiopoda, where the embryonic coelom is derived in enterocoelous fashion, as it is in typical deuterostomes. A variation of the enterocoel theory is that of *cyclomerism,* which assumes that the enteric pouches closed off in an originally radially symmetrical coelenteratelike ancestor and became bilaterally symmetrical. The *fission (corm) theory* assumes that body segments result from suppression of asexual fission after a degree of initiation during ontogeny. The asexual reproductive phase in which reproduction occurs by fission in Metazoa is common in the coelenterates, flatworms, nemerteans, and occasionally occurs in some annelids. The asexual phase produces new individuals which may have an obligatory sexual reproduction through gamete production (jellyfish and tapeworms), or sexual reproduction occurs in asexually produced individuals only under favorable environmental conditions (some flatworms). The alternation of sexual and asexual generations, or metagenesis, is a well established phenomenon, although not necessarily obligatory, also in the unicellular Protozoa, algae, and nearly all the plants. The phenomenon is so common as to suggest that metagenesis was a characteristic of ancient simple organisms, and became suppressed in the

higher animals. Only traces of the phenomenon remain in the metameric protostome coelomates, united in the Cuvieran taxon Articulata. As used here,
Articulata is a supraphylum taxon not to be confused with the brachiopod
class Articulata. This group is represented today by the Annelida,
Onychophora, (?) Tardigrada, (?) Pentastomida, and Arthropoda. Whether
the Mollusca should be included in the Articulata, as might be indicated by the
apparent segmentation of the Monoplacophora, is not discussed here. The
Tardigrada and Pentastomida are enigmatic groups whose relationship to the
arthropods is difficult to establish.

None of the theories of metamerism satisfactorily accounts for all types of
body segmentation. Each seems to have merit when applied to a limited
group of animals, suggesting that body segmentation arose several times independently among animals. The fission theory is best applied to scyphozoan
coelenterates, cestode flatworms, and the Articulata. In the Scyphozoa the
polyp produces medusae asexually by fission (really a budding process), and
generally the medusae produce the polyp stage sexually. The developing larval medusae (ephyrae) are usually seen in a graded sequence, the oldest
dropping off the oral end and swimming away, while new ephyrae are initiated at a zone of growth lower down. For a while the developing future
medusae are stacked in a series, and the series only suggests segmentation.
Since the Scyphozoa are generally not considered to be primitive, the details
of this type of metagenesis seem to have originated in this class independently. In the tapeworms, the sexually produced larvae ultimately mature
(the scolex) to bud off a graded series of asexually formed sexual reproductive
units (the proglottids). The mature units farthest from the budding zone usually become detached, and release fertilized eggs by rupture of the unit. This
type of segmentation also appears to have originated in the ancestral
cestodes, and is probably another independent evolutionary event, correlated
with parasitism.

The fission theory, when applied to the Articulata, may be sustained in the
phenomenon of teloblastic growth of segments. A zone of growth in the larva
(metatrochophore of polychaetes), or in the embryo of oligochaetes,
onychophores, and to some extent in some members of all subphyla of arthropods, is formed immediately ahead of a caudal mass of tissue associated
with the anus. The growth of new segments proceeds in such a way that the
oldest segment is anteriormost, the younger ones nearest the zone of growth.
If the growth of segments is initiated by an ancient genetic code for asexual fission, the execution of the code is quickly suppressed by a new genetic system
peculiar to the Articulata, and the functional body becomes composed of a
thoroughly integrated series of segments. A metamere can then be defined as
the homologue of the sexual phase of a metagenetic metazoan only in terms
of the persistent genes that barely initiate the fission of incipient individuals.

The metamere of today falls far short of being an individual. The mutations causing suppression of fission beyond an early stage, and integration of the buds into a whole animal are restricted to the Articulata, and therefore this type of segmentation may represent a third independent origin of segmentation.

Many students of articulate phylogeny distinguish between primary and secondary metameres. They are called primary if they result from nearly simultaneous segmentation of a pair of anterior pieces of embryonic mesoderm into coelomeres. Subsequent metameres, generated through successive teloblastic growth at the posterior end, whether during embryogenesis or as a postembryonic event, are called secondary. Great significance is attributed to this distinction for phylogenetic reasons. There is no way to be absolutely certain that agreement in the number of primary segments indicates common ancestry, such as has been proposed as the chief basis for relating the trilobites with early chelicerates (Ivanov, 1933). Trilobites share derived characteristics in common with the Crustacea and not found in chelicerates, as discussed later. The arthropods are generously endowed with scattered similarities between obviously distantly related forms, which must be interpreted as convergences.

In Onychophora there are no primary segments. All the metameres, and some future head muscles originate from teloblastically generated cells in the embryo. In many arthropods, a varying number of primary segments are formed from the mesodermal germ band before hatching, and in most insects all the somites originate from the germ band without the addition of teloblastic growth. It would appear that the genetic system controlling the generation of mesoderm becomes turned on at various times in different groups. Later the genetic system(s) controlling segmentation of the mesoderm and the formation of coelomeres becomes activated.

The notion that primary segments are fundamentally different from secondary segments is reinforced, by those who so believe, by statements that no gonads ever form in primary segments. However, in only a few annelids are there gonads in all the so-called secondary segments, and in the Archiannelida (Dinophilus and Nerilla) all the segments are primary, including the genital segment. The localization of gonads in any metamere should be regarded as the peculiar property of the genetic system involved. There is little reason to believe that the number of primary segments is anything more than a larval or embryonic adaptation (Fauchald, 1974).

Still a fourth independent origin of segmentation may be considered as leading to the enterocoelous segmented deuterostomes. In this case the segmentation is in no way comparable to metagenesis. The segments have evolved apparently in connection with lateral undulatory swimming motions (Clark, 1964).

THE ARTICULATA

While the origin of the coelom and of the metamerism of annelids, arthropods, and onychophorans must be considered as uncertain, there is practically total agreement that these groups of animals most probably arose from a marine aquatic stock that had evolved into an elongate bilateral wormlike metameric animal, the traces of which have not been preserved in the fossil record. The reconstruction of a hypothetical ancestor of the Articulata must be based on apparent primitive features retained in recent forms, and therefore can only vaguely suggest its nature.

All the Articulata have a cephalic sensory front end anterior to the mouth, as do most bilateral Metazoa. The nervous system of the cephalic sense organs (the prostomial brain) arises from neuroblasts originating from preoral blastomeres. In annelids, these form in the upper part of the embryonic prostomium. In arthropods and onychophorans the protocerebrum similarly arises dorsally in an embryonic nonmetameric cephalic lobe known as the acron. Sensory nerves for simple or elaborate eyes and for tactile and chemosensory palps or antennules lead to a simple or elaborate acronal brain. Association centers called corpora pedunculata of similar construction occur in the simple brains of annelids and onychophorans, and in the protocerebrum of arthropods. The embryonic acron of onychophorans and arthropods and the annelid prostomium appear to be to some extent homologous, so the original articulate ancestor may be presumed to have elaborated a primitive sort of head largely devoted to sensory functions, and positioned ahead of the mouth.

The trunk of the hypothetical ancestor most probably consisted of a series of segments. Each trunk segment may be called a somite, or metamere, to distinguish this type of segment from the cephalic lobe. Each metamere probably bore a pair of lateral coelomic cavities surrounding the intestine and lined with an epithelium, or peritoneum. The only trace of the peritoneum left in modern arthropods may be the very thin cellular basement membrane lining parts of the haemocoel. The hind wall of one coelomic pair and the front wall of the coelom in the next rear somite formed segmental septa probably containing muscle fibers, a condition that still occurs in annelids. In arthropods, such septa begin to form in some embryos, but do not persist. The dorsoventral muscles of arthropods may be remnants of such septa. Each adult coelomic pouch probably communicated with the outside via a coelomoduct penetrating the body wall, which served the excretory-osmoregulatory nephridium, and also provided an exit for gametes, as is the case today in most of the Polychaeta and in Onychophora. Coelomoduct remnants are retained in arthropods only in a few metameres, but are modified into the mesodermal gonoducts and the coxal glands. It is not possible to decide if gametes were produced in all the somites, or if gametes were formed only in one or a few somites in the first articulates. There is only one gamete- producing somite in the Archiannelida, Onychophora, and most Arthropoda, suggesting the latter alternative. In many polychaetes a large number of posterior metameres bear the gonads. The seemingly metameric distribution of gonads in pycnogonids and entognathous insects may be suggestive of the first alternative.

Apart from the lining epithelium, the mesal wall of the paired coelomic pouches produced the muscles and supporting tissues of the intestine and the vessels of the circulatory system. At least the dorsal vessel produced peristaltic waves pumping the blood forward, as is the case still in all the Articulata (except a few forms, such as the Acarina and Pauropoda, which have lost all their blood vessels). The apposed mesal walls of the coelom above and below the intestine formed vertical sheets of dorsal and ventral mesentery, as still occurs in most annelids.

In modern articulates, all have body wall muscles consisting at least of bundles of longitudinal fibers, most often arranged in paired dorsolateral and ventrolateral sections or sheets. This common pattern suggests that the ancestral articulate probably possessed such longitudinal body wall muscles. Circular muscles surrounding the body outside the longitudinal series occur in Annelida and Onychophora. The occurrence of circular and longitudinal as well as dorsoventral muscles may have arisen before metamerism, as might be inferred from the presence of such similarly arranged muscles in nonmetameric turbellarian worms. Metamerism in a coelomate animal permitted the evolution of more efficient creeping locomotion. Forward going waves of local extension of the body occurred through the narrowing effect of contraction of circular muscles in short sections of the body. Extension was by the forward and rearward pressure exerted by the coelomic fluid. Alternately, waves of adjacent local contraction of sections of longitudinal muscles caused local shortening and increase in diameter of the body. Each somite must have had its own neural paired ganglia derived by migration of neuroblasts from the embryonic ventral blastoderm, since the double ventral chain of the soma arises in this fashion in all the Articulata.

The primeval articulate probably was covered with a continuous secreted cuticle containing protein. It is not clear whether the protein was collagenous, as in Annelida, or of a different sort, as in Arthropoda or Onychophora, but the cuticle was probably unsclerotized, and to some extent limited the amount of stretching produced locally by the internal pressures developed in locomotion. Chitin probably was produced in such a cuticle, but it is restricted today in annelids to setae and the stomodaeal lining, while it is a constant ingredient complexed with protein in arthropod integument. Annelid chitin and ar-thropod chitin are not identical according to X-ray diffraction studies (Rudall, 1963).

Metameric appendages may have had their origin with mutations that affected the fate of simple lobelike outgrowths, such as the sensory organs found on the cephalic part of many bilateral Metazoa. The parapodia of polychaetes and the arthropod appendages have been variously homologized, but it is more probable that the ancestral articulate had neither parapodia nor appendages as they appear in Recent forms. Annelid parapodia and arthropod appendages are homologous only to the extent of possibly being initiated as outgrowths of the body through the execution of a simple genetic code causing outpocketing. The evolution of parapodia followed a different path in Annelida than did the evolution of locomotor appendages in the line which led to Onychophora and Arthropoda. The annelid parapodium can be characterized as a chaetopodium, bearing protrusible and retractile setae containing chitin and sclerotized protein. Parapodial muscles appear to be derived from the circular muscle layer because in polychaetes bearing well developed parapodia, circular muscles are not well developed (Fauchald, 1974). The lobopodium of the Onychophora and Arthropoda differs in lacking such setae, and in having developed extrinsic as well as intrinsic muscles making bending motions of the leg itself possible. The lobopodia of the Onychophora bear both circular muscles and extrinsic muscles originating inside the body.

A teloblastic zone of growth just ahead of a nonmetameric ring of tissue surrounding the anus was probably the source of all the metameres of the ancestral articulate. Such a zone of growth occurs in some members of all the present articulate classes, producing the paired mesodermal germ bands in characteristic fashion in each group.

The anal segment is not a part of the serial metameric segmentation, since it never bears a coelomere, appendages, or ganglia. It has variously been called pygidium, telson, or periproct. However, the telson may not separate from the last metamere in some arthropods. An extreme case occurs in the Symphyla, in which postembryonic generation of segments occurs just ahead of the ultimate metamere that bears the vestige of a pair of limbs in the form of trichobothria, along with neuromeres.

MONOPHYLY VERSUS POLYPHYLY IN THE ARTICULATA

If the hypothetical ancestral articulate was properly envisioned above, application of the principles of phylogeny suggest a parting of the ways, with the Annelida evolving as a sister group of the Onychophora-Arthropoda assemblage. The annelid line developed the lateral appendages into parapodia bearing movable setae. The sister group developed lateral appendages into musculated lobes directly movable by extrinsic and intrinsic muscles, useful in crawling-stepping locomotion rather than as paddles. On this basis, the annelid line may be referred to as the chaetopod branch, and the Onychophora-Arthropoda as the lobopod branch (Snodgrass, 1938). The chaetopods further restricted their chitin and sclerotized protein to the setae, the pharyngeal teeth, and the lining of the gizzard (Rudall, 1955), while the lobopods acquired a chitin-protein complex as the main ingredient of the whole epidermal cuticle. The chaetopods bear collagen as the main cuticular protein. Collagen when present in lobopods is always internal to the epidermis. The cuticular proteins of lobopods are remarkably similar in amino acid composition (Hackman and Goldberg, 1975). The chaetopods variously elaborated the vascular, muscular, reproductive, and nervous systems of the primitive ancestor, and in parallel with the lobopods united some metameres into functional units (tagmata) such as the peristome, a simple headlike region, or the reproductive epitoke of some polychaetes. The lobopods developed an open vascular system (hemocoel, mixocoel). Although paired coelomeres occur early in embryonic lobopods, the cavities of the coelomeres persist only in gonads and coelomoduct remnants. Otherwise, the coelomeral cavities merge with each other and with other cavities (e.g., epineural sinus) without developing septa, forming a definitive body cavity of mixed origin. The dorsal blood vessel became provided with segmental ostia guarded by valves, through which the blood (hemolymph) from the hemocoel enters and is driven by the dorsal vessel to distant parts of the body. The dorsal vessel has become the functional heart, and may pump blood into arteries, but there are no capillaries or veins returning blood back to the ostia. The lobopods of all the classes have developed various anterior limbs into gnathal appendages. Another advanced feature apparently acquired by the early lobopods was the yolk-laden shell-covered egg, in which cleavage is meroblastic. The lobopod zygote nucleus produces, primitively, free nuclei, most of which move to the surface of the egg. There, the nuclei become organized into a cellular blastoderm. The nuclei remaining in the yolk are generally considered to function in processing the yolk for nourishment of the embryo. Holoblastic cleavage may occur in some lobopods, such as in viviparous onychophorans, collembolans, parasitic hymenopterans, pauropodans, symphylans, many crustaceans, and pycnogonids. Such instances of total cleavage are best regarded as simplifications from the primitive yolk-rich condition, because they are exceptional, and although occasionally cleavage may initially be total, a blastoderm eventually forms enclosing the yolk so that most of the embryo grows from a ventral germ band, as happens in the meroblastic yolk-rich eggs of most lobopods.

The semen of the chaetopods is emitted as a fluid from the male genital system. In the lobopods, the semen is usually enclosed in a membranous spermatophore. The spermatophores are transferred to the female in various fashions characteristic of the various lines. The terrestrial Onychophora, Arachnida, and Atelocerata generally produce spermatophores. In Crustacea, the copepods and a few decapods produce spermatophores. However, xiphosurids, pycnogonids, and most crustaceans deposit fluid semen without forming spermatophores, and the evolution of spermatophore production appears to be polyphyletic.

The distinctness of the annelids and of the lobopods is such that one may have confidence that neither evolutionary line could have been ancestral to the other. Their derived characters are of different sets of features, strongly suggesting that the ancestral articulates evolved along two separate lines. The evolution of the lobopod branch has become the subject of great controversy. The central question has become that of whether arthropodization resulted only once from a sequence of mutations that resulted in the emergence of an ancestral arthropod becoming the progenitor of all members of the phylum Arthropoda. A sister group of the Arthropoda could be the phylum Onychophora. Onychophoran evolutionary advances were of a sort which set these animals apart from the arthropods before early Paleozoic times. The opinon that evolution of arthropods was monophyletic has been widely held for a long time, but is being vigorously opposed recently by supporters of the notion that parallel and convergent evolution has produced at least three distinct lines from evolving lobopod worms (Manton, 1973a). The polyphyletic origin of arthropods is seen as having come about through selection for special needs to perfect particular habits. Evolution designed to serve functional needs is repeatedly invoked as a supreme guiding principle for interpreting differences as evidence of lack of evolutionary relationships. The structural differences correlated with function which are seen in the detailed study of representatives of Recent species of most types of arthropods are emphasized in support of the assumption of the independent origin, from soft lobopod worms, of the chelicerates and trilobites as one line, the crustaceans as another line, and the phylum Uniramia as a third line. The Uniramia includes Onychophora, insecta, and Myriapoda. All the arthropod features these three groups share are considered to have arisen independently in each line: paired jointed locomotor appendages; compound eyes; hemocoel; ostiate dorsal heart; a cuticle containing chitin and protein (noncollagenous), typically locally sclerotized by tanning of the outer layer of protein-chitin; periodic growth during molting of cuticle; typically yolk-rich eggs undergoing meroblastic cleavage; all muscles striated; skeletal muscles attached to cuticle through epidermis; presence of intersegmental tendons; suppression of motile cilia. The polyphyletic viewpoint denies that the above characteristics, although found in many members of all the arthropod classes, in any way are evidence of a monophyletic origin of arthropods. The similarities are glossed over as being only superficial or roughly similar, and considered to have evolved by different means. The convergencies of compound eyes, biting jaws, jointed limbs, and so on are explained as examples of the limited morphological detail capable of serving common needs.

It is characteristic that when one has strongly held opinions, evidence interpreted as supporting such opinions is emphasized while conflicting evidence is minimized. Phylogenetic studies must still employ evidence whose usefulness or validity is a matter of opinion. The testing of phylogenetic hypotheses for predictability cannot be done in the fashion of a controlled laboratory experiment. As long as the evidence employed in constructing phylogenetic hypotheses can be interpreted in more than one way, there will arise conflicting hypotheses. Whenever there can be serious sustained argument about such viewpoints, it may well be that all sides represent viewpoints that are not entirely correct, rather than that one side is wrong and the other is right. The conclusions presented in the following pages are in favor of a concept of monophyly, based in part on the same evidence that might favor a concept of polyphyly among members of the phylum Arthropoda.

CHAPTER 3

The Lobopods

In order to investigate the evolutionary relationships of arthropods, it is best to begin with an attempt to envision a hypothetical ancestral species in some detail. In the previous chapter this species was characterized as a segmented wormlike creature. It presumably bore a sensory presomatic segment at its anterior end. This sensory segment may be homologized with the annelid prostomium and probably had its origin in a metozoan ancestor older than the first metameric worm. The concentration of sensory functions in the front end of animals in general seems to be an evolutionary event of prime importance early in the history of animals, for obvious reasons. The anterior sensory segment or acron has been retained as a more or less separate segment or prostomium, in the annelids, but in no lobopod, fossil or recent, is such a separate segment known in the adult stage, except in the mystacocarid crustacean *Derocheilocaris*, which remains neotenous as an adult. Its presence is indicated in the embryos of arthropods, where the embryonic acron (blastocephalon) is always the site of origin of the visual center of the brain (protocerebrum). In common with the brain of annelids, the protocerebrum also encloses presumably association neurons in a pair of structures known as corpora pedunculata.

Another segment probably existing in the adult ancestral lobopod was a ring surrounding the anus, called periproct in annelids, and variously seen in arthropods, where if present it is usually called telson. This postsomatic segment appears in some members of all classes of articulate animals and is located behind a zone of growth that is the source of tissue that generates somatic body segments, or metameres. The telson, or trace of it, is retained in adult arthropods as a tail spine, such as in scorpions and xiposurids, a more or less flat structure in malacostracan crustacea, or a distinct segmentlike ring in proturans.

Between the acron and the telson, the body of the ancestral lobopod externally appeared as a series of another type of segment, the metameres. These

26

segments differ from the acron and telson essentially in that each metamere probably contributed to the main body functions: locomotion, nutrition, excretion, reproduction. The homology of the lobopod metamere with the annelid metamere is largely suggested by the repetition of organs, and by the roughly similar embryonic origin of segments suggested by the widespread occurrence of teloblastic generation of new segments. In the more advanced arthropods of each major division, teloblastic growth is faintly discernible in the sequence of organization of segments in a ventral germ band. The band is produced along the length of the ventral blastoderm, more or less all at once. The segments generally form from front to back, thus suggesting that the anterior are chronologically older than the posterior segments (Figure 2).

The concept of a linear series of repeated metameres between the acron and periproct has become so well established that most students of arthropods do not question that the first lobopods probably had a uniformly segmented body. Each metamere supposedly bore ventrolaterally a pair of musculated lobelike protrusions useful in creeping locomotion, which ultimately became a crawling locomotion as the legs elongated and lifted the venter from the substrate. A great deal of argument has taken place concerning whether the first two or three metameres originally might have been of a different sort. The cyclomerism theory of metamerism assumes that such was the case. Associated with such arguments are others concerning whether there can be a common genetic origin for annelid prostomial palps and ar-

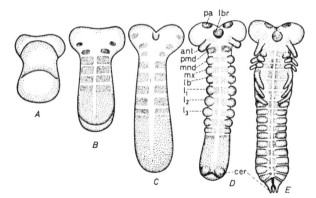

Figure 2. Successive steps in the development of the germ band of *Lepisma saccharina*, Insecta, Thysanura. ant, antennal rudiment; cer, rudiment of cercus; lb, labial rudiment; lbr, labral rudiment; l_1, l_2, l_3, thoracic leg rudiments; mnd, mandibular rudiment; mx, maxillary rudiment; pa, preantennal coelomere. From Sharov: BASIC ARTHROPODAN STOCK, 1966, with permission of Pergamon Press, Ltd.

thropod antennae, or whether the finding in various arthropod embryos of paired preoral coelomeres, corresponding paired rudiments of ganglia, and fleeting paired protrusions suggesting metameric appendages, all of which occur in the embryonic arthropod acronal lobe, must be interpreted as evidence that the acron consists of the homologue of the annelid prostomium on to which has become merged former anterior somatic metameres, such as a preantennal and an antennal segment.

If the mouth is considered originally to have been just ahead of the first metamere, behind the primeval head, or acron, and if a pair of transversely joined ventral ganglia were part of the original organs of a metamere, then the transverse nerve fibers of the commissure of the metameric ganglia should all occur in a postoral position. Since the transverse commissures of the antennal ganglia (deutocerebrum of antennate arthropods) are always preoral, this fact has been a strong argument against the view that there is at least an antennal segment, no matter how obscure, joined to the acron of lobopods. Other arguments against the segmental origin of antennae point out that these organs never have the structure of a leg, and are filamentous in the oldest fossil arthropods, the trilobites. On the other hand, in Onychophora the musculature of the antennalike palps develops from the first teloblastically generated embryonic coelomeres, which are pushed forward past the embryonic mouth into the anterior region past the eyes. While the musculature is postoral in origin, the ganglia originate from a pair of ventral rudiments situated in front of the mouth, and their transverse commissure is preoral. The events of embryology can thus be interpreted for or against the idea of an original sensory function and acronal origin of the various lobopod anterior organs commonly called antennae. Perhaps there is too much emphasis on the use of embryonic sequences of morphogenesis as indicators of primitive states. There is no reason to assume that there has been less evolution in the processes of cellular differentiation than there has been in the nature of the adult organs. For instance, in two fairly closely related groups of myriapods, the pauropods and symphylans, the muscles of the midgut arise embryonically from entirely separate sources. In the symphylan *Hanseniella agilis* the midgut muscles differentiate from the splanchnic walls of some of the segmental coelomeres. In the pauropod *Pauropus silvaticus*, the midgut muscles differentiate from cells migrating backward around the midgut from a source near the stomodaeal invagination, entirely independent of any coelomeres (Tiegs, 1940, 1947). In both cases, these myoblasts are typified as mesodermal. The digestive epithelium of the midgut and the gonads in each of the two species also have different embryonic origins. In *H. agilis* the midgut epithelium forms from an axial column of yolk cells budded from the blastoderm, while in *Pauropus* it forms from a single cell entering the central cavity of the very early embryo. More strangely, the midgut epithelium of an

Indochinese species of *Hanseniella* forms by a typical invagination of the classic "archenteron," although the "blastopore" is not the future mouth (Dawydoff, 1943). The gonad tissue in *H. agilis* is derived from pieces of coelomeres, while in *Pauropus* the gonads arise from a median cord of cells which never were associated with coelomeres. Thus closely related species have evolved remarkably different pathways of differentiation to produce organs ordinarily considered homologous.

It is evident then, that the true nature of the antennae (antennules of crustaceans) must remain conjectural, except in the minds of those whose opinions are firm. There is no indisputable evidence that can settle the question whether one or more metameres must be included in the embryonic acron of lobopod animals. At any rate, no known arthropod antennae are ever used by its owner in the manner of a crawling locomotor appendage so typical of the metameric legs, nor is there any fossil evidence that such antennae could have been employed in this manner. All fossil arthropods bearing antennae appear to have employed them for sensory functions only, or secondarily used them in swimming. For discussion purposes, in this book the first metamere is taken to be the first one whose paired ganglia are joined transversely by a postoral commissure. Until the true nature of preoral coelomeres is settled (preantennal, antennal), these will be considered to be part of the acron, and therefore nonmetameric.

THE PHYLUM ONYCHOPHORA

These soft-bodied wormlike animals are represented today by only 65 or so species, distributed discontinuously in the tropics and the Southern Hemisphere. They seem to have evolved very slowly and have not produced a great variety of forms. Only two extant families are recognized, and a third is known as a fossil, represented by *Aysheaia pedunculata* (Wolcott, 1911b). These animals appear to have retained relatively primitive features of the presumed lobopod ancestor. The cuticle contains chitin, but the protein component of the cuticle is heavily sclerotized only in the terminal claws of the legs. All the muscles except those of the jaw are unstriated. Functional ciliated nephridial segmental organs occur ventrally, opening on the bases of the walking legs. Cilia also occur in the gonoducts. Except for the epidermis, nervous system, and digestive tract, most tissues form through cellular differentiation of segmental paired clumps of cells (coelomeres) growing from a teloblastic zone of growth just anterior to the anus. There is a single pair of simple eyes. In addition to the dorsal, lateral, and ventral longitudinal muscles, the outer muscle layer is circular in orientation, under which there are two layers of crossed oblique muscle sheets. Because of some of these characteristics, they have been regarded as representing a stage intermediate between annelids and arthropods. It is better to regard the Onychophora as having evolved from a lobopod segmented ancestral form coming from a more remote progenitor of both annelids and lobopods. They share with arthropods a number of derived characters that are not at all annelidan.

The onychophoran leg is a typical walking leg in that it is lifted, set down, moved forward and backward principally through the action of extrinsic leg muscles inserted on the limb base, and the limbs are variously bent and shortened employing mostly muscles contained within the leg (Figure 3). Unlike arthropod legs, they are not sclerotized or jointed, and they can be lengthened or shortened individually. The lengthening apparently is possible using hemocoelic pressure exerted locally through the intrinsic circular and

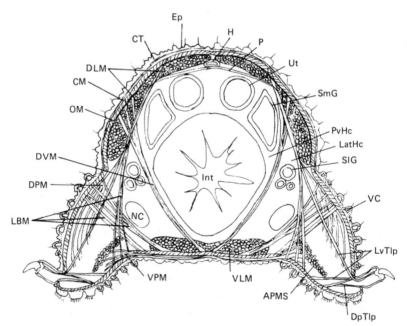

Figure 3. *Peripatoides novozealandiae.* Diagram of main musculature of an onychophoran. Cross-section through legs at about the third quarter of the body. APMS, anterio-posterior muscular septum of leg; CM, circular muscle layer; CT, connective tissue; DLM, dorsal longitudinal muscle; DPM, dorsal promotor muscle of leg; DpTlp, depressor of telopodite; DVM, dorsoventral muscle (lateral diaphragm); Ep, epidermis; H, heart; Int, intestine; LatHc, lateral hemocoel; LBM, leg base muscles (which isolate the lateral hemocoel from the leg hemocoel); LvTlp, levators of telopodite; NC, nerve cord; OM, oblique muscle layers; P, pericardial floor; PvHc, perivisceral hemocoel; S1G, salivary gland; SmG, slime gland; Ut, uterus; VC, vascular channel; VLM, ventral longitudinal muscle; VPM, ventral promotor muscle. Original.

transverse muscles. In walking, the legs of each pair are raised together in metachronal waves passing forward, together with a wave of body contraction, similar to the crawling action of a caterpillar.

The body cavity is a hemocoel, rather than a coelom as is typical of annelids, and forms as is typical in arthropods through the merging of parts of the embryonic coelom with spaces formed apart from the embryonic coelom. The heart is perforated by paired ostiate inlet valves arranged segmentally. The blood fills the spaces of the hemocoel and is propelled forward by the heart after entering via the ostia.

Although most onychophorans have small, yolk-poor eggs, this must be considered to be a derived state associated with viviparity. The embryonic pattern of development of viviparous and ovoviviparous species is easily seen to be derived from the typical meroblastic pattern found in the large yolky shell-covered eggs of oviparous onychophorans, and typical of arthropods in general. The embryo mostly forms from the ventral blastoderm, with the dorsal blastoderm contributing not much more than the dorsal epidermis to the developing animal.

As in arthropods with yolky eggs that develop a blastoderm over the yolk, the ventral blastoderm or germ band is the source of cells that produce nearly all the internal organs except the tracheae. In the viviparous yolk-poor species, in parallel with other yolk-poor arthropods such as the smaller highly evolved crustaceans, there is, along with the loss of yolk, a return toward the invagination of blastoderm to form the digestive tract, more typical of protostomes in general.

The feeding organs of lobopods seem in all cases to be the result of modification of appendages homologous with the walking legs. In onychophorans, the principal gnathal appendage is a clawlike pair of sclerotized structures commonly regarded as the terminal claws of the limbs (Figure 4, I). These jaws, therefore, are only functionally equivalent to the mandibles of crustaceans. In the latter case the biting surface is undoubtedly an enlarged median lobe of the limb base, rather than the limb tip. The feeding claws of onychophorans develop from the first metamere, as defined above, since their ganglia comprise the first postoral pair that is joined by a postoral commissure. The nature of the antennalike palps is discussed below with respect to fossil onychophores.

Judging from the nature of the specializations of onychophorans, it appears that their lineage originated before any arthropod. Their outstanding specialization is the elaboration of the use of hemocoelic pressure for producing local rigidity and for extension of the body and legs in locomotion. This can be characterized as a fluid skeleton, in contrast to the typical arthropod skeleton consisting of various hardened portions of the cuticle. The leverage of skeletal muscles of arthropods is largely exerted through the rigidity of parts of the cuticle articulated in various ways.

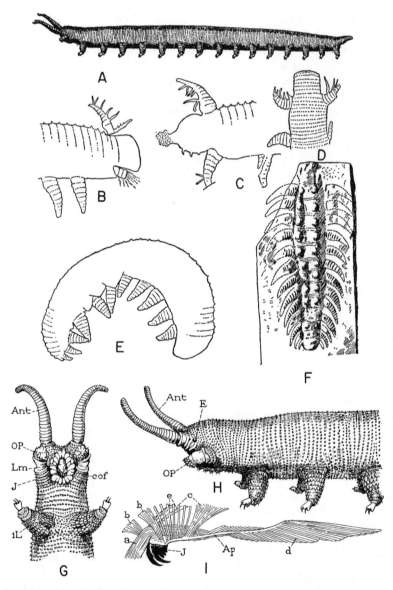

Figure 4. Onychophora. A, *Peripatoides novozealandiae*, modern; B, C, D, E, *Aysheaia pedunculata*, Cambrian; F, *Xenusion auserwaldae*, Algonkian; G, ventral view, H, lateral view of head end of *P. novozealandiae*; I, mouth hook and associated muscles; a-e, jaw muscles; Ant, antenna; Ap, jaw apodeme; cof, circumoral fold; E, eye; J, mouth hook (jaw); 1L, first leg; Lm, labrum; OP, oral papilla. A-F, from Snodgrass, 1958; G-I from Snodgrass, 1938.

In harmony with the usefulness of a fluid skeleton, adult onychophorans have eliminated the ringlike metameric segmentation of the body. The longitudinal body muscles extend unbroken the length of the body, not even being superficially attached to the relatively prominent collagenous connective tissues underlying the epidermis (Figure 5). In arthropods the longitudinal muscles are discrete segmental muscles extending individually, typically from the intersegmental lines. The dorsoventral muscles of onychophorans consist of two continuous sheets whose dorsal attachment to the body wall separates the dorsal longitudinal muscles into dorsal and lateral components, and whose ventral attachment goes medially between the widely separated nerve cords and penetrates the ventral longitudinal muscle. These sheets form a pair of long diaphragms that divide the hemocoel into a perivisceral space and a pair of lateral hemocoels. Pores in the diaphragm control the passage of blood from one space to the other. In arthropods, the dorsoventral muscles never form such diaphragms, but appear as individual muscles variously extending dorsoventrally lateral to the longitudinal muscles and the nerve cords.

Another feature known only in onychophorans is the presence of numerous vascular channels of the body wall (Manton, 1967), which outwardly appear as small ridges that encircle the body and the legs (Figure 5). These channels appear to be important in the functioning of a fluid skeleton as a means of achieving dimensional rigidity associated with ability to bend, shorten, and lengthen, analogous to the pleats of an accordion or the spiral wire in a vacuum cleaner hose. These vascular channels also provide for the passage of blood from the lateral hemocoel to the pericardial space around the heart. These vascular rings were interpreted by Sharov (1966a) as a primitive incipient segmentation of a lobopodium into podomeres. He apparently did not know the true nature of these rings.

Since the Cambrian onychophoran fossils are found in marine sediments associated with other marine animals, and modern onychophorans are terrestrial, the group is one of several that have independently come out of the water on the land. Others that have members that made this transition are chelicerates, crustaceans, and the insect-myriapod group. As in the other groups, onychophorans breathe air into invaginations of the epidermis, the tracheae. Onychophoran tracheae are unicellular and form in large numbers as short tubular tiny invaginations in the furrows between the superficial vascular rings. They are lined internally with cuticle. Before leaving the body wall, each trachea ends with two or three bundles of fine unbranched tracheoles, also lined with an extremely fine cuticle. Each tracheole forms in a cell whose protoplasm covers the tube. These penetrate to the internal organs (Cuenot, 1949). It is believed that cuticular respiration may occur also on the general body surface and through the thin walls of the fluid-filled eversible sacs on the legs of some species.

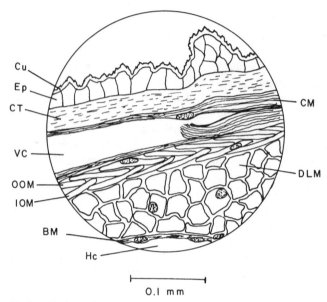

Figure 5. Body wall of *Peripatoides novozealandiae*, sectioned obliquely through a vascular channel (body ring). BM, basement membrane; CM, circular muscle; CT, connective tissue; Cu, cuticle; DLM, dorsal longitudinal muscle; Ep, epidermis; Hc, hemocoel; IOM, inner oblique muscle; OOM, outer oblique muscle; VC, vascular channel. Original.

The antiquity of the Onychophora is suggested by the Precambrian fossil *Xenusion auerswaldae* Pompeckj, and the middle Cambrian *Aysheaia pedunculata* Walcott (Figure 4). *Xenusion* may be a preonychophoran because individual body segments are indicated on the preserved ventral side. The appendages appear ringed, however, as they are in the legs of *Aysheaia* and modern onychophorans. The body segmentation is not evident in *Aysheaia*, and it appears that the rings or ridges over superficial vascular channels of the body were fewer than in present forms. Eyes and antennae are not certainly apparent on the *Aysheaia* specimens. The species was originally thought to be an annelid with parapodia, and an annelid protomium with small palps and large eyes were figured by Walcott (1911b) by retouching the photograph of the type specimen. It appears that in attempting reconstructions from fossils, a great deal of imagination has been used and structures have been inferred to occur when they are not at all apparent on the fossil simply because in the mind of the author such structures ought to be there. The actual fossil specimens of *Aysheaia* bear no trace whatever of any prostomium, eyes, or palps (Snodgrass, 1958) other than an extension of the body ahead of the first pair of legs (Figure 4). On the type specimen, the first, fourth, fifth, sixth, and eighth legs bear slender filamentlike structures that have been interpreted as respiratory organs. Other specimens bear no trace of such filaments, suggesting that these may have been retractable vesicles. Modern onychophorans retain only one retractable vesicle per leg on some of the walking legs in some genera. These vesicles seem to be useful as water absorbing organs in present species.

Since the fossil *Aysheaia* seemed to lack antennal palps, the nature of such organs on recent onychophorans might be considered. In *Aysheaia*, the body ahead of the first legs protruded so as to suggest the presence of a cephalic structure. No eyes are evident, but possibly the animal had visual organs. The existence of jaws is not evident in the fossil. But these, if present, must have been somewhere anterior to the first walking legs, as in recent species. If this is the case, then the oral papillae of recent forms are homologous with the first legs of *Asyheaia*. The oral papillae develop from the limb buds of the segment immediately following the jaw segment. This suggests that the antennal palp of recent species is a new structure in the Onychophora, if the evidence from paleontology is reliable. It appears, then, that the antennal palp of onychophorans may not be homologous with the antenna (antennule) of arthropods, because such antennae originated in the ancestors of trilobites before the Precambrian, while those of Onychophora appear to have been acquired after the Midcambrian, when the Onychophora were already highly specialized animals which as such cannot have been ancestral to any Arthropod.

The preceding discussion is based on the assumption that *Aysheaia pedunculata* was essentially typical of all the onychophorans of its time. This assumption is not warranted. *Aysheaia* may just as well have been representative of a separate line of onychophoran evolution which had suppressed its antennal palp, and the line which gave rise to modern onychophorans bore such palps, but no fossils of this type have yet been discovered. Another possibility is that the imprints of *Aysheaia* are such that the palps are not perceptible because of their orientation in the matrix. Whatever is the true situation regarding the presence or absence of palps in *Aysheaia*, the acronal palps of modern onychophorans do not necessarily belong to a cephalized metamere merely because a pair of coelomeres having a posterior embryonic origin are responsible for the initiation of the embryonic growth of the palps, and some of the coelomeres cells become papal muscle.

In their embryonic development, the onychophoran antennal palps take their origin from two sources (Anderson, 1973). Blastodermal cells behind the future proctodaeum bud off two "mesodermal" bands which grow forward below the yolk. Groups of cells are pinched off in succession from front to back as the bands grow forward, and each clump of cells becomes hollow, forming pairs of coelomeres. The anteriormost pair moves forward past the mouth into the embryonic head. There, these cells give rise to the antennal, head and stomodaeal muscles. As elsewhere in the body, the presence of these coelomeres induces cells in the ventral blastoderm ("ectoderm") to differentiate into a pair of ganglionic masses which become the antennal ganglia, analogous to the deutocerebrum of mandibulate arthropods. Lateral to the developing ganglia, the blastoderm is induced to evaginate to form the surface epithelium of the antennal palp. There are no claws on this organ. It is clear then that the antennae are composed of cells originating in part from the posterior end of the embryo and in part from the preoral head (acron). The antennal ganglia take their origin ventrally in the acron. The ontogeny of the onychophoran antennal palp has been interpreted to indicate that the palps, their ganglia and muscles represent the anteriormost originally postoral metamere which has become cephalized. This is in direct conflict with the embryonic evidence. If the first metamere really did become cephalized, all of its components should arise postorally and migrate forward. Embryonic development consists of a sequential induction of the expression of genes. One induction event evokes another. The position of the induction event varies in different species and is not tied down by man-made rules. But all embryonic cells must be presumed to have the same genetic makeup, if we can believe the implications of the process of mitosis. The induction of the expression of any gene system depends on the production of the inducing agent (organizer), which itself must be genetically controlled. Therefore, we must expect that mutations will occur which affect not only the adult form, but also the timing and sequences of events leading to other events during embryonic development.

Another peculiar onychophoran character is the slime gland. Each oral papilla is invaginated terminally into a long tubular gland extending into the perivisceral hemocoel. Prey, such as soil arthropods, is captured by emission of a thick fluid which entangles it.

Since the onychophorans share with arthropods characters which are obviously ancestral or derived from a common lobopod ancestor, and since onychophorans have so many derived characters of their own which seem to have originated early in the Paleozoic, it appears that this group of animals is the present remnant of a phyletic line which very early went its own evolutionary path apart from the arthropods. The phylum Onychophora then would be a sister group to the phylum Arthropoda. The phyla together may be regarded as a sister group to the annelids.

Much has recently been written in support of the theory that the onychophorans, insects, and myriapods comprise an independently evolved monophyletic assembly, phylum Uniramia (Manton, 1972). It is pointed out that the chelicerates, crustaceans and trilobites, basically "arthropods with biramous limbs," are additional independent groups which are only convergently similar through "arthropodization." The Uniramia are assumed to share four features distinguishing them from other lobopods: (1) primitively uniramous limbs, (2) a wholelimb jaw or mandible, biting with the limb tip, (3) a long midgut without digestive diverticula, and enclosing particulate food, and (4) embryological characters not present in Crustacea and Chelicerata.

There is reason to suspect that ancestral lobopod animals probably originally possessed simple unjointed limbs, as suggested by the Precambrian fossil *Xenusion auerswaldae.* The Midcambrian fossils of *Ayshaeia* might suggest, as Sharov (1966a) inferred, that exite lobes had their origin in common with the protrusions on some of the limbs of these fossils, and that in the ancestral lobopod these protrusions formed a starting point on which arthropods with biramous limbs developed the epipodites as specialized aquatic respiratory structures. The usefulness of thin-walled protruding respiratory structures on limbs of terrestrial arthropods is limited to those species which can protect them from drying out. Such is the case in some terrestrial species inhabiting damp environments, exemplified by some species of Peripatidae, most symphylans, the pauropods, machilids, amblypygids, and a few millipedes and centipedes, whose eversible sacs can be withdrawn into the suppressed limb base by a retractor muscle and protruded by hemocoelic pressure. If such eversible vesicles originally were respiratory in the ancestors of the above, today they appear to be useful primarily as an organ for maintaining water balance, since none of these terrestrial arthropods have respiratory pigments in their blood. Instead, they have developed tracheal systems as respiratory structures. However, there is no good reason to invoke a common ancestry for the tracheae of onychophorans and those of the insects and myriapods. Convergently evolved tracheal systems also arose among terrestrial arachnids and crustaceans. At any rate, the primitive uniqueness of the uniramous limb of the insect-myriapod group can seriously be questioned. The shaft of the leg of all arthropods is so dissimilar to any exite structures such as epipodites and exopodites, and is so basically similar in all arthropods, that a uniramous limb was most probably the ancestral state. So-called biramous limbs are the results of specializations.

The second alleged unique character of the phylum Uniramia, that of mandibles consisting of whole limbs, with the tip serving the biting function, can also be seriously questioned. Only the onychophorans undoubtedly employ the entire limb of the first postoral segment to bite. The rest of the Uniramia employ a structure on the modified limb of the second postoral metamere. The concept that the mandible of the insect-myriapod assembly is a whole limb (superclass Atelocerata Heymons, 1901), arises from several observations. First, the embryonic rudiment of the jaw never possesses a palp or telopodite in any atelocerate, such as is found on the mandible of the nauplius larva of many crustaceans. The crustacean mandible clearly consists of a biting surface derived from a coxal endite, or gnathobase. Second, the mandible of centipedes, millipedes, and symphylans is segmented, suggesting the

38

The Lobopods

presence of podomeres. Third, the mechanism of biting in the myriapods is said to be unique, and impossible to be derived from any mechanism in any other arthropod. Myriapods open their jaws through the swinging action of movable anterior tentorial arms pushing the jaws apart (Manton, 1964). However, the flexing of the jointed mandible is always mainly through the pull of a large tergal muscle inserted on the gnathal lobe. This dorsal coxal promotor muscle of arthropod legs is general, as will be discussed in the next chapter. The "segments" of the myriapod mandible are totally unlike podomeres, in that these "segments" are not cylindrical. The inner faces of all segments are open to the hemocoel, unlike any known podomere other than the coxa. The opposing view that atelocerate mandibles are of coxal origin, with suppression of the mandibular palp, is denied because of the lack of any recapitulation of such a palp in atelocerate embryos. The absence of embryonic recapitulation cannot be proof against the presumed former existence of a structure. If that were the case, the insect orders Mallophaga, Anoplura, and Grylloblattaria would have to join the apterygote insects as creatures whose ancestors never were winged.

The third uniramian character of a long midgut lacking digestive diverticula and enclosing particulate food can also be interpreted in a fashion other than suggesting unity of the Onychophora with the Atelocerata. While there is no evidence that diverticula were part of the original intestine of onychophorans, it seems that early arthropods were so equipped. Diverticula are characteristic of chelicerates, trilobites, and crustaceans. Intracellular digestion is a specialization of the terrestrial chelicerates. Xiphosurans take particulate food, and digestion is mostly accomplished in the lumen of the long midgut as well as in the diverticula. Various crustaceans have a long tubular midgut, which may even be coiled (Cladocera, Cumacea), and lack diverticula (Meglitsch, 1967). All free-living crustaceans take particulate food. The midgut and diverticula of Crustacea have the same cell types as are found in the midgut of insects and millipedes. The epithelium consists of secretory cells which become disrupted, and are replaced by mitotic growth. The absence of digestive diverticula in the Atelocerata may be viewed as a specialization in parallel with their suppression in many crustaceans.

Anderson (1973) emphasized some embryonic similarities of the Uniramia not found in chelicerates and crustaceans. He considered total cleavage to be primitive in these two latter groups, and yolky eggs with intralecithal cleavage to be specializations, whereas the yolky eggs with intralecithal cleavage are considered primitive for the Uniramia. Anderson likened the formation of anterior midgut in Onychophora [from the bilateral invasion of the edges of the blastoderm after its separation along the midventral blastoderm, leaving the yolk temporarily exposed, with migration of sheets of blastoderm ("endoderm") to surround and absorb the yolk] to the similar migration of ventral cells mitotically produced and detached from the blastoderm in centipedes. He did not comment on just as great similarities in the formation of the midgut of arthropods in general. The anterior midgut is not distinct from the posterior midgut in epimorphic centipedes. The entire midgut forms from cells detached from the blastoderm the whole length of the germ band, surrounding the yolk. In onychophorans, crustaceans, xiphosurans, symphylans, anamorphic centipedes, and pauropods, the anterior midgut is closely associated with the digestion of yolk. The posterior midgut forms from blastoderm cells associated with the posterior end of the embryo, forming a yolk-free tube of posterior midgut (Anderson, 1973).

It is easy to find interpretations of almost any kind of evidence to support opinions. It is also easy to ignore some evidence which does not support a viewpoint. The theory of a polyphyletic origin of arthropods from different stocks of lobopod worms not only ignores evidence suggesting that arthropods in general share many fundamental characters or that some ancestral characters have changed in various evolutionary sequences, but the theory assumes that a vast amount of independently derived convergent features became remarkably so similar as to suggest homology. The polyphylists argue that the huge differences in detail found in the functioning of modern arthropods are fundamental and deep seated, arising in response to functional needs serving specialized habits. The polyphylists do not admit that evolutionary novelties may be the basis for new functions and habits. It also seems that there is too much persistence in seeing ancestral stages in modern embryos, overlooking the distinct probability that novelties of evolutionary importance occur in the embryonic stage as well as in the adult.

The tables that follow are arranged to illustrate the fundamental unity as well as the diversity that accounts for the evolution of the various groups of metameric animals.

The members of the superphylum Articulata share ancestral characters as follows:

1. Schizocoelous metamerism between the preoral acron (prostomium) and the nonmetameric telson (periproct).
2. Acronal protocerebrum serving the eyes, and containing an association center associated with pedunculate bodies.
3. Double ventral somatic nerve cord.
4. Dorsal and ventral longitudinal muscles.
5. Coelomoducts, their vestiges or derivatives.
6. Dorsal blood vessel with forward-going peristalsis.
7. Simple eyes.
8. Prostomial palps or antennae.

The members of the phylum Annelida have acquired derived characters of their own as follows:

1. Locomotor-respiratory chaetopodia with protrusible and retractile setae.
2. A cuticle bearing collagen but no chitin except in setae and stomodaeum.
3. Anterior tagmosis (in parallel with lobopods).
4. Annelid nephridial system.

The annelids have retained primitive articulate characters that in the lobopods have become altered:

1. Septate coelom.
2. Coelomic fluid useful in aiding locomotion as a fluid skeleton.
3. Closed vascular system with blood vessels.
4. Holoblastic cleavage.
5. Distinct prostomium and telson.

Primitive characters retained by onychophorans and annelids, but not retained by arthropods are as follows:

1. Ciliated coelomoducts (gonoducts in annelids, gonoducts, and excretory structures in onychophorans).
2. Smooth muscle fibers. Onychophoran jaw muscles are striated.
3. Circular muscles in body wall.

The Onychophora and the Arthropoda share derived characters that have not been acquired by the Annelida, segregating the annelids as a sister group to the lobopod animals:

1. Locomotor lobopodia (coxae) with extrinsic and intrinsic muscles, and with a sclerotized terminal structure on each (telopodites).
2. Ostiate dorsal blood vessel, open vascular system.
3. Nonseptate hemocoel (mixocoel).
4. Chitinous cuticle over entire body, secreted by the epidermis.
5. Noncollagenous protein incorporated in cuticle.
6. Unicellular sensory setae in the epidermis.
7. Coelomic excretory organs or their vestiges.
8. Yolky eggs with centrolecithal cleavage and an egg shell.

9. Cleavage always results in the formation of a blastoderm enclosing a yolk mass, except in species whose cleavage has become secondarily total, correlated with loss of yolk, viviparity, or precocious hatching into specialized larvae.

10. The embryo develops typically from cells arising by mitosis in a ventrally situated germ band. The remainder of the blastoderm typically is extraembryonic and serves the embryo in various ways, but does not contribute to adult organs other than dorsal epidermis in a few groups.

11. Development of a deutocerebrum as a sensory center for palps or antennae.

The Onychophora possess derived or totally new characters of their own, acquired independently from their sister group, the Arthropoda. These are as follows:

1. Suppression of external body segmentation.

2. Wide separation of ventral nerve cords.

3. Compartmentalization of hemocoel into longitudinal pericardial, perivisceral, and lateral spaces, separated by longitudinal muscular diaphragms.

4. Ringlike vascular channels between circular and oblique muscle layers of body wall and legs, serving a fluid skeleton.

5. Stretching of body limited by extensive collagenous connective tissue under epidermis, external to muscle layers.

6. Two oblique crossed layers of muscle of body wall, under circular muscle layer.

7. Jaws from appendage I.

8. Slime gland, appendages II.

In addition, the Onychophora have other new characters convergently acquired with analogous characters in arthropods.

1. Viviparity in most species, some with placental nutrition of the embryo.

2. Yolk-poor eggs undergoing total cleavage in some viviparous species.

3. Production of a spermatophore, enclosing the semen.

4. Numerous fine unicellular tracheae, in epidermal furrows between vascular rings.

5. Terrestrial habit.

CHAPTER 4

The Arthropoda

The phylum Arthropoda (Siebold and Stannius, 1848) consists of the largest number of species known in any phylum, so many that the total number exceeds the number of species known in all the rest of the animal phyla combined. In spite of such a great variety of species, they all share certain characteristics in a combination found in no other phylum. Some of these characteristics occur in long extinct species dating from the beginning of the Paleozoic, in species already highly specialized so that they may be separable into class-rank taxa (Trilobita, Chelicerata). This suggests that their evolutionary origin as arthropods probably occurred a considerable time before the Paleozoic.

In spite of the many arthropodan characters found in all classes of arthropods, there is still uncertainty concerning whether arthropodization was an evolutionary process that produced a single ancestral arthropodized line that was the precursor of all arthropods, or whether arthropodization arose perhaps four times so as to produce independently four evolving lines in parallel, each of which not only acquired its own evolutionary specializations, but also independently evolved many arthropod characters in common with other lines: trilobites, chelicerates, crustaceans, and atelocerates.

The fossil record furnishes no definite evidence for either possibility. The oldest known fossil representatives of each of the four main groups are clearly identifiable as already specialized forms, none of which appear to be candidates for ancestors of species in another of the main groups. The evolutionary relationships among arthropods, therefore, must be deduced largely through the study of existing species.

Vigorous champions of the hypothesis of a polyphyletic origin for arthropods follow the lead of a group that may be called the Manton school, because the evidence believed to support the hypothesis has been widely discussed by S. M. Manton in a series of investigations devoted to elucidation of the functional anatomy of various arthropod types. Her studies have resulted in the publication of a large amount of new information on the anatomy of the locomotor and mandibular mechanisms of arthropods. How arthropods employ their appendages in the process of living is described in intimate detail, and the specialization in structure and habit typical of each of the classes of arthropods is revealed for the first time. Because in each class the specializations employed in performing many actions are so different from the specializations found in other classes, in which similar actions are found, it is argued that none of the specializations of any one class could be the starting point for specializations in any other class. The theory concludes that gross structural resemblances among members of different classes arose convergently to serve common needs dictated by habit. The assumption that convergence can be unlimited follows from the admission that convergence is possible. The conclusion that convergence explains similarities shared by two taxa is inescapable when such similarities are the exception among the dissimilarities that separate various groups. For example, tracheal respiration seems to have arisen independently at least four times, assuming that the ancestors of land arthropods and onychophores were aquatic and that these did not evolve tracheal respiration until some time after the transition to terrestrial habits. These assumptions must be made because the older fossil representatives of each class (except for insects and myriapods) bear evidence of employing blood gill respiration and living in water long before life on land was possible, and some classes still contain thoroughly aquatic gill-breathing species. But more important, although tracheal respiration occurs in some arachnids, isopods, all modern onychophorans, and atelocerates, in each group the tracheae are fundamentally nonhomologous by virtue of their different positional distribution. In arachnids the tracheae invaginate from lung books, in onychophorans from highly scattered positions in the grooves of the numerous body "rings," in insects and myriapods from the pleural areas, and in isopods they are surface pseudotracheae that do not penetrate into the body. Nevertheless, if convergence can be admitted in this case, it must be possible in others. The Manton school assumes this possibility to an unlimited extent: All the similarities of arthropods can be convergent, and it is their differences in modern forms that speak for their separate origin.

Among those who favor a theory of monophyletic origin of arthropods there is thorough agreement that specialization within a group can only give rise to further specialization. A highly specialized modern or extinct animal cannot represent the ancestral condition of a group whose specializations are greatly different. When different taxa share many features, each similarity of which is explainable on a basis of homology and can be derived from a probably ancestral state, the probability of convergent evolution is lessened. The fundamental similarities exhibited by arthropods are so numerous that the probability of convergence in all these similarities seems much less than does the probability of common ancestry. In a monophyletic theory, the differences between groups then become explainable on the basis of unlimited divergence.

In Chapter 3, a list of eight characters shared by all Articulata suggests that in their ancestral state these characters were to be found in the common ancestor of annelids, onychophorans, and arthropods. Various derived states of these characters can be seen in most members of Recent classes. The absence of all traces of some of the characters can be regarded as a secondary state. For example, pauropods and mites lack a dorsal vessel, and many arthropods lack eyes. Schizocoelous metamerism in the lobopods is different from that of annelids in that in the lobopods the coelom is only incipient in the embryo, and there is never an adult coelom as such. No one doubts the unity of the articulates, with their schizocoelous metamerism, similar forebrains, double ladderlike nerve cord, dorsal vessel, coelomoducts, basically similar simple eyes, and so on.

The list of 11 characters shared by arthropods and onychophorans similarly suggests that the common ancestor of these two groups possessed the ancestral state of each of these characters. If they were ancestral states in the lobopod common ancestor when compared to the descendants, these same character states are to be considered as derived from the condition in the ancestral articulate. The derived states in modern arthropods and onychophorans may include suppression as a character state. In every class, for instance, the typical centrolecithal cleavage is suppressed in the egg of species whose eggs undergo total initial cleavage, correlated with small size and paucity of yolk material, such as in some mites, in pauropods, symphylans, many highly specialized small crustaceans, and collembolans. The deutocerebrum is lacking in most chelicerates.

Following is a list of 17 characters whose origin may be presumed to have been in a lobopod line leading to the arthropods, independently of the onychophorans. By contrast with the 11 lobopod characters referred to in the preceeding paragraph, these are either derived or new. They are considered as ancestral states, or ground plan characters, for arthropods. Each arthropod group has evolutionarily modified the ancestral states in its own fashion. Some characters may have remained with much less change in one group than in another. These basic characters are found together only in arthropods. If these are convergent and have arisen independently in each of Manton's four arthropod phyla, it should not be possible to trace their origins from a common ancestral source. The rest of this chapter attempts to show that these characters may not be convergent, and that the phylum Arthropoda may be a natural monophyletic taxon.

All arthropods share the following synapomorphic characters or their variously derived states:

1. Epidermal cuticle consisting of chitin and protein.
2. Sclerotization of the cuticle through the tanning of protein, complexed with chitin, into metamerically arranged plates.
3. Typical periodic molting of the cuticle in a well characterized sequence controlled by hormones.
4. Ecdysial glands in anterior end producing alpha- and/or beta- ecdysone, a molting hormone.
5. Typically segmented metameric locomotor appendages, consisting of a coxal base followed by a distal segmented telopodite.
6. In some metameric appendage some sort of endite lobe on the coxal part, related to feeding (gnathobase).
7. Exite lobes on coxa, useful or not. Absence of exite lobes interpreted as suppression.
8. Variously modified intersegmental tendon system.
9. Tonofibrillae penetrating epidermis, for muscle attachment to cuticle.
10. Longitudinal dorsal and ventral muscles segmented metamerically.
11. All muscles striated.
12. Deutocerebrum developed as antennal ganglia (suppressed in most chelicerates), protocerebrum developed as an ocular center.
13. Compound eyes in some members of all major groups.
14. Suppression of all circular body wall muscle.
15. Expansion of lateral edges of dorsum into lateral tergal folds (paranotal lobes).
16. Cephalic tagmosis including at least one metamere.
17. Suppression of all motile cilia except in spermatozoa.

CUTICULAR SCLEROTIZATION AND MOLTING

Arthropod cuticles are remarkably similar in their structure and chemical content. Chitin and protein are always incorporated in the cuticle, which is secreted by the cells of the epidermis. The sclerotization of the cuticle consists of the chemical bonding of protein molecules to one another in the outer layers, where a complex is formed with chitin, which is at the same time light and strong. The chemical bonding, or tanning process, of the protein occurs at points along the protein chain where a phenolic compound attaches the protein chains together through amino side groups of one of the amino acids, usually lysine. This process occurs in all arthropods. There is the possibility that cysteine or cystine in the cuticles of chelicerates, crustaceans, and insects takes part in the bonding process involving their sulfhydryl groups (Lafon, 1943; Krishnan, 1953; Neville, 1975).

In the deeper layers below the sclerotized portion of the cuticle (endocuticle), and throughout in flexible arthrodial membranes, the tanning does not occur, and there is relatively less protein. In some arthropods, particularly in some crustaceans and millipedes, and apparently in trilobites, additional hardness is achieved through mineral deposition in the cuticle. Barnacles have such a thick and hard mineralized cuticle that they were originally thought to be molluscan.

Wherever it has been studied, it has been found that the processes of discarding an old cuticle, its replacement by a new cuticle, and a spurt of epidermal growth activity (molting) are remarkably similar in all arthropods. The process involves a steroid hormone, ecdysone, produced by ecdysial glands associated with the back part of the head in mandibulates, or in the prosoma of chelicerates (Herman, 1967). The ecdysial glands of insects have shifted into the prothorax. Insects produce two types of ecdysone, differing only in the presence or absence of an (-OH) group replacing a (-H). Crustaceans produce mostly the type containing (-OH) (beta-ecdysone). The ecdysones induce the beginning of molting when artificially introduced into xiphosurans, arachnids, crustaceans, and insects (Krisnakumaran and Schneiderman, 1970). Molting typically begins with the enlargement of epidermal cells, followed by mitosis. The old cuticle detaches from the epidermis (apolysis) and a new cuticle is secreted by the active epidermis, while a molting fluid dissolves the inner part of the old cuticle. When the epidermis finishes growing and a new cuticle is laid down, the thin outer remnant of the old cuticle is shed (ecdysis) as the animal increases its volume by ingesting air or water. The epidermis returns to its resting state after the sclerotization of the new cuticle. The physiology of molting is so typical in arthropods that it appears to be a homologous process in all groups, and probably is not to be regarded as convergence (Figure 6).

THE METAMERIC LOCOMOTOR APPENDAGES, SKELETAL SYSTEM, AND TENDONS

The oldest fossil sclerotized arthropod limbs were essentially fully segmented into eight podomeres. Whether the segmentation came about all at once, or gradually increased from no segments to the fully segmented state, or through another process, cannot be determined from the fossil record. If the

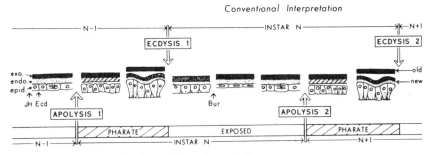

Figure 6. Diagrammatic representation of the molting cycle. Bur, time of action of bursicon (hardening hormone); Ecd, time of action of ecdysone; endo., endocuticle; epid., epidermis; exo., exocuticle plus epicuticle; JH, time of action of juvenile hormone. From CSIRO, 1970, *The Insects of Australia* by permission of Melbourne University Press.

original lobopod was able to perform stepping motions with unsclerotized limbs, the mechanism is suggested by the onychophoran method. Promotion, remotion, and up and down motions employ both extrinsic and intrinsic muscles, and lengthening and shortening occur through the operation of intrinsic muscles working or relaxing against hemocoelic pressure. As each leg is moved off the ground and brought forward, hemocoelic pressure extends the leg until it touches the ground. As the leg bears the weight and is moved rearward, the plantar surface in contact with the ground remains in contact at one point, the leg shortening during its swing backwards until it attains 90°, then lengthening for the rest of the propulsive move. As each pair of legs is lifted, the forward swing is a ventrolateral motion, aided by local shortening of the body through action of the longitudinal muscles, adding a parasagittal axis (transverse-horizontal) to the leg motion. Judging from the stepping motion of legs of modern arthropods, if a leg is sclerotized it should have at least three articulations and three podomeres. The coxa-body articulation, whether condylar or membranous, provides movement forward and rearward around a roughly vertical axis. A coxa-telopodite articulation, usually dicondylar, provides movement away from and toward the substrate, permitting lifting the leg after the backstroke and depressing the leg at the beginning of the backstroke. As the leg touches the ground, starting the backstroke, the tip remains in contact at one point as the body moves forward. The backstroke involves bending the telopodite at a minimum of one "knee" joint since the point of contact is a fixed distance lateral to the body during the backstroke. In modern arthropods leg base movements in a parasagittal plane around a transverse-horizontal axis is provided by various ways of articulating the leg base (Manton, 1950 through 1977).

That a three segmented leg is a workable locomotor apparatus is clearly shown in the Collembola (Manton, 1972). The coxa-body articulation is entirely membranous, with no coxa-body condyles. The coxa-trochanter articulation is a dicondylar joint providing for most of the levation and depression of the telopodite. A poorly flexible, occasionally absent trochanter-femur articulation provides for only slight bending at that point. The femur-tibiotarsal joint provides for flexion and extension at a knee joint. The tibia and tarsus fail to separate in the Collembola, remaining as a single podomere, followed by the small pretarsus. Flexion at the knee joint is provided by a large femoral muscle inserted into the base of the tibiotarsus, and by a major portion of the pretarsal flexor muscle originating in the femur (Figure 7).

However, there is no evidence that the functionally three-segmented leg condition of Collembola is a primitive state. Rather, it appears that such a state in the Collembola is traceable to an ancestral state in which the telopodite consisted of additional podomeres. The selective advantage of a multisegmented telopodite is not clear. Since the oldest fossil arthropods apparently possessed a telopodite with seven podomeres, it may be that this

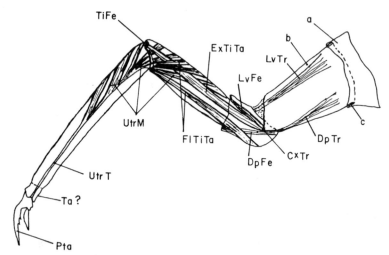

Figure 7. Middle leg of *Tomocerus flavescens* (Collembola), illustrating a functional arthropod leg with three main joints, the acondylar coxa-body joint, the dicondylar coxa-trochanter joint (CxTr), and the monocondylar femur-tibia joint (TiFe). a, upper (catapleural) coxomere; b, lower coxomere; c, undeveloped coxa-body joint. In this species the functional coxa-body joint is between the catapleurite and the body. DpFe, depressor of femur; DpTr, depressor of trochanter; ExTiTa, extensor of tibiotarsus; F1TiTa, extensor of tibiotarsus; LvFe, levator of femur; LvTr, levator of trochanter; Pta, pretarsus; Ta?, vestige of tarsus; UtrM, unguitractor muscles; UtrT, unguitractor tendon. Original.

condition was established at the time that cuticular sclerotization occurred in the lobopod ancestor. If there originally were more podomeres than now occur in arthropod limbs, it seems that the number most appropriate for efficient locomotion was established long ago with eight as a basic number. Subsegmentation of the tarsus apparently occurred several times in parallel. A subsegmented nonmusculated tarsus occurs in pycnogonids, arachnids, myriapods, and most insects. The tarsus is simple in trilobites, xiphosurids, eurypterids, arthropleurids, crustaceans, and entognathous apterygote insects.

If the evolution of a jointed limb was begun in an ancestral arthropod as a one-time event, the process has resulted in the characteristic limbs of each class. In each class there are differences in detail of musculature, relative lengths of podomeres, and types of articulation. The specializations are such that there is little agreement among students of arthropods on the homologies especially of the basal podomeres of the various types of legs. Two main features of modern arthropod legs can be found in all the classes: (1) the limb base (coxa, coxopodite) always bears the important extrinsic leg muscles whose origins are on the dorsal and ventral body wall, and whose function is mainly the bringing forward (promotion) and backward (remotion) of the entire limb, and (2) the coxa-trochanter joint between the base and the rest of the limb (telopodite) bears two condyles, which permit motion in a vertical plane. Exceptions to these two basic features are found in highly specialized forms. For instance, in the arachnids the coxa has lost most of its ability to swing on the body, and the main promotion and remotion of the walking legs is at the modified coxa-trochanter or other joint, about a roughly vertical axis. Elevation and depression of the rest of the leg is at more distal articulations (Manton, 1973b).

The coxa with its extrinsic muscles is so commonplace among arthropods that there is a strong implication that this segment is homologous in all arthropods. The coxa-body joint, however, is of various sorts, more or less typical in each class. The variations are such that it is difficult to infer a primitive condylar joint. It appears, rather, that originally the coxa-body joint consisted mostly of flexible membrane all around the coxa, with perhaps only a narrow membranous union of the upper coxal base and the lower edge of the tergum. This is suggested by the apparent absence of recognizable condyles on trilobite coxae. The xiphosuran coxa is completely surrounded by soft cuticle except for a small sclerite in the membrane near its dorsal end. A membranous articulation of the coxa is also the case in other species, such as in the Cephalocarida among the Crustacea and the Collembola among the insects. Such membranous coxal articulations attest to the workability of a presumed noncondylar ancestral coxa. It is not to be supposed that the condition in Collembola and Cephalocarida is primitive. The absence of coxal condyles is probably a specialization correlated with small size in these two groups. But an ancestral acondylar coxa-body joint serves as a satisfactory beginning for the various ways that the coxa is articulated with the body in the various arthropod groups, each evolved independently of the others. A single dorsal coxal condyle occurs in Xiphosurida, and among insects in the Machilidae, Odonata, and Ephemerida. A dorsal and a ventral condyle occurs among the Crustacea Malacostraca, and in insects in the Protura and most pterygotes. A single ventral condyle is typical of the myriapods and among insects in the Diplura. The similarities in coxal articulations of various insect groups with those of other classes are to be interpreted as convergencies. In pycnogonids, the coxa-body joint is dicondylar, but the condyles are anterior and posterior, allowing only up and down movement at that point, while the promotor-remotor movement occurs at the coxa-trochanter joint, as in arachnids.

The telopodite is that part of an arthropod leg beyond the coxa. It is tempting to homologize the telopodite with the small narrow portion of the tip of the onychophoran leg that bears the terminal claws. It has become standard practice to define the segments of the telopodite (podomeres) by the presence of a set of muscles originating in the proximal segment, which operates the next podomere. A podomere is said to be subsegmented if the annuli are free of muscles as is often the case in the tarsus. By this definition, the largest number of podomeres in the telopodite of modern species is seven in the legs of pycnogonids, acarina, and solpugids. There are apparently seven podomeres in the telopodites of extinct trilobites, arthropleurids, and eurypterids, but it is impossible to be certain because the muscles have not been preserved in these fossil forms.

The podomeres of the arthropod leg have been given names that suggest that there is a typical arthropod homologous series. Thus we have the coxa (coxopodite), trochanter (basipodite, first trochanter), prefemur (ischiopodite, second trochanter), femur (meropodite), patella, tibia (carpopodite), tarsus (propodite), and pretarsus (dactylopodite). The basic pattern of eight podomeres per leg has been variously modified. The second trochanter appears to have been suppressed independently several times. Among the chelicerates, the Xiphosurida and most arachnid orders have no second trochanter. Insects, pauropods, and symphylids bear only one trochanter. The patella has been eliminated in all recent arthropods except the chelicerates. Judging from the possible presence of a patella in the extinct *Cheloniellon* (sister group to the Crustacea) and Arthropleurida (sister group to Myriapoda) the patella may have been eliminated twice independently in the Mandibulata. A patellalike subsegment of the tibia occurs in the peculiar diplopod *Polyxenus,* (designated as postfemur by Manton, 1956). By our man-made definition of a podomere, the postfemur of *Polyxenus* is part of the tibia (see Figure 50), judging from the absence of muscle from it to the tibia. The tarsus appears to have become subsegmented several times in the evolution of arthropods. The tarsus is simple in recent xiphosurids, and the legs of fossils of this group are insufficiently known to indicate whether the simple tarsus is an ancestral state. If it is, then subsegmentation of the tarsus occurred at least twice in the Chelicerata. The pycnogonids bear a subsegmented tarsus, and they are generally regarded to be a distant relative (sister group) of other chelicerates. Among the latter, xiphosurids bear a simple tarsus, and most arachnids bear a subsegmented tarsus. The possibility that the extinct eurypterids also had a subsegmented tarsus was discussed and discarded (Snodgrass, 1952) in favor of a possibly subsegmented trochanter, but in the absence of preserved muscles there is still uncertainty. A simple tarsus appears to be the ancestral state in the Mandibulata. It is simple in Crustacea, Arthropleurida, Symphyla, the entognathous insects, and most diplopods. There are tarsomeres in the higher insects, pauropods, chilopods, and one group of millipedes, the Pselaphognatha, suggesting four more independent origins of tarsal subsegmentation. Thus we have the good possibility of a five-time elimination of a trochanter, a two-time elimination of the patella, and a six-time acquisition of subsegments in the tarsus. It is no wonder that students of the polyphyletic theory of arthropod origin can embrace a policy of unlimited convergence.

The common dicondylic articulation at the coxa-telopodite joint is so widely distributed that it is quite possible that this joint was established in the ancestral arthropod. If this is so, the rotation to a vertical axis at this joint in pycnogonids and arachnids must be interpreted as secondary, and it probably happened more than once. The nature of the joints between podomeres in the telopodie vary from being membranous, as in most palpi and between tarsomeres, to monocondylar or dicondylar. There is no demonstrable uniform pattern except within classes or lower categories. It appears that each type evolved its typical condylar articulations independently of other types from ancestral membranous joints.

The fossil record supports the assumption that the first arthropods bore various outgrowths on their appendages, traces of which are to be found in some appendage of nearly all extinct and living classes: exite and endite lobes of the basal podomeres. In existing arthropods, endite lobes are called gnathobases when they occur on appendages near the mouth, and hence are useful in feeding. Endite lobes on gnathal coxae are employed in such diverse ways in feeding that it suggests that their use in feeding has been exploited by arthropods after such endites had become acquired by the primeval arthropod, and that originally their main function may not have been related to feeding. In the trilobites, one of the generalized features is the lack of specialization in their legs. All along the body the legs were similar, and the coxae all possessed weak to well developed endite lobes. It appears that on the trunk, their tips were too far apart to permit their action as food handling organs (Størmer, 1944). In the ancestral arthropod such endites may have been useful extensions of the tip of the coxa, acting as bearing surfaces on the substrate before the telopodite evolved as a means of lifting the body during creeping locomotion. In the Onychophora such is the case. The ventral tip of the large lobelike coxa rests on the ground, and the poorly developed telopodite generally does not bear weight. In arthropods, subsequent evolution of the telopodite replaced the coxal endite as a bearing surface. Through natural selection the anterior coxal endites variously became useful in feeding and became gnathobases.

In the Xiphosurida, legs 2 through 6 bear large coxae which are freely movable in all directions (see Figure 11). Each coxa bears gnathobasic lobes with spines pointing dorsomedially. The jawlike motion of the coxae in feeding involves not only a transverse abduction-adduction (Manton, 1964), but the coxae also move alternately in a dorsoventral fashion so that food is literally shoved into the centrally located mouth by the slanting spines of the paired coxae, alternating their ventrally directed protraction with a dorsally directed retraction. It appears that protraction of some coxae results from hemocoelic pressure exerted on them by the retraction of other coxae made possible by the flexible membrane surrounding the coxae. This mechanism of feeding probably was employed by the trilobites, using the usually well developed gnathobases of the first three pairs of appendages. The venter of trilobites has not been preserved, suggesting that these had not acquired a sclerotized sternum. The trilobite leg base did not appear to be articulated by firm condyles, since these have not been observed. Therefore, the leg base was probably not only freely movable, but protrusible by localized hemocoelic pressure.

In the mandibulate arthropods, the mandibles are fundamentally similar in their actions and musculature, but each class has evolved differences in detail of structure, action, and musculature. The detailed differences among mandibles have been described by Manton (1964). The fundamental similarities were explained as convergent, because of the belief that the insect-myriapod mandible is a whole limb, with tip-biting, rather than a modified coxa with strong gnathal endites.

Before discussing mandibles, a consideration of the basic segmental and extrinsic leg musculature of arthropods must be made (Figure 8). In any arthropod one may study, there can be found some variation of an apparent basic plan. The figure is a hypothetical construction suggesting the minimum operational musculature of a metamere. From this plan may be derived all the variations found in arthropods. Such variations are shifts in muscle attachments, additions of new or loss of muscles, enlargement, reduction, fusion or elimination of the transverse tendons, and so on. This hypothetical segment differs from that envisioned by Snodgrass (1935) mainly in incorporating the transverse tendon system.

In any class of arthropod, usually in the relatively unspecialized species, and especially in the mandibulates, the paired dorsal and ventral longitudinal muscles (DLM, VLM) are intrasegmental, in that the muscles are broken up into segments whose ends are attached to collagenous tendonlike masses of tissue situated at the intersegmental lines (DTT, VTT). These transverse tendons, both dorsal and ventral, are also anchor points for the dorsoventral muscles (DVM) and for the extrinsic muscles of the legs. In the more highly evolved species, there is a tendency in all classes for the intersegmental transverse tendons to be reduced, to the point that in most cases the muscles are attached to the cuticle by way of apparent vestiges of such tendons, in the form of tonofibrillae which penetrate the epidermis.

Dorsal transverse tendons between the ends of the dorsal longitudinal muscles have been retained by many species of various classes. The dorsal transverse tendons are most highly restricted in the chelicerates, where they are known only as dorsally paired tendons at four intersegmental grooves in

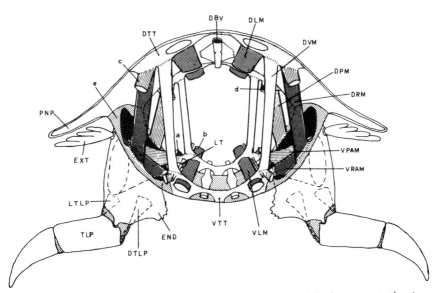

Figure 8. Reconstruction of hypothetical primitive arthropod body segment showing minimum operational somatic and extrinsic leg muscles, viewed from posterior aspect. a, stump of VRAM at anterior intersegment; b, stump of VLM at anterior intersegment; c, stump of DPM at posterior intersegment; e, unsclerotized pleuro-sternum. For other legends see p. 79. Figures 8 through 32 reprinted from Boudreaux, in *Arthropod Phylogeny*, A. P. Gupta, Ed., 1979, by permission from Van Nostrand Reinhold Company, New York.

the opisthosoma of the genus *Liphistius* (Figure 9A), a somewhat primitive spider (Millot, 1949b). They are present in the almost typical hypothetical state in the head and thorax of *Hutchinsoniella macracantha* (Figure 10), a presumably relict primitive crustacean, and in other Crustacea as well (Hessler, 1964). In the higher Crustacea there are various degrees of suppression of dorsal tendons. In the myriapods, dorsal tendons have mostly been eliminated, except in the Scutigeromorpha. In insects, dorsal tendons are commonplace in the thorax and abdomen of Diplura, Collembola, and Machilidae. The pterygote insects lack dorsal tendons. In their thorax the dorsal intersegmental groove is usually deeply inflected forming a large plate (phragma) on the surfaces of which the powerful indirect wing muscles attach.

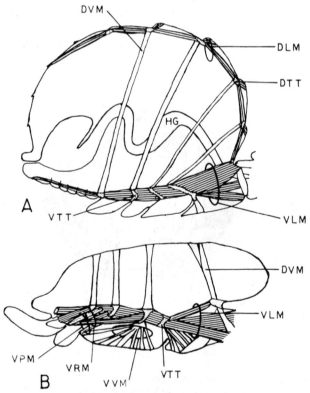

Figure 9. Longitudinal sections of abdomens of spiders (Arachnida). A, *Liphistius batuensis*, with persistent dorsal and ventral intersegmental tendons; B, *Uroctea durandi*, with only ventral tendons persisting. VPM and VRM operate spinnerets. For legends see p. 79. From Boudreaux, 1979.

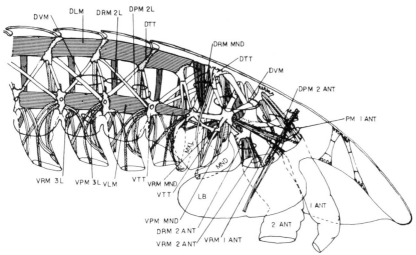

Figure 10. *Hutchinsoniella macracantha* (Crustacea, Cephalocarida). Section showing longitudinal view of head and first three trunk segments. First leg is sometimes considered to be second maxilla, and is articulated on head. For legends see p. 79. From Boudreaux, 1979.

The ventral transverse tendon system is more widespread, in that all classes of arthropods except the pycnogonids retain some form of ventral tendons. In other chelicerates the ventral tendon system of the prosoma is usually detached form the venter, and the segmental tendons are united in the form of a single plate on which the ventral muscles of all the legs take their origins. The tendinous plate is suspended below the alimentary canal by suspensory muscles, the remants of the dorsoventral muscle system (Figure 11). This structure is called endosternum, but it is not apodemal or sclerotized, as is the endosternum of the malacostracan Crustacea. The solfugids, ricinulids, and some mites appear to be the only arachnids that have eliminated the endosternal plate. In the opisthosoma, separate ventral tendons occur in the first four segments of spiders, in which the tendons serve as attachment points of the ventral longitudinal muscles and the dorsoventral muscles on each side.

The ventral trunk tendons are mostly suppressed in the Pauropoda (Figure 12), Symphyla (Figure 13), and Diplopoda. The ventral trunk tendons of pselaphognathid millipedes (Figure 14) are retained, but they are highly modified in that the tendons contain chitinous rods (Manton, 1956). However, the ventral muscles of the mandibles in these three classes, as in the mandibles of many Crustacea, Chilopoda, and the apterygote insects, originate on a large persistent ventral transverse tendon (Manton, 1964.) Ventral tendons provide attachment of dorsoventral muscles, ventral

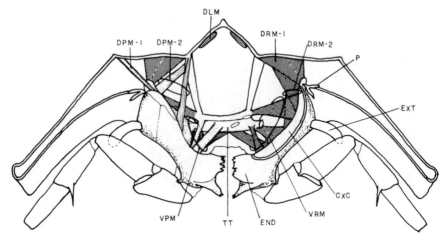

Figure 11. Cross section of *Limulus polyphemus* (Xiphosurida), viewed from anterior aspect. Front edge of left coxa is cut away to show remotor muscles inserting on posterior border of coxa. DPM-1 is segment of DPM converted to abduction of coxa. DRM-1 is segment of DRM converted to abduction of coxa. Clear oval areas are insertion points of left VPM. For legends see p. 79. From Boudreaux, 1979.

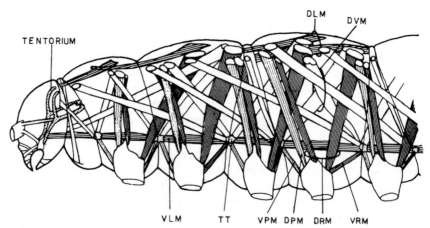

Figure 12. *Pauropus sylvaticus* (Pauropoda). Lateral view of trunk and limb musculature. Only ventral transverse tendons remain. For legends see p. 79. From Boudreaux, 1979.

56

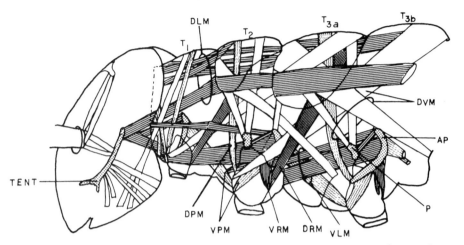

Figure 13. *Hanseniella agilis* (Symphyla). Lateral view of head and first three trunk segments. Transverse tendons are suppressed, and muscles attach ventrally on pleuron (P) or sternal apodemes (AP). Third tergum consists of two hemitergites (T_{3a} and T_{3b}). For legends see p. 79. From Boudreaux, 1979.

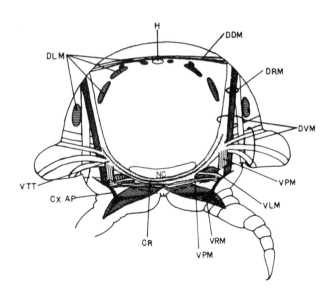

Figure 14. *Polyxenus lagurus* (Diplopoda, Pselaphognatha). Cross section of trunk showing musculature. Note that dorsal tendons are suppressed, and that a cuticular rod (CR) is embedded in ventral transverse tendon. For legends see p. 79. From Boudreaux, 1979.

longitudinal muscles, and ventral leg muscles in the centipedes (Figures 15 and 16), apterygote insects (Figures 17-20), and crustaceans of all the subclasses. In some orders of malacostracan crustaceans (for example, the Decapoda) the ventral tendons are not developed, and the ventral muscles mostly originate on sternal apodemes, as they do in parallel in the pterygote insects. Cisne (1974) demonstrated that the musculature and tendons of the trilobite *Triarthrus eatoni* (Figure 21) appeared to be similar to the Cephalocarida (Figure 10).

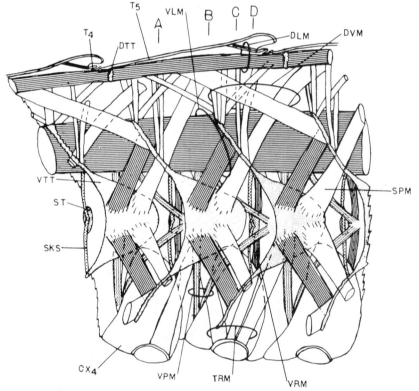

Figure 15. *Scutigera coleoptrata* (Chilopoda, Scutigeromorpha). Dorsal view of body musculature. Fourth, fifth, and sixth body segments are shown as if opened dorsally and spread out. Somatic muscles were omitted in lower (left) side to reveal some of the leg muscles. Note the reduced terga of fourth (T_4) and sixth segments. A, B, C, and D indicate level of sections shown in Figure 16. For legends see p. 79. From Boudreaux, 1979.

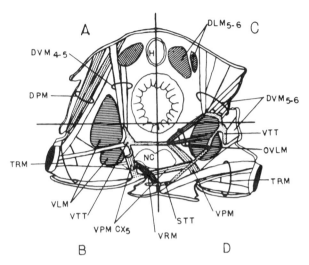

Figure 16. *Scutigera coleoptrata*. Cross sections through body segment 5 at approximate levels A, B, C, and D of Figure 15. Each quadrant represents one-fourth of respective section. For legends see p. 79. From Boudreaux, 1979.

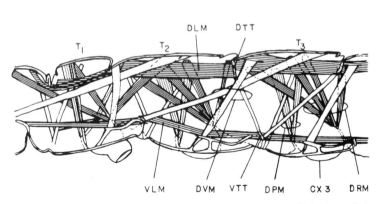

Figure 17. *Campodea* sp. (Insecta, Diplura). Internal view of right half of thorax. Only small intersegmental tendons persist dorsally, and ventrally tendons are more extensive than shown. Origins of DVM, DRM, and DPM shifted to cuticle dorsally. Insertion of vertical DVM shifted to endosternal apodeme (AP). For legends see p. 79. From Boudreaux, 1979.

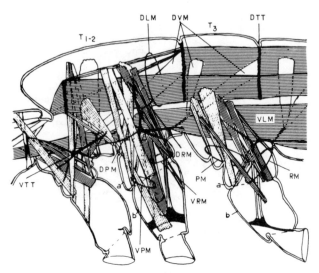

Figure 18. *Tomocerus longicornis* (Insecta, Collembola). Thoracic musculature and tendons. Origins of DRM, DPM, and vertical DVM shifted from dorsal tendon to tergum. Ventral transverse tendons elaborated. T_{1-2}, combined prothoracic and mesothoracic terga; a, upper coxomere (catapleurite); b, lower coxomere (definitive coxa of other insects). For legends see p. 79. From Boudreaux, 1979.

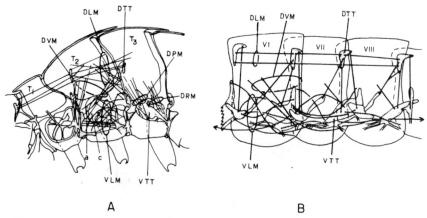

A   B

Figure 19. A, *Lepismachilis y-signata* (Insecta, Microcoryphia). Thoracic musculature and tendons. No muscles are shown in prothorax except for some dorsal muscles. Somatic muscles are shown in mesothorax, and extrinsic coxal muscles in metathorax. Anapleurite (a) and catapleurite (c) are shown in mesothorax. Muscles so numerous that each is represented by a line. B, *Trigoniophthalmus alternatus* (Insecta, Microcoryphia). Schematic representation of abdominal musculature and tendons in sixth, seventh, and eighth abdominal segments. Note twisted ventral muscles (VLM). Dorsal and ventral tendons are separated medially and appear paired in each segment. For legends see p. 79. From Boudreaux, 1979.

60

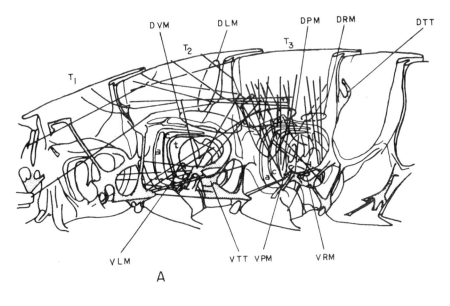

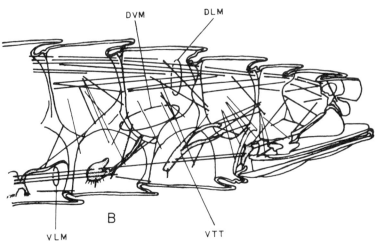

Figure 20. A, *Lepisma saccharina* (Insecta, Thysanura). Schematic representation of thoracic musculature and skeleton. Prothoracic muscles not shown. Longitudinal musculature (DLM, VLM) and dorsoventral muscles (DVM) only shown in mesothorax. Extrinsic leg muscles (DPM, DRM, VPM, VRM) only shown in metathorax. Ventral thoracic tendons are well developed, but dorsal tendons have been reduced and separated into lateral fragments. B, *Thermobia domestica* (Insecta, Thysanura). Schematic representation of musculature and associated structures of sixth through eleventh abdominal segments. Dorsal tendons have been suppressed, and ventral tendons are reduced and laterally separated. In both A and B, the muscles are represented by lines. For legends see p. 79. From Boudreaux, 1979.

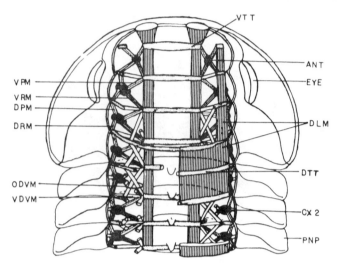

Figure 21. *Triarthrus eatoni* (Trilobita), Ordovician. Dorsal view of head and first three thoracic segments. About 1 cm of animal is shown schematically. For legends see p. 79. From Boudreaux, 1979.

In the pterygote insects, only tiny fragments of ventral intersegmental tendons may persist, above the spinae of the thoracic sterna. The larvae of *Corydalus cornutus* (Megaloptera) (Figure 22) were shown to have traces of ventral tendons in the thorax (Barlet, 1977).

The homology of the mandible of the Mandibulata is suggested by several features of all mandibles. It is always the modified appendage of the second metamere. In all mandibulate classes the extrinsic muscles comparable to a basic plan. The gnathal surface is a modified endite lobe. Some members of all classes retain the transverse tendon on which the ventral extrinsic muscles originate. Figures 23-29 illustrate the variations of the common plan exhibited by the mandibles of various mandibulates. The muscles are labeled in accordance with those of Figures 8 through 22, indicating their homologies.

The segmentation of symphylid, chilopod, and diplopod mandibles can be interpreted as a secondary flexibility acquired in the coxal part of the appendage. The lack of a mandibular palp (telopodite) on these forms represents a suppression of that part of the appendage.

The subsegmentation of a coxa is allowed because in general, other podomeres of arthropods may be subsegmented. The most commonly subsegmented podomere is the tarsus, as was discussed above. The trochanter is divided in dragonflies, and the femur is basally subsegmented in some Hymenoptera (Snodgrass, 1935). Solpugid arachnids also bear subsegmented femora on their two last legs. The tibia is subsegmented in pselaphognathid millipedes (postfemur, Manton, 1956), and in the walking

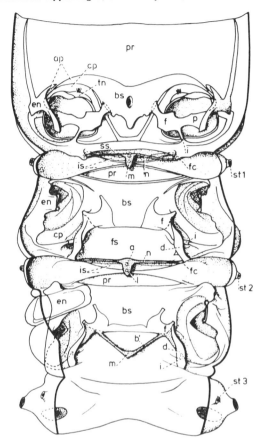

Figure 22. *Corydalus cornutus* (Insecta, Megaloptera). Dorsal view of ventral side of larval thorax, showing persistent fragments of ventral transverse tendons (m, n, b') at the intersegments associated with the spinae (a, 1). a, first spinal attachment; ap, anapleural arc; b', lateral arm of intersegmental transverse tendon of metathorax; bs, basisternite; cp, catapleural arc; d, postanapleural attachment to furca; en, endopleurite; f, cuticular furca; fc, furcilla; fs, furcasternite; i, pleural attachment; is, intersternite; l, second spinal attachment; n, lateral intersegmental tendon of pro- and mesothorax; p, prothoracic pleural process; pr, presternite; ss, spinasternite; st, spiracle; tn, trochantin. From Boudreaux, 1979.

legs of amblypygids (Millot, 1949a). The coxa is subsegmented in the middle and hind thoracic legs of Collembola (Manton, 1972). The insect maxilla is a large subsegmented coxa bearing a small telopodite. In *Limulus*, the coxae of the second, third, and fourth walking legs each bear a flexible endite lobe movable by an intrinsic coxal muscle. A flexible lacinia mobilis is also found on some mayfly and dipluran mandibles. There is no compelling reason to regard the segments of myriapod mandibles as podomeres.

The simplest segmented mandible is found in the Symphyla (Figure 23). The base is articulated with the head by a single dorsal ball and socket joint. The large coxa bears a flexibly articulated gnathal endite lobe. The chief flexor of the gnathal lobe consists of three parts of the dorsal promotor muscle (DPM) that pull on a cuticular apodeme of the gnathal lobe and on the forward edge of the mandible, causing a slight promotor roll. The ancestral dorsal remotor muscle was eliminated in Symphyla. Remotion is possible through muscles originating on the anterior tentorium and inserting on the proximal edge of the mandible (VRM). These appear to be muscles originally part of the ventral complex that have shifted their origin from the transverse tendon. The transverse tendon of the mandible extends between the tentorial arms, as in machilids (VTT-1). The ventral muscles remaining also originate on the tentorial apodeme and insert into the body of the mandible, acting as adductors (VRM). One segment of the promotor complex (VPM) has shifted its insertion to the inner border of the mandibular condyle, and acts as a weak direct abductor of the mandible. The main abduction (opening) action is provided by the movement of the swinging anterior tentorial arms (AT) pushing the mandibles apart.

The diplopod mandibular coxa (Figure 24) is subsegmented into two parts called cardo and stipes, by analogy with the insect maxilla, and the endite is

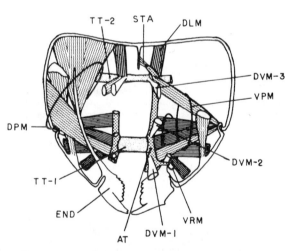

Figure 23. *Scutigerella immaculata* (Symphyla). Section through head showing mandibular musculature. DRM suppressed. Left DPM excluded to show other muscles. DVM-1 and DVM-2 operate swinging tentorium in abduction of MND. TT-1 is mandibular tendon and TT-2 is maxillary tendon. Upper sector of VPM may act as an abductor muscle of MND. For legends see p. 79. From Boudreaux, 1979.

articulated on the stipes. The former dorsal promotor muscle (DPM) is in two parts, one inserting on the forward edge of the stipes and the other inserting via a strong apodeme on the endite lobe. The ventral muscles (VRM, VPM) are in several parts which originate on the transverse tendon and insert on the cardo, stipes, and endite lobe. A pair of dorsoventral muscles (DVM-4) originate on the cranium and insert on the transverse tendon. As in the Symphyla, these muscles all provide a strong biting action to the mandible. As in the myriapods in general, opening of the jaw is accomplished by a pair of specialized swinging apodemes, the anterior tentorium, which are swung outward by several muscles (DVM-1, DVM-2) and which then push the mandibles open. The tentorium is retracted by DVM-3.

The jaw of centipedes is also flexible (Figure 25), and the main biting muscles are arranged roughly as in millipedes. The cranial muscles insert on the dorsal (DPM) and ventral (DRM) (morphologically, anterior and poster-

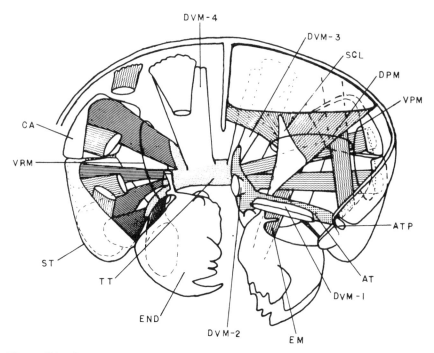

Figure 24. *Poratophilus punctatus* (Diplopoda, Juliformia). Cross section of head, anterior view. Promotor muscles of right side are not drawn in, to show remotor muscles. Most DPM and DRM operate as abductors. DVM-1 and DVM-2 insert on movable tentorium, and cause indirect abduction of mandibles. DVM-3 and DVM-4 aid in abduction. DRM suppressed. For legends see p. 79. From Boudreaux, 1979.

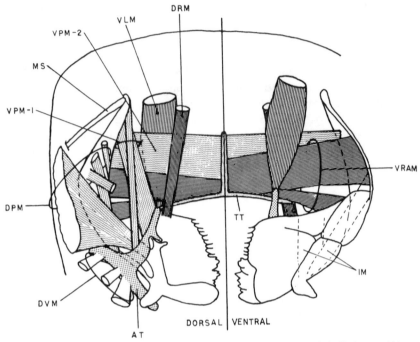

Figure 25. *Cormocephalus nitidosus* (Chilopoda, Scolopendromorpha). Right mandible and associated muscles from ventral and dorsal views. VPM-1 represents sectors that transferred insertions to tentorium, acting as retractors of tentorium, aided by VLM. Muscles DVM insert on tentorium instead of tendon, and aid in abduction. For legends see p. 79. From Boudreaux, 1979.

ior, since the jaw is "hanging" in a forward direction) edges of the distal part of the mandible, and ventral muscles (VRAM, VPM-2) originating on the transverse tendon insert inside the hollow mandible. Centipedes have acquired additional dorsal muscles (VPM-1) that help to protrude the mandible while it is being swung open by the movements of the tentorial apodemes, and slender segments of DPM help retract the mandible. The pauropod mandible is not segmented, and it usually is fragile, elongate, and pointed. But in some primitive pauropods (*Millitauropus*) the mandible is robust and bears a strong biting endite (Remy, 1950).

In most of the myriapod mandibles other than in Symphyla there is not a firm monocondylar articulation at the base of the mandible on the cranium. It is either a membranous broad union, as in millipedes, or the mandible is attached to an articular rod that permits protraction and retraction as in centipedes and pauropods. The myriapod mandible has changed its motion from the primitive promotor-remotor movement of the coxa about a roughly vertical axis to transverse adduction. Insects and crustaceans also have acquired a transverse biting action, but in these the originally vertical axis (Figures 26 and 29) of the promotor-remotor swing has changed to a tilted axis, (Figures 27, 28, and 30). The former promotor action has become the action of opening the mandibles (abduction), and the remotor movement has become the closing action (adduction). In both, a dorsal coxal articulation permits the rotation of the coxa against the head. In the more advanced crustaceans and insects, in each case apparently independently, an additional secondary condyle has developed on the formerly anterior lower border of the coxa (now essentially dorsal) to articulate with the head, and the transverse tendon was replaced by the tentorium (insects, Figures 30 and 31) or ventral head apodemes (isopods, Figure 28).

The entognathous insects are aberrant among insects in having an oral fold enclosing the mouthparts. The mandibular musculature, however, is derivable from the usual extrinsic coxal muscles. In the Collembola maxillary and mandibular transverse tendons have persisted, but some of the ventral muscles (VPM, Figure 32) have shifted their origin to the anterior tentorium.

In the chelicerates and trilobites, the gnathal appendages all bear a well developed leglike telopodite that is secondarily reduced in some arachnids (pedipalps). In the mandibulates, the gnathal telopodites are always reduced and often are suppressed. The mandibular and maxillary telopodites are not usually leglike, and that of the mandible has no more than three podomeres, as in the Crustacea. The retention of a small palp in crustacean mandibles is correlated with the usefulness of the naupliar appendages as specialized swimming organs—the antennules, antennae, and mandibles. Suppression of the gnathal telopodites has gone farthest in the myriapods, especially in millipedes, pauropods, and symphylans. A small palplike structure was retained on the gnathochilarium by pselaphognathid (palp-jaw) millipedes. At best, in other millipedes and in symphylans maxillary palps are reduced to pegs. Palps on the first and second maxillae are retained to varying extents by centipedes and insects. In centipedes the palps of the first maxilla are generally much reduced, and the coxae tend to be fused together.

From the above discussion, it is not necessary to suppose that mandibles of myriapods and insects are "whole-limbs" that bite with their tips. Rather, the structure and musculature are consistent with the idea that these are modified coxae, and that the telopodites have been suppressed. It is also apparent that in each group of mandibulates the further evolution of the mandible from a primitive coxa has occurred, each following its own evolutionary path (Boudreaux, 1979).

The question of whether the uniramous limbs of the Atelocerata (Uniramia, in part) became so with the advent of terrestrial life through the suppression of

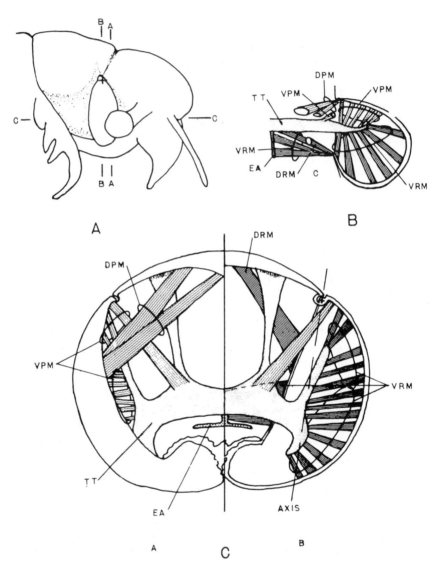

Figure 26. *Chirocephalus diaphanus* (Crustacea, Anostraca). A, lateral view of head showing levels of sections A. B. and C: B, right mandible as seen in section C-C. viewed dorsally; C, cross section through mandible, anterior view. At left, section A-A is near anterior border, while at right section B-B is through axis. Symbol + indicates primary dorsal articulation of rolling mandible in parts (A) and (C). For legends see p. 79. From Boudreaux, 1979.

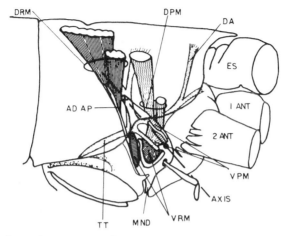

Figure 27. *Anaspides tasmaniae* (Crustacea, Anaspidacea). Lateral view of head as if transparent. Note inclined axis of mandible, with a secondary articulation ventrally indicated by solid black circle. For legends see p. 79. From Boudreaux, 1979.

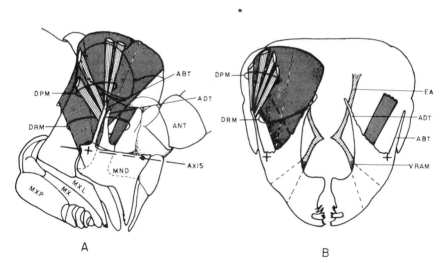

A   B

Figure 28. *Ligia oceanea* (Crustacea, Isopoda). A, head shown laterally as if transparent; mandibular axis is almost horizontal, and is dicondylic; B, posterior view of head in cross section posterior to mandibles; ventral muscles of MND are small, and insert on a sternal apodeme (EA) rather than on a ventral tendon, which is suppressed. For legends see p. 79. From Boudreaux, 1979.

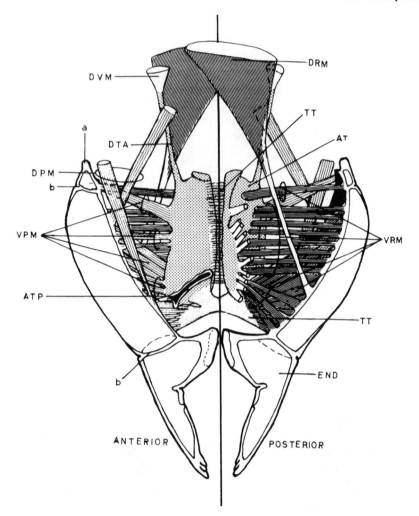

Figure 29. "Machilis" *strenua* (Insecta, Microcoryphia). Right mandible and associated structures in anterior and posterior views. There is only one mandibular condyle (a). Internal sclerotic
ridges (b) simulate cardino-stipital and stipito-lacinial unions. Remnants of VTT persist between
and below anterior tentorial arms (TT). Unshaded stumps on rear of tentorium at right are maxillary muscles. For legends see p. 79. From Boudreaux, 1979.

a presumed coxal epipodite, or whether there never were epipodites on the
limbs of these animals cannot be satisfactorily resolved. There is no doubt that
the earliest arthropods, such as trilobites and trilobitelike forms, did have
outgrowths on the coxa of all the legs, and that these epipodites were not

equivalent to the telopodite, but rather were specialized exite lobes. According to the Manton school, the uniramous legs of the Atelocerata never bore epipodites, and were modeled on the lobelike legs of an ancient onychophoranlike ancestral species. Yet, it appears that outgrowths, perhaps respiratory, did occur on the legs of ancient Onychophora, as in *Aysheaia pedunculata*. It is generally supposed that the booklike epipodite gills of aquatic chelicerates became enclosed, and evolved into the book lungs of modern terrestrial arachnids. On some of these, tracheae form in the invaginated book lungs. The only suggestion of an epipodite in the xiphosuran prosoma is the flabellum on the coxa of the last leg. However, all the legs of an Asian species of xiphosurid bear evanescent epipoditelike structures on the embryonic limb buds (Kishinouye, 1891), suggesting that epipodites have been suppressed in the prosomal legs of chelicerates.

In modern Crustacea, respiratory epipodites are commonly found on the coxae of thoracic legs, but in many crustaceans these are entirely suppressed, and the exopodite or endopodite of the reduced abdominal limbs have developed accessory respiratory structures in the form of filamentous outgrowths (isopods and stomatopods). The suppression of epipodites has occurred repeatedly in crustaceans, and therefore the suppression of

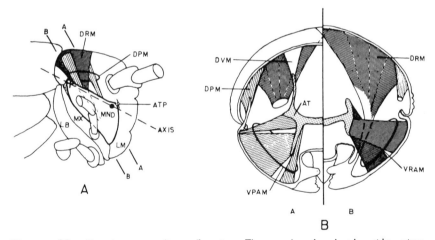

Figure 30. *Ctenolepisma ciliata* (Insecta, Thysanura). A, head, side view, represented as a transparent object; sections A-A and B-B are shown in part B; B, semi-vertical cross sections through mandible, with left side sectioned near front (A) and right side more posteriorly (B). There are no more head transverse tendons. Ventral muscles (VPAM, VRAM) originate on anterior tentorium (AT), and serve as adductor muscles, while dorsal muscles (DPM, DRM) serve as strong adductors and abductors. Mandibular axis is tilted nearly horizontally, and there is a secondary mandibular articulation shown by a solid black circle in part A. For legends see p. 79. From Boudreaux, 1979.

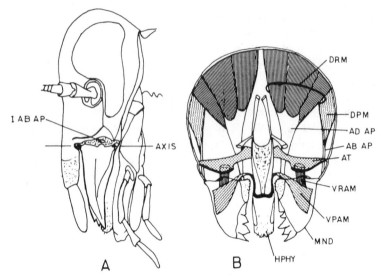

Figure 31. *Periplaneta americana* (Insecta, Blattaria). A, side view of head showing dicondylic horizontal mandibular axis and insertion of abductor muscle (I AB AP) which is converted from mandibular DPM; B, cross section through mandible. As in *Ctenolepisma*, ventral muscles (VRAM, VPAM) have shifted their origins to tentorium and hypopharynx, respectively, in absence of ventral transverse tendon, and are now adductors only. Huge DRM serves as main adductor. For legends see p. 79. From Boudreaux, 1979.

epipodites on the ancestral terrestrial atelocerate is a distinct possibility. The monophyletic theory of arthropod evolution requires that the atelocerates have evolved from an ancient unknown aquatic species that possessed a biramous limb in the form of a leg, bearing a respiratory epipodite.

If one looks for traces of vestiges of respiratory epipodites in terrestrial arthropods, the most suitable candidates are in the form of eversible thin-walled vesicles that can be everted by hemocoelic pressure and retracted by muscles. These are paired structures on the limbless abdomens of some insects such as the collophores of collembolans, abdominal coxal sacs of thysanurans, gill-bearing vesicles of megalopteran larvae. In the chelicerates, the Amblypygi (Pedipalpi) bear a pair of such eversible vesicles on the third opisthosomal venter (Millot, 1949a). Pycnogonid protonymphon larvae develop evanescent paired ventral epidermal invaginations which are lost at metamorphosis (Fage, 1949b). These develop opposite the ventral ganglia, as do the ventral organs of embryonic onychophorans and Symphyla (Tiegs, 1940). The symphylan embryonic ventral organs later develop into the paired coxal eversible sacs of the adult. Eversible coxal vesicles are known in some diplopods and chilopods, and on the venter of the collum segment of pauropods. If such eversible vesicles of terrestrial arthropods are homologous, it is only to the extent that perhaps the ancestral genetic code that caused initiation of exite lobe growth in aquatic chelicerates, trilobites, and crustaceans was retained in the terrestrial arthropods in a suppressed state, and later became derepressed in the lower insects, myriapods, and amblypygids. Additional independent mutations in each group were established by natural selection, providing for retraction of the thin-walled evaginations into a protected internal state for life on land. Their function in terrestrial species appears to be that of providing a means of regulating the internal water balance through transepidermal absorption from the environment.

The biramous limb as a fundamental arthropod type was imagined long ago when the annelid parapodium with its two lobes was envisioned as the beginning of the arthropod limb. Actually there is no such biramous limb in any arthropod, consisting of two identical branches each containing podomeres. All arthropod limbs are essentially uniramous. The crustacean biramous limb consists of the usual coxa-telopodite structure, with a specialized exite lobe (exopodite) borne by the trochanter (basipodite).

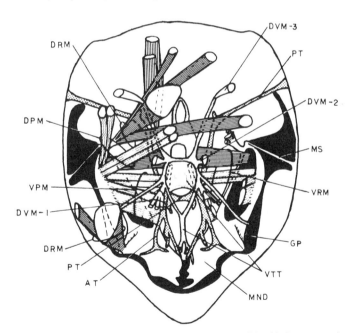

Figure 32. *Tomocerus longicornis* (Insecta, Collembola). Mandibular musculature from above. Some muscles on left of head omitted to reveal underlying mandibular muscles. Mandibular VPM origins shifted from mandibular VTT to anterior tentorium. DVM serve as suspensors of the VTT. VLM insert on transverse tendons VTT. For legends see p. 79. From Boudreaux, 1979.

COMPOUND EYES

The antiquity of arthropod compound eyes is well established. The earliest fossils of trilobites bore well developed compound eyes whose facets were close-packed as they are in modern crustacean and insect compound eyes (Hupe, 1953). Such close-packed ommatidia are characterized as holochroal. Later in the Devonian, especially in the phacopidoid trilobites, the ommatidia became fewer and farther apart, forming schizochroal eyes similar to the poorly developed compound (or agglomerate) eyes of some myriapods and lower insects, in which facets are not closely packed. Several other superfamilies included eyeless forms. A number of mid-Cambrian and later trilobitomorph crustaceanlike fossil forms possessed well developed compound eyes: *Cheloniellon, Burgessia, Waptia, Hymenocaris.*

Compound eyes were characteristic of the early chelicerates also, but they have been retained only in the Recent Xiphosurida. Xiphosurid compound eyes are intermediate between the holochroal eye in which lenses and ommatidia are closely packed, and schizochroal eyes in which lenses are ommatidia are separated by some space. Their lenses are not separated. The cuticle is smooth over the eyes and forms lenslike internal bulges over each ommatidium. The ommatidia are simple and separated from one another. The xiphosurid compound eye is here regarded as being derived from the primitive holochroal ancestral arthropod eye.

The compound eye of modern *Limulus* seems to be almost useless as an organ of vision. Rather than producing useful images of objects, the *Limulus* eye appears to be useful mainly as a means of distinguishing degrees of illumination. This is borne out by the simple structure of the individual visual units and by the behavior of a captive *Limulus polyphemus.* The specimen I observed spent most of its time buried in the sand, moving about by burrowing. It emerged infrequently, swimming in an inverted position in an aimless fashion, bumping into pieces of coral and shell in the aquarium. Most of the swimming activity occurred at night, when there would be minimal sight of objects. When it came out and swam during the daylight, it would recognize food in the form of worms or fragments of oyster or shrimp only if the legs came in contact with the food. It appears, then that modern *Limulus* makes almost no use of its visual apparatus to see objects, and can find food by recognizing its prey using other senses. The presence of large nonfunctional compound eyes in *Limulus* is explainable on the basis that this is a burrowing animal that has retained compound eyes largely because mutations for their suppression have not taken place. The limulid compound eye has no selective disadvantage in a burrowing animal, and may have its best use in conjunction with the habit of swimming at night, when predators with vision would not be likely to find *Limulus* as prey. But distinguishing night from day is possible with much fewer visual units than found in *Limulus*. Here is a case where "disuse" has not brought about suppression, but has permitted some simplification.

The extreme correspondence in most details between the compound eyes of malacostracan crustaceans and of insects (including the primitive machilids) is "one of the most disconcerting problems of arthropod phylogeny" (Tiegs and Manton, 1958) to those who maintain the separate origins of these two groups and thereby admit of extremely detailed convergence of these organs. In spite of contrary evidence, Tiegs and Manton regarded as most improbable that various simplifications of compound eyes could have come about, as appears to be the case in trilobites, arachnids, myriapods, and insects. The lower Devonian myriapod, *Kampecaris forfarensis* and the Carboniferous *Pleurojulus levis* possessed well developed compound eyes. Of recent myriapods, only the scutigeromorph centipedes have retained compound eyes. Whether compound eyes are fully developed or reduced to only a few facets, the visual center is always in the protocerebrum of the brain. The viewpoint of the monophyletic nature of arthropods also assumes extreme convergence, but a convergence in the trend towards simplification of the compound eye correlated either with burrowing as in millipedes and xiphosurids, or with very small body size as in the small crustacea and apterygote insects or with less dependence on vision, as in web-spinning spiders and most other arachnids.

THE DEUTOCEREBRUM AS AN ARTHROPOD STRUCTURE

Wherever antennae occur in modern arthropods, their nerve center is a specialized part of the brain called the deutocerebrum. The deutocerebral ganglia are found not only in arthropods, but also in Onychophora, in which these innervate the antennal palps. Even in *Limulus*, a pair of small deutocerebral centers connected by a preoral commissure can be found in the brain ahead of the postoral cheliceral ganglia (Fage, 1949a). The remnants of a deutocerebrum in a chelicerate imply that the ancestral chelicerate probably bore antennae which were suppressed early in the evolution of this group.

CIRCULAR BODY WALL MUSCLES

Among the lobopods, only the Onychophora have retained and even elaborated the circular muscles occurring just beneath the epidermis. It has been supposed (Snodgrass, 1948b) that the dorsoventral muscles of arthropods are remnants of primitive circular muscles. However, in primitive arthropods that have a well developed intersegmental tendon system, the dorsoventral muscles are typically attached to the intersegmental tendons at the intersegmental lines. Further, dorsoventral muscles are present in the Onychophora, in which there are also circular muscles, and in annelids the intersegmental septa contain some dorsoventrally oriented muscle fibers. The arthropod dorsoventral muscles then probably are more nearly homologous with septal muscles than with circular body wall muscles. The circular muscles seem to have become entirely suppressed in an early proarthropod ancestor, and the absence of circular muscles is another monophyletic character of arthropods.

THE LATERAL TERGAL FOLDS (PARANOTA)

If the polyphyletic theory of arthropod evolution is valid, lateral tergal folds in the form of canopies overlying the ventrolateral limbs have originated separately in each of several arthropod lines. These are to be found in fossil and recent Xiphosurida. In the fossil xiphosurids, paranota were apparent on the still separate terga of the opisthosoma. In *Limulus,* all the terga, prosomal, and opisthosomal, fail to separate and remain as two dorsal plates which are expanded laterally. These lateral expansions are commonly homologized with unseparated paranota. Another line, the trilobites and their trilobitoid relatives, clearly possessed paranota. The main body portion is usually called the rachis, and the lateral paranota are called pleura. The apparent separation of the body into a central and two lateral longitudinal lobes was the basis for naming these trilobites. It is now known that only the rachis represents the dorsal main part of the trunk. The trilobite pleura are not equivalent to the lateral surface of the body of myriapods and insects, also called pleura.

Paranota are found also in a third evolutionary line, the Crustacea. They are typically present on all the trunk segments of isopods. In most Crustacea, they have been retained variously in restricted parts of the trunk. In decapods, paranota are restricted to the abdominal segments. In the nonmalacostran crustaceans, paranota may be essentially absent such as in barnacles, branchipods, and copepods. Some fossil crustaceans bore many paranota, such as in *Lepidocaris rhyniensis* Scourfield of the mid-Devonian. Thoracic paranota are not common in modern Malacostraca, but in a Carboniferous anaspidacean, *Pleuronectes annulatus* Calman (Roger, 1953) there appear to be distinct thoracic paranota. Recent anaspidaceans have separate thoracic segments, but there are no paranota except on the gnathal segments and the abdomen (see Figure 43, p. 101).

A fourth evolutionary line terminating in peculiar crustaceanlike forms was the Cheloniellida. The only known representative is the lower Devonian *Cheloniellon calmani* Broili, whose only tagmosis was the union of the acron with the segment of the second antennae, such as is found in most Crustacea, and the union of the next two metameres. The paranota were quite extended laterally, so that the animal appeared as broad as long, although the main body of about 13 metameres was somewhat slender (see Figure 41, p. 97).

A fifth evolutionary line with paranota is represented by the carboniferous myriapodlike giant Arthropleurida (see Figure 44, p. 103). These apparently terrestrial animals are known mostly from trunk segments, but each segment from above was trilobitelike in that lateral paranota extended over the legs of each segment (Rolfe and Ingham, 1967).

Among the myriapods, paranota are weakly or not at all developed in pauropods, symphylans, centipedes, and juliform millipedes. Most other millipedes bear recognizable paranota, and the Carboniferous *Pleurojulus levis* Fritsch, a fossil millipede, bore paranota clearly marked from the body as they are in *Arthropleura*.

The insects are a seventh evolutionary line in which paranota may or may not be found. They are essentially absent in Collembola, Diplura, and Protura, although the thoracic terga of the first two weakly protrude beyond the pleura. In the Thysanura, Microcoryphia, and Monura, paraterga occur on the trunk segments. The fossil Paleodictyoptera and related insects also bore thoracic and abdominal paranota, but those on the meso- and metathorax were expanded into wings, as they are in all the pterygote insects which have retained wings. Prothoracic paranota still occur in modern insects, especially in the Orthoptera where these cover the prothoracic leg bases.

Proponents of the monophyletic viewpoint regard the widespread occurence of paranota among arthropods as an attribute whose origin occurred in an unknown ancestral arthropod. In some cases, tergal paranota seem to have been suppressed in some lines, and then later derepressed. Among the Acarina, the absence of paranota is an inheritance from the ancestral arachnid mutation which caused their suppression, probably before the differentiation of arachnids into their respective orders. Among the oribatid mites, several families appear to have redeveloped paranotumlike structures independently. Prothoracic paranota seem to have been redeveloped among some tropical mantids, and in the insect family Peloridiidae of the Homoptera. Such cases can be interpreted as instances of reverse mutations, or as derepressions of a formerly repressed but still persistent genetic code whose expression is in the form of paranota, and whose origin was established in the first arthropod. The process of evolution has resulted in all the variations in size and structure of paranota that occur in arthropods in general.

CEPHALIC TAGMOSIS

All known arthropods (except the neotinic crustacean *Derocheilocaris*) bear an anterior cephalic tagma formed from the union of at least one undoubted metamere with the primarily sensory acron. In *Derocheilocaris*, the first two naupliar stages bear an unsegmented head tagma including the antennae, antennules, and mandibles fully developed. The third stage first exhibits the separation of the acronal antennular lobe from the head, which at this state bears the antennae, mandibles, maxillules, and maxillae. The head is unsegmented (Deboutteville, 1954). An acronal head-lobe entirely separate from all postoral metameres was postulated for the enigmatic fossil *Opabinia regalis* Walcott, from the mid-Cambrian (Sharov, 1966a). The state of preservation of this fossil does not reveal the nature of the head, and Sharov's reconstruction cannot be trusted because it is based on Walcott's retouched photographs and a great deal of speculation. Hutchinson (1930) restudied the available fossils of *Opabinia*, and interpreted the head to include the first metamere united with the acron. He thought he could see traces of a tiny antennule, and the base of the second antenna on a head that bears large stalked compound eyes. The "proboscis" of Sharov, or frontal process, is suggestive of the male copulatory organ of some branchiopods, in which the endopodites of the second antennae are partially fused in the male. Whittington (1975) did not consider *Opabinia* to be either a trilobitid or an annelid. He showed that the cuticle probably was not heavily sclerotized, that there were five stalked compound eyes, that the gills were above the lateral body flaps, and that there were no antennae or legs discernible. The true nature of *Opabinia* remains unknown.

Cephalic tagmosis in the Chelicerata has resulted from the absence of separation of six or seven metameres, best displayed in the Xiphosurida and Eurypterida, in which six or seven pairs of appendages are evident. The first pair, the chelicerae, are always at most three-segmented and are not locomotory. Their ganglia are the first of the series of postoral nerve centers whose transverse commissures lie behind the mouth, so they are evidently homologous with the crustacean second antennae. The cephalic tagma of chelicerates is known as the prosoma. It is functionally a cephalothorax rather than a head, because the appendages are primarily locomotor, although in *Limulus* the second through the sixth are at the same time gnathal appendages with well developed gnathobases used in feeding. Tagmosis of the prosoma in the Pycnogonida appears to have come about beginning with the fusion of the acron with the cheliceral, pedipalpal, oviger, and first leg somites, forming a sort of head (Figure 34). The somites of legs 5, 6, and 7 in most pycnogonids are largely unfused, although together with the head they form a distinct prosoma. Fusion of the prosomal segments into a single unsegmented unit occurs in a few advanced species, suggesting that prosomal fusion in pycnogonids was in parallel with this process in other chelicerates. In all other arthropods at least the first three postoral metameres bear appendages which are not primarily locomotory. The first is never locomotory (except in trilobites) and is either in the form of an antenna (second antenna) typical of Crustacea, or is suppressed as in the Atelocerata. In trilobites the first apparent postoral appendage appears to be leglike, and bears a strong gnathobase as do the other two pairs of head appendages, all of which are on the head tagma consisting of three postoral metameres united with the acron (Cisne, 1974).

Considering that the first metamere is united with the acron, and that a variable number of gnathal metameres joins this protocephalon in different evolutionary lines of mandibulates, the formation of a protocephalon seems to be the ancestral state in the original arthropod. The union of gnathal segments with the protocephalon to form a head capsule independently evolved in several arthropod evolutionary lines, such as chelicerates, trilobites, and mandibulates, and within the mandibulates perhaps also independently in the Crustacea and Atelocerata. A version of the evolution of the various arthropod head types is discussed in Chapter 5, based on phylogenetic principles discussed in Chapter 2.

Legends Used in Figures 8 Through 32

AB AP	Abductor apodeme of MND
AD AP	Adductor apodeme of MND
ANT	Antenna
AP	Sternal apodeme
AT	Anterior tentorial arm
ATP	Anterior tentorial pit
CA	Cardo of MND
CR	Cuticular rod of transverse tendon
CX	Coxa
CX AP	Coxal apodeme
CXC	Coxal cavity
DA	Dorsal arm of transverse tendon
DBV	Dorsal blood vessel

DDM	Dorsal diaphragm muscle
DLM	Dorsal longitudinal muscle
DPM	Dorsal promotor muscle
DRM	Dorsal remotor muscle
DTLP	Depressor muscle of telopodite
DTT	Dorsal transverse tendon
DVM	Dorsoventral muscle
EA	Endosternal apodeme
EM	Adductor muscle of endite (segment of DPM)
END	Endite lobe of coxa
ES	Eye stalk
EXT	Exite lobe of coxa
GP	Gnathal pouch
H	Heart
HG	Hind gut
HPHY	Hypopharynx
I AB Ap	Insertion of abductor apodeme of MND
IM	Intrinsic muscle of MND
L	Leg base
LB	Labium
LM	Labrum
LT	Longitudinal ventral tendon
LTLP	Levator muscle of telopodite
MND	Mandible
MS	Mandibular suspensory rod
MX	Maxilla
MXL	Maxillule
MXP	Maxilliped
NC	Nerve cord
ODVM	Oblique dorsoventral muscle
OVLM	Oblique ventral longitudinal muscle
P	Pleurite
PM	Promotor muscles
PNP	Paranotal process
PT	Posterior tentorial arm
RM	Remotor muscles
SCL	Adductor sclerite of MND
SKS	Skeletal strut
SPM	Sternopleural segment of DVM
STA	Suspensory Tendon of tentorium
ST	Stipes of MND
STT	Sternal fragment of VTT
T	Trunk terga
TENT	Anterior tentorium

TLP	Telepodite
TRM	Trochanteral muscles
TT	Transverse tendon
VDVM	Vertical dorsoventral muscle
VLM	Ventral longitudinal muscle
VPAM	Ventral promotor-adductor muscle
VPM	Ventral promotor muscle
VRAM	Ventral remotor-adductor muscle
VRM	Ventral remotor muscle
VTT	Ventral transverse tendon
VVM	Vestigial ventral muscles of former appendages

CHAPTER 5

Phylogeny of the Arthropod Classes

In this chapter a classification is proposed that is constructed to avoid paraphyletic or polyphyletic taxa. Each taxon represents a line in which ancestral features are assumed to have been modified so as to produce one of two sister groups. Each group is characterized by changes that are presumed to have occurred within the group, independently of changes that occurred in the companion group. The actual ancestral species of each line remains unknown, and the presumed nonsurviving radiations that probably evolved along any single line also remain largely unknown. The phylogenetic classification can be represented as an outline, or as a family tree. Either representation depicts the minimum number of branching sequences. The family tree (Figure 33) covers down to the class categories, and nothing is implied other than deduced splits of the various lines. All splits at the same level are taxonomic, representing the same taxonomic rank. The family tree or the outline makes no attempt to indicate probable ages or geological time of origin of the splits, and the degree of difference between groups is not graphically indicated. New taxa proposed by Boudreaux (1979) are indicated in the outline in italics. No attempt was made to correlate name endings with rank. The descending ranks of divisions of common categories are indicated, respectively, by the prefixes *sub-*, *infra-* and *subter-*. The prefix *subter-* is introduced to avoid creating a new category, or to avoid employing little-used categories such as division, legion, phalanx, series, and so on whose categorical ranks are either not apparent, or the terms are used at various rank levels by different authors. The phylum Onychophora is listed as a separate taxon for the reasons discussed in Chapter 3.

    I. Phylum Onychophora Grube, 1853.
   II. Phylum Arthropoda Siebold and Stannius, 1848.

A. Subphylum *Cheliceromorpha* Boudreaux, 1979.
  1. Infraphylum Pycnogonida Latreille, 1810.
  2. Infraphylum Chelicerata Heymons, 1901.
    (1) Superclass Xiphosurida Latreille, 1802.
    (2) Superclass *Cryptopneustida* Boudreaux, 1979.
      (a) Class Eurypterida Burmeister, 1843.
      (b) Class Arachnida Latreille, 1810.
B. Subphylum *Gnathomorpha* Boudreaux, 1979.
  1. Infraphylum Trilobitomorpha Størmer, 1944.
    (a) Class Trilobita Walch, 1771.
    (b) Class Trilobitodea Størmer, 1959.
  2. Infraphylum Mandibulata Snodgrass, 1938 [not Mandibulata Clairville, 1798].
    a. Subterphylum *Crustaciformia* Boudreaux, 1979.
    (a) Class Cheloniellida Broili, 1933.
    (b) Class Crustacea Pennant, 1777.
    b. Subterphylum *Myriapodomorpha* Boudreaux, 1979.
    (1) Superclass Arthropleurida Waterlot, 1934.
    (2) Superclass Atelocerata Heymons, 1901.
      (a) Class Myriapoda Leach, 1814.
        i. Subclass *Collifera* Boudreaux, 1979.
          (i) Infraclass Pauropoda Lubbock, 1866.
          (ii) Infraclass Diplopoda Blainville-Gervais, 1844.
        ii. Subclass *Atelopoda* Boudreaux, 1979.
          (i) Infraclass Symphyla Ryder, 1880.
          (ii) Infraclass Chilopoda Leach, 1814.
      (b) Class Insecta Linnaeus, 1758.

SUBPHYLUM CHELICEROMORPHA

There are several advanced features that are shared by the members of this group, which seem not to have evolved in precisely the same fashion in the Gnathomorpha. The suppression of acronal antennules was accompanied by extreme reduction of the deutocerebrum. The appendages of the first metamere were reduced to typical three-segmented organs known as chelicerae. The chelicerae are appendages of the first metamere because their origin is on the first postoral segment, and their ganglia are joined by the first postoral commissure. This pair of ganglia is homologous with the tritocerebrum of mandibulates.

Respiratory epipodites were probably borne on all postcheliceral appen-

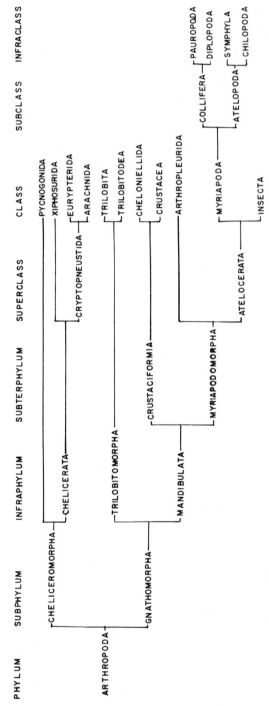

Figure 33 Cladogram illustrating proposed phylogenetic classification of the Arthropoda.

dages, but these are not known in any fossil or recent form on any leg of the first five metameres. A trace of an epipodite on leg 5 (sixth metamere) remains on the coxa of Limulus as the flabellum, and perhaps the chilaria of metamere 7 are persistent vestiges of epipodites. Transient rudiments of epipodites were described on the five prosomal legs in the embryos of the xiphosurid Tachypleus tridentatus by Kishinouye (1891), but they persist in the adult only on the last legs of the prosoma.

In this group, another derived character not found elsewhere is the restriction of typical locomotor appendages to the prosoma, a tagma consisting of the first six or seven metameres. In this line initial anterior tagmosis, consisting of the solid union of the first four metameres with the acron, was retained only in Pycnogonida. All others developed a solid prosomal tagma without intersegmental grooves or flexibility. The posterior tagma, or opisthosoma, never bears walking legs. Vestiges only of the telopodites of the first six opisthosomal appendages (metameres 8 through 13) were retained by xiphosurids. Another characteristic of the cheliceromorphs is intracellular digestion of some or most of the food, in vacuoles of the epithelial cells of the midintestine. The process occurs mostly in the diverticula. This peculiarity of digestion is modified in most terrestrial arachnids in that secretory epithelial cells produce digestive enzymes that partly liquify food that has been taken in before it is passed to the diverticula for final digestion. Assimilated digestive products generally are stored as lipid or other substances in interstitial midgut cells. In pycnogonids and some acarina the epithelial cells containing digestive vacuoles become detached and fall into the lumen of the diverticulum. This group of arthropods has retained a characteristic patella in some of their walking legs.

Probably before the Paleozoic this group became separated into two sister groups, represented by the Pycnogonida as one division and the Chelicerata as the other. Pycnogonids bear derived features of their own, and have apparently retained some ancestral characters. A characteristic of Recent species is the almost total suppression of the opisthosoma. The young hatch into a larva (protonymphon) bearing the chelicerae, palps, and ovigerous legs, each with their ventral ganglia, plus an incipient fourth segment. With further molts, new segments form teloblastically, but segment generation stops on completion of the prosoma. In some species vestiges of opisthosomal segments, in the form of one or two tiny pairs of transient embryonic ganglia, briefly exist. The opisthosoma therefore at most may include the telson and two undeveloped segments. It is always small in Recent species, but in the Devonian fossil Palaeoisopus problematicus Broili (Figure 34), the opisthosoma was narrow, consisting of four legless segments and a telson (Lehman, 1959).

The pycnogonid prosoma typically includes an unsegmented cephalic

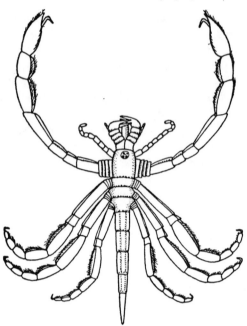

Figure 34   Reconstruction of Devonian *Palaeoisopus problematicus* Broili, a pycnogonid with a segmented opisthosoma. From Sharov: *Basic Arthropodan Stock*, 1966, with permission of Pergamon Press, Ltd.

tagma composed of the acron plus four metameres and a segmented portion consisting of three metameres. The first metamere bears the chelicerae, the second limb is modified into a palp (pedipalp) and the third is a slender leglike oviger, which the males employ for holding the eggs until hatching. The fourth somite bears the first walking legs. The usual three following pairs of walking legs are on a series of three articulated segments, but species in several families appear independently to have acquired the ability to generate extra prosomatic segments, and some end up with five, others with six pairs of walking legs. The ancestral state of the locomotor section of the prosoma is generally thought to be segmented and flexible. But the total fusion of the entire prosoma appears in a few species of different families suggesting independent loss of somatic joints (Figure 35). Another specialization of the pycnogonids is the elongate proboscis extending forward on most species. The proboscis has been compared with the polychaete eversible pharnyx, but there is no evidence that the proboscis is anything other than the elongated acron.

   If endites and exites were original in the ancestral arthropod, these are entirely suppressed in pycnogonids. Also suppressed are the compound eyes,

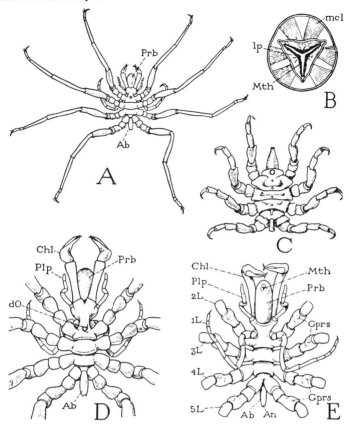

Figure 35   Pycnogonida. A, *Nymphon hirtipes* Bell, female, dorsal; B, same, cross section of proboscis near mouth; C, *Pycnogonum littorale* Strom., female, dorsal; D, *Nymphon hirtipes*, female, body and legs, dorsal; E, same ventral. Ab, abdomen; An, anus; Chl, chelicera; dO, dorsal ocelli; Gprs, gonopores; L, leg; lp, lip; mcl, muscle; Mth, mouth; Plp, palpus; Prb, proboscis. Reprinted from R.E. Snodgrass: *A Textbook of Arthropod Anatomy*. Copyright © 1952 by Cornell University. Used by permission of Cornell University Press.

paranotal lobes, and the intersegmental tendons. An apparently ancestral character is the presence of openings of the gonads on the trochanter of each walking leg, and another is the typical presence of well developed appendages on all seven prosomatic segments.

The infraphylum Chelicerata is the sister group of pycnogonids. The derived characters in this group are fairly typical. In all chelicerates the prosoma is usually an unsegmented tagma consisting of the first six or seven metameres. The gonopores are on the coxae (or equivalent) of the appendages of the eighth metamere. The ventral prosomal intersegmental tendons have united to form a tendinous endosternum, on which the extrinsic limb muscles originate. The endosternum is suspended over the ventral prosomal nervous system by persistent dorsoventral muscles. Opisthosomal transverse tendons join the lateral ventral muscles, and lie below the nerve cord. Although in ticks and xiphosurids there is some quinone bonding in the tanned cuticle, there is additionally a method of bonding involving sulfhydryl groups of cysteine, which is the typical tanning method in the cuticle of the terrestrial arachnids. No chelicerate is known bearing appendages on the last five segments of the 11 or 12-segmented opisthosoma, in those orders in which the full segmentation persisted.

The Chelicerata evolved into two sister groups. One of these is the superclass Xiphosurida, the other is the superclass Cryptopneustida. The xiphosurids are represented in the oldest Cambrian rocks by the Aglaspids, which bore an 11-segmented opisthosoma. Present species do not develop more than six opisthosomal segments, all united into a solid tagma (Figure 36). The appendages of the first opisthosomal segment (metamere 8) are broad and flat, and bear the gonopores. The next five segments bear limbs modified into flat plates, the epipodites of which are in the form of stacked lamellae, serving as gills. The flat opisthosomal appendages are powerful swimming organs, and also serve to flush suspended debris stirred up by the legs during burrowing. These free appendages contrast to the essentially non-movable gill or book lung covers of the Cryptopneustida. The prosoma of xiphosurids includes seven anterior metameres and a portion of the eighth, or genital, segment. The hinge between the prosoma and the opisthosoma is within the eighth segment.

The xiphosurids appear not to have evolved a subsegmented tarsus, and there is no second trochanter in any leg. The ancestral paranota together with the forward and laterally extended head shield fused into a structure similar to the head shield of trilobites. They have not firmly articulated their legs with the body. The coxae are freely movable in all directions, including motions of protrusion and retraction, a probable ancestral state.

The superclass Cryptopneustida is so named because the respiratory lamellae are enclosed by vestiges of opisthosomal coxae that form fixed gill covers in the extinct eurypterids, and in the terrestrial arachnids the respiratory lamellae have become sunken into respiratory atria in the form of "book lungs." The seventh metamere is devoid of any trace of appendages and the seventh sternum generally is lacking. This, the pregenital segment, is part of the prosoma of xiphosurids and bears short vestigial appendages, the chilaria. In the cryptopneustids, the tergum of the seventh somite is large and seems to form the first part of the opisthosoma in eurypterids, but is reduced in most arachnids to form a narrow pedicel, or appears only in the embryo of scorpions.

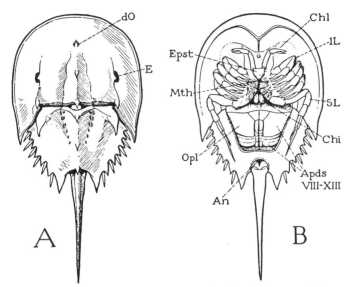

Figure 36 *Limulus polyphemus* L., Xiphosurida. A, dorsal; B, ventral; An, anus; Apds, opisthosomatic appendages; Chi, chilarium; dO, simple dorsal eye; E, lateral compound eye; Epst, epistome; L, leg; Mth, mouth; Opl, operculum. Reprinted from R.E. Snodgrass; *A Textbook of Arthropod Anatomy. Copyright © 1952 by Cornell University, Used by permission of Cornell University Press.*

The paranota have mostly been suppressed in this group. Subsegmentation of the tarsus has occurred, probably in parallel with pycnogonids, since tarsal segmentation is unknown in xiphosurids. The coxae of the cryptopneustids are more firmly fixed to the body, and in modern forms are barely movable.

It has been customary to combine the xiphosurids with the eurypterids as the class Merostomata, because in both the mouth is presumably displaced rearward to lie among the leg-bases. But in the eurypterids, as in scorpions, it is rather that the coxal endite lobes converge forward toward the mouth, and although the mouth may be near several coxal lobes, it is not in a real rear position as it is in xiphosurids. As discussed above, the eurypterids share derived characters with arachnids that are not found in xiphosurids. This is consistent with the belief of arachnologists that the first land animals, the scorpions, evolved from a eurypteridlike stock.

The Cryptopneustida includes two sister groups, the Eurypterida and the Arachnida. Eurypterids have only a few of their own derived characters, such as the enlargement and flattening of the last legs into swimming paddles and the generally gradually increasing length of successive legs from front to rear.

Some species have secondarily enlarged second legs, as in the scorpionlike *Mixopterus kiaeri* Størmer (Figure 37). The eurypterids have retained some ancestral characters, such as the compound eyes, the fully segmented opisthosoma, a small "doublure," or forward and lateral expansion of the prosoma, and a nearly normal tergum of the seventh metamere (first opisthosomal).

The class Arachnida includes all the terrestrial air-breathing chelicerates (Figure 38). The ancestral book gills are enclosed in respiratory pockets, or atria, and serve as lungs. Their metameric distribution varies in the various orders, from absence of book lungs in several orders to four pairs in scorpions. Some species in several orders have secondarily acquired tracheae as respiratory organs. Compound eyes are entirely suppressed in arachnids, and the tarsus is mostly subsegmented. The cuticle is tanned through sulfur bonding in addition to phenolic bonding, by linkages between cysteine units in the protein chains. A partial change toward adoption of sulfur tanning occurred in the xiphosurids, but arachnids have not eliminated the phenolic tanning process (Neville, 1975). The reduction of the seventh metamere to a narrow structure between the prosoma and opisthosoma is typical of arachnids, and the seventh metamere occurs only in the embryo, in scorpions. Another specialization of the arachnids is the production of the highly insoluble guanine as a final nitrogenous excretory product, in midintestinal interstitial cells and "malpighian tubules" formed from the midintestine but applied close to the cells surrounding the caeca. The various coxal glands, which seem to develop from rudimentary embryonic coelomoducts, generally excrete urate nitrogen and guanine.

Many arachnids enclose the semen in a spermatophore, which is inserted into various types of seminal receptacles of the female. Since the xiphosurids and pycnogonids produce fluid semen, and the eggs are fertilized externally, it is most probable that the arachnid spermatophore evolved as a convergent evolutionary process with spermatophore formation in Onychophora and mandibulate arthropods.

SUBPHYLUM GNATHOMORPHA

This subphylum includes the descendants of an ancestral species in which the main advanced character was a functional separation of cephalic functions from locomotor functions. The cephalic functions may be divided into sensory and feeding functions. In the chelicerates the cephalic and locomotor functions are performed on a specialized tagma, the prosoma. This prosoma is the best candidate for bearing the name cephalothorax, if thorax is a term for a specialized locomotor tagma. The so-called cephalothorax of crustaceans is not comparable to that of chelicerates. In many crustaceans, the

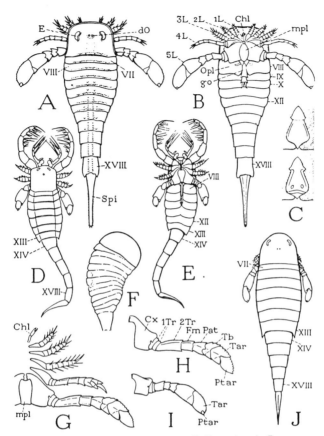

Figure 37 Eurypterida. A, *Eurypterus remipes* DeKay, dorsal; B, same, ventral; C, *Pterygotus rhenaniae* Jaekel, opercular appendage of female, dorsal and ventral; D, *Mixopterus kiaeri* Størmer, dorsal; E, same, ventral; F, *Strabops thatcheri* Beecher, Upper Carboniferous; G, *Dolichopterus macrochirus* Hall, left appendages, ventral; H, Same hind leg; I, *Eurypterus remipes*, hind leg; J. *Hughmilleria norvegica* (Kiaer), dorsal. Roman numerals indicate body segments beginning with cheliceral segment. Chl, chelicera; Cx, coxa; dO dorsal ocellus; E, Compound eye; Fm femur; go, genital organ; L, leg; mpl, metastomal plate; Opl, operculum; Pat, patella; Ptar, pretarsus; Spi, tail spine; Tar, tarsus; Tb, tibia; 1Tr, first trochanter; tr, second trochanter. Reprinted from R.E. Snodgrass: *A Textbook of Arthropod Anatomy.* Copyright © 1952 by Cornell University. Used by permission of Cornell University Press.

posterior and lateral edges of the maxillary segment grow backward to form a carapace over the functional thorax, giving the impression that there is a single cephalothoracic tagma (Snodgrass, 1958). The first ancestral change toward tagmosis in the arthropods may have been the union of the first metamere with the acron, as is suggested by the mid-Cambrian genera *Cheloniellon* and perhaps *Opabinia*. The protocephalon composed of the acron and the first metamere was retained, though variously modified, in the Crustacea.

Some time before the Cambrian the gnathomorph line split into two sister groups, the Trilobitomorpha and the Mandibulata. There appeared to be very little initial specialization in the trilobitomorph line. One such was the union of two more metameres with the protocephalon to form a head, on which three pairs of metameric appendages differed from the trunk appendages primarily

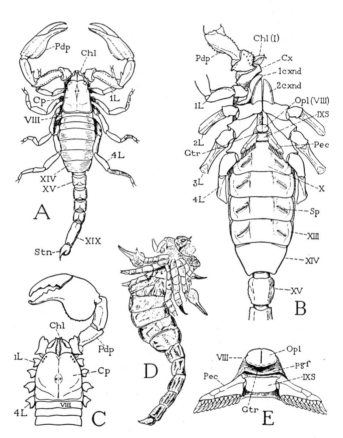

FIGURE 38   Scorpions, Arachnida. A, *Chactus vanbenedeni* Gervais, Chactidae, Colombia; B, *Pandinus* sp., Scorpionidae, Congo, ventral surface; C, same, dorsal; D, *Palaeophonus hunteri* Pocock, Scottish Silurian; E, *Pandinus sp.*, genital region of segment VIII,and pectine of segment IX, ventral. Roman numerals indicate body segments beginning with cheliceral segment. Chl, chelicera; Cp, carapace; Cx, coxa; cxnd, coxal endites; Gtr, gonotreme; L, leg; Opl, operculum; Pdp, pedipalp; Pec, pecten; pgf, pregenital fold; S, sternum; Sp, spiracle; Stn, sting. reprinted from R. E. Snodgrass: *A Textbook of Arthropod Anatomy.* Copyright © 1952 by Cornell University. Used by permission of Cornell University Press.

in that the gnathobases on the coxae were larger on the head than on the trunk (Cisne, 1974). These three pairs of head appendages were probably the gnathal appendages. In addition to the three pairs of gnathal appendages, the acronal antennules generally became well developed into long filamentous structures. The ancestral trilobitomorph probably was essentially homonomously segmented on the trunk, bearing typical walking legs, each of which bore a lateral coxal respiratory epipodite.

It was probably a pretrilobitomorph that gave rise to the chelicerate line. It is commonplace to suppose that trilobites were the ancestors of chelicerates. This supposition is purportedly indicated by the superficial similarity between the broad head of trilobites with the similarly broad prosoma of xiphosurids, and by the hatching of limulids into a so-called trilobite larva. The resemblances are only superficial. The xiphosurid prosoma from the beginning seemed to consist of at least 7 metameres (Cambrian aglaspids) followed by 11 trunk segments, with fully developed limbs only on the prosoma. The trilobites at no time exhibited any tendency to use head appendages as the primary locomotor organs, or to eliminate locomotor appendages on any part of the trunk.

The newly hatched trilobite was a true larva, metamorphosing gradually by teloblastic generation of new metameres at molting time. The newly hatched trilobite consisted of the future head, already a single tagma, plus the periproctal zone of generation of future segments (telson). The newly hatched trilobite larva of xiphosurids issues from the egg as a fully segmented individual able to swim upside down as do all other instars, employing mainly the flat opisthosomal appendages. It seems more likely, therefore, that one cannot find the ancestor of one line in the other line, because each line was already specialized in its own manner by the time they appeared in the Cambrian.

The specialization in the Trilobita seemed to be the production of a heavily sclerotized dorsum with prominent lateral paraterga on all trunk segments (Figure 39). The trunk segments bore anterior and posterior lateral articular condyles so that the segments were firmly hinged with one another such as to prevent lateral bending, but the segments could move at the articulations around a horizontal transverse axis. This action permitted some trilobites to roll up by bending the body ventrally, the head touching the tail. The earliest trilobites (Olenellidae) from the lower Cambrian had a series of freely jointed trunk segments, but within the trilobite line there developed a tendency for the posterior segments to fail to separate into moveable structures. The result was the formation of a solid tagma, called the pygidium, characteristic of the higher orders of trilobites. However, the pygidium apparently retained fully formed and functional legs.

The sister group of the Trilobita, the Trilobitodea, became specialized main-

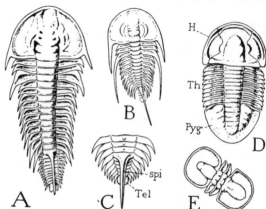

Figure 39   Examples of trilobite forms. A, *Olenellus vermontanus* Hall; B, *Olenellus gilberti* Meek; C, *Schmidtiellus mickwitzi* Schmidt; D, *Asaphicus wheeleri* Meek. E, *Peronopsis montis* (Matthew). A, B and C are early forms, with segmented pygidium. D is a later species, with a solid pygidium. E has a reduced thoracic segmentation. H, head or prosoma; Pyg. pygidium; spi. spine; Tel. telson; Th. thorax, reprinted from R.E. Snodgrass; *A Textbook of Arthropod Anatomy*. Copyright © 1952 by Cornell University. Used by permission of Cornell University Press.

ly in the development of the head appendages into large structures variously adapted to seizing food. The representatives are known from a few genera found together in mid-Cambrian Burgess shales (Figure 40). All bear filamentous antennules and three or less pairs of gnathal appendages on the head. The trunk was more or less uniformly segmented, but in *Naraoia* the dorsum of the trunk appeared to be a single unit. As far as is known, the trunk appendages consisted of the usual limb with the respiratory epipodite, typical of trilobites. The genera here included, all described by Walcott (1911a), are *Emereldella, Molaria, Naraoia, Sidneyia, Leanchoilia, Burgessia, Waptia,* and *Marella*. These have been considered as forms which suggest intermediate stages in the evolution of merostomes from trilobites (Størmer, 1944). They are too highly specialized to be any other than unsuccessful evolutionary lines related to the trilobites. The genera *Cheloniellon, Opabinia,* and *Yohoia* have been included in the Trilobitodea. They are not here included in this group because these three genera have a simple protocephalic head, more typical of the ancestral arthropod line.

The name Mandibulata is used here because it is a well known name. As a sister group of the Trilobilomorpha, however, the name is not descriptive of the specializations that appear to be fundamental and that probably occurred in the ancestral line leading to extant mandibulates. The union of the first metamere with the acron, forming a protocephalon separate from the following segments, is an ancestral arthropod character. The protocephalon per-

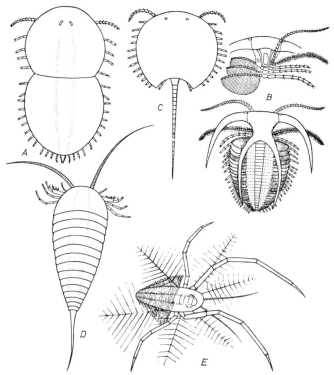

Figure 40   Trilobitodea. A, middle Cambrian *Naraoia compacta* Walcott; head from below, and dorsal view; C, middle Cambrian *Burgessia bella* Walcott; B, middle Cambrian *Marella splendens* Walcott; D, middle Cambrian *Emereldella brocki Walcott; E, lower Devonian Mimetaster hexagonalis* Gurich. From Sharov; *Basic Arthropodan Stock,* 1966, with permission of Pergamon Press, Ltd.

sisted as a separate tagma among the Crustaciformia. The appendages of the first metamere were changed from primarily locomotor organs to largely sensory organs (second antennae) and their epipodites were suppressed, or the appendages were entirely suppressed. The ganglia of that segment moved forward to lie next to the deutocerebrum, and became the tritocerebrum. In the species in which the tritocerebral appendage persisted, its sensory function is correlated with its form, in that it is antennalike. The tritocerebrum is the only adult vestige of the first metamere of the Atelocerata. Embryonic transient rudiments of the appendage appear briefly as short buds in some species of various classes.

The second, third, and fourth metameres bear primarily feeding appendages, and are often borne on a gnathocephalic tagma. The gnathocephalon frequently is united with the protocephalon, and occasionally with one or two posterior segments to form a head capsule. But although the heads of various groups of mandibulates may be formed of the same number of metameres, it is not likely that the head capsules of mandibulates are of a single origin. It appears rather that in addition to the origin of a head in the trilobite line, the head capsule formed again independently some time during crustacean evolution, and once more at least in the line leading to insects and myriapods. The evolution of head capsules of various groups will be discussed later.

In addition to the suppression of epipodites on the second antennae, it appears that these were suppressed early also on the appendage of the second metamere, which became the mandible of the mandibulates. The most nearly primitive mandible known is that of the Devonian *Cheloniellon calmani* Broili, which consists of a four-segmented limblike structure. The large coxa bears a prominent gnathobasic biting endite. The telopodite has only three podomeres, the same as the maximum number of palpal podomeres on crustacean mandibles. Although Broili (1933) in his reconstruction figured a small epipodite on the mandible, the excellent photographs of the ventral side of the second specimen exhibit no trace of such, and he described the mandible as uniramous or questionably biramous.

Mandibulates evolved a different type of digestive process than that of intracellular digestion of the chelicerates. The epithelium of the digestive diverticula (Crustacea) and of the midgut (Atelocerata) consists of two main types of cells. One type produces digestive enzymes, and releases them through rupture and death of the cell. Another type remains mitotically competent and serves to regenerate new digestive cells lost by rupture. Particulate food is digested in the lumen of the intestine, as a rule.

Mandibulata parted into two sister groups, the Crustaciformia and the Myriapodomorpha, probably very early in the Paleozoic. The Crustaciformia advanced along lines which favored aquatic life, and the other line eventually assumed a fully terrestial existence. In the Crustaciformia the coxal gland of the second antennae become an osmoregulatory and excretory organ (green gland).

Two sister groups of crustaceanlike forms produced the Cheloniellida (Figure 41) and the Crustacea. The specializations of the Cheloniellida were very few. Strong gnathobases developed on the mandible, first maxilla, second maxilla, and maxilliped. On these appendages there seemed to be only one trochanter and no patella. The second antennae acquired numerous sensory bristles, and as in the Crustacea. The coxal gland of the second antenna became the main excretory organ. The orifice of the antennal coxal gland is prominent in *Cheloniellon*. Endites are not apparent on the trunk limbs, which were of the trilobite type, and they may have lacked a patella. Although *Cheloniellon* expanded the tergal paranota to the extent that the

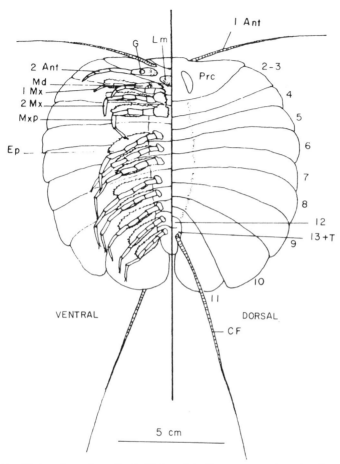

Figure 41 *Cheloniellon calmani* Broili, Chelionellida. Reconstruction from the photographs in Broili, 1933. Ant, antennae; CF, caudal filament; Ep, epipodite (gill); G, opening of green gland; Lm, labrum; Md, mandible; Mx maxillae; Mxp, maxilliped; Prc, protocephalon (acrom plus first metamere); T, telson (periproct); arabic figures at right designate body segments.

body seems to be a disc, it remained largely primitive in its tagmosis. The mandibular and first maxillary metameres fused into a single structure, and the last two metameres lacked paranota. A pair of slender jointed tail filaments of unknown nature dangled from the telson. The mid-Cambrian *Opabinia regalis* Walcott and *Yohoia tenuis* Walcott are tentatively regarded as species which might belong among the Cheloniellida. The only reasons for including them are that they seem to have a separate protocephalon, and their paranota are well developed except on two or more rear end metameres. These two species could just as well be aberrant crustaceans. The nature of their mouthparts and limbs is unknown. Although *Opabinia* appears to have branched epipodites consisting of flat comblike plates as in *Cheloniellon*, there are no recognizable appendages (Whittington, 1975).

It was the Crustacea which became the successful line of this lineage. Advanced features most typical of the Crustacea are the development of a prominent exite on the basipodite (first trochanter), the elaboration of the second antennae into specialized sensory organs, and the nauplius larva. The trochanteral exite is so well developed in the legs of many species that it was the basis for the original concept that the biramous appendage was the original type for all arthropods. The exite was named exopodite and the remainder of the telopodite became the endopodite. The biramous limb as an ancestral limb became dogma. This concept can no longer be maintained. The crustacean limb is fundamentally a uniramous limb which has become highly specialized by the development of trochanteral exites which are not at all limblike.

The crustacean second antenna, so typical of the group, is the modified limb of the first metamere. The organ typically is first apparent in the nauplius larva, in which it is useful as a swimming organ. As such it usually bears a well developed exopodite. As the larva metamorphoses, the exopodite generally does not remain as well developed, but most second antennae retain some vestige of the exopodite in the adult. The exopodite is retained as a major component of the second antennae in Cephalocarida and Mystacocarida. In most Crustacea, the second antennae have shifted forward ahead of the mouth, and are borne on a sclerite of the head that also bears the eyes and the first antennae, forming the so-called protocephalon. The unit is frequently movable as a unit separate from the gnathal region, but the protocephalon is not separated from the head in copepods, *Bathynella*, isopods, amphipods, cephalocarids, and mystacocarids. It is uncertain whether the crustacean protocephalon was acquired as a secondary separation from the head, or was part of the fundamental nature of the ancestral crustacean. In favor of the latter alternative is the presence of a distinct protocephalon in the Cheloniellida. However, the nauplius larva hatches without a distinct protocephalon, a point in favor of the first alternative. Furthermore, although the crustacean second antennae are commonly borne on a protocephalon, their muscles originate in the head capsule. In *Nebalia bipes* (Leptostraca) only the eyes and the first antennae are borne on an acronal structure. The second antennae of *Nebalia* articulate, with the gnathal appendages, on the definitive head. A lobe bearing the first antennae separates from the nauplius front end after several molts in the eyeless *Derocheilocaris* (Mystacocarida) simulating a protocephalon.

The nauplius larva stage is a unique character of the Crustacea (Figure 42). No other arthropod group hatches into such a larva. The protonymphon

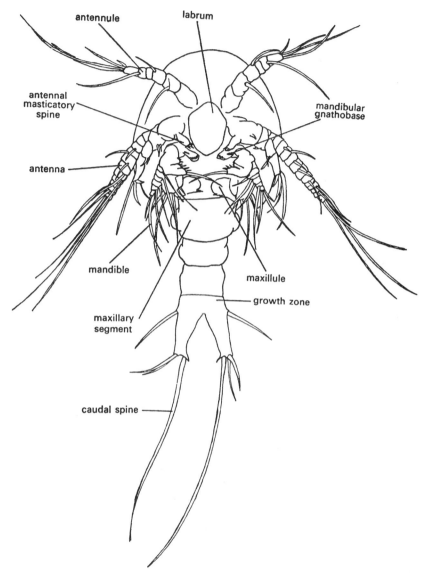

Figure 42 First stage nauplius larva of *Hutchinsoniella macracantha* Sanders (Cephalocarida), ventral view. From Anderson: *Embryology and Phylogeny in Annelids and Arthropods*, 1973, after Sanders, 1963, by permission of Connecticut Academy of Sciences and Pergamon Press, Ltd.

parasitic larva of pycnogonids resembles a nauplius superficially, but its three pairs of appendages are part of the first three metameres. While the nauplius larva also bears three pairs of appendages, the first appendages are the antennules (first antennae) of the acron, and therefore not metameric. The nauplius larva results from the precocious hatching of the embryo. Growth of the larva is anamorphic with the addition of body segments occurring at a teloblastic zone of growth just ahead of the telson. Anamorphic posthatching growth is generally the rule in arthropod species whose eggs are tiny and bear little yolk. The Crustacea whose embryos develop fully before hatching also grow by teloblastic generation of posterior segments, but are able to complete development before hatching. There is no nauplius larva in the malacostracan groups Leptostraca, Syncarida, and Peracarida, and these bear relatively large yolky eggs whose cleavage is essentially meroblastic.

The crustacean head typically consists of the protocephalon and a gnathal tagma consisting of the second, third, and fourth metameres, whose appendages are respectively the mandibles, first maxillae (maxillules), and second maxillae. The gnathocephalon sometimes includes another metamere, whose appendages are called maxillipeds. When the maxilliped somite is united with the gnathocephalon, the protocephalon also generally is united, to form a syncephalon (amphipods and isopods). Crustacean heads so frequently are extended laterally and backward to form a carapace that it appears that the ancestral crustacean may have evolved a carapace, to be suppressed or reduced in several lines. Amphipods, isopods, anostracan branchiopods, mystacocarids, and anaspidaceans do not develop a carapace. The carapace is a rearward and lateral expansion of the posterior tergum of the somite of the second maxillary segment. In the Malacostraca the maxillary somite itself often grows posteriorly on the dorsum such as to crowd the median section of terga of the thoracic segments essentially out of existence (Eucarida). The carapace sides hang down over the epipodites as gill covers, and the result is a seeming cephalothorax (Snodgrass, 1958), not at all morphologically comparable to the chelicerate cephalothorax, or prosoma. The heads of carapace-bearing forms and of anostracan branchiopods and *Anaspides* (Figure 43) frequently bear a dorsal more or less transverse groove, the cephalic groove. This is said to be a persistent intersegmental groove between the mandibular and the maxillulary terga (Snodgrass, 1952). There is never any flexibility at this groove. The sclerotized cuticle is continuous on the head, and the groove first appears in a later naupliar stage, or at hatching in *Anaspides*, which hatches fully formed. Whether the groove is truly intersegmental, its presence provides mechanical rigidity resisting the strong pull of mandibular muscles.

Still another derived character of the Crustacea is the tagmosis of the trunk behind the head into a thoracic, or locomotory section, on which the locomotor appendages are borne, and an abdominal section whose appen-

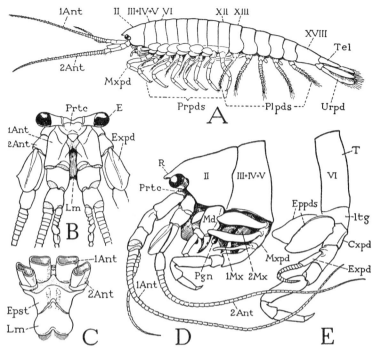

Figure 43 *Anaspides tasmaniae* Thompson. Crustacea. A. entire animal with left appendages:
B. head. anterior: C. epistome and labrum. with bases of antennae. ventral: D. head and
anterior part of body: E. Body segment of first legs (first free segment). Roman numerals
designate body segments. beginning with segment of second antenna. Ant. antennae: Cxpd.
coxopodite: E. compound eye: Eppds. epipodites: Epst. epistome: Expd. exopodite: Lm.
labrum: ltg. laterotergite: Md. mandible: Mx. maxillae: Mxpd. maxilliped: Pgn. paragnath:
Plpds. pleopods. Prpds. pareiopods: Prtc. protocephalon: R. rostrum: Tel. telson: Urpd.
uropod. Reprinted from R.E. Snodgrass: *A Textbook of Arthropod Anatomy.* Copyright ©
1952 by Cornell University. Used by permission of Cornell University Press.

dages are reduced or modified for functions other than primary locomotion.
The internal organs in the abdomen are generally restricted to muscles, hind
intestine, nerves, and vascular spaces.

The crustacean telopodite lacks a patellar segment. However, there
seemed to have been a patella in the trunk limbs of Cheloniellida if the bristly
tips of the legs represent the pretarsus, or dactylopodite. It is equally possible
that the bristles are nothing more than terminal spurs, in which case the
cheloniellid limb is similar to that of crustaceans in lacking the patella. The loss
of the patella then would be a feature of the Crustaciformia, in parallel with its
loss in the Atelocerata, because the myriapodlike Arthropleurida retained the
patella.

The spermatophore produced by a few Crustacea, such as copepods and some decapods, is unusual in this group. Most crustaceans produce a fluid semen which is introduced into the female genital system. The spermatophore in this class appears to have arisen independently of that of Onychophora, Arachnida, and Atelocerata. Evolution within the Crustacea has produced such a variety of specialized forms that it is not possible to find a truly primitive modern living species. The diversity is such that some authors would rank the Crustacea as a phylum.

The subterphylum Myriapodomorpha is the sister group to the Crustaciformia. Specialization in the ancestral myriapodomorph included adaptation to terrestrial air-breathing life. It is assumed that the ancestral transition from aquatic to terrestrial living occurred as a monophyletic event in this line mainly because the myriapods and insects share so many advanced characters not found anywhere else. But the aquatic-terrestrial transition phenomenon, even considering only the arthropods, appears to have occurred several times independently: Onychophora, Arachnida, Isopoda, Amphipoda, land crabs, and Atelocerata. Isopods, crabs, and amphipods are essentially aquatic animals. Some isopods became freshwater living, others remain on the shore near marine habitats, and the superfamily Oniscoidea has become essentially terrestrial from fresh water. Only a few species of amphipods live in moist soil.

The nature of the ancestral myriapodomorph is suggested by the extinct Carboniferous Arthropleurida. There are no known fossil arthropods that can suggest what the aquatic progenitor was like. The fossil record for this group begins in the Upper Silurian-Lower Devonian with *Archidesmus longanensis* Peach, and *Kampecaris forfarensis* Page, which are myriapodlike in bearing a head capsule with antennules and compound eyes, and a series of trunk segments each bearing a pair of simple legs. They could be ancestral pauropods, symphylans, or chilopods, or may belong to unsuccessful extinct lines.

The Myriapodomorpha as terrestrial animals suppressed the coxal epipodite so common in the aquatic arthropods. It is only a matter of opinion whether the coxal eversible vesicles of myriapods and primitive insects are vestiges of coxal exites converted to organs of water regulation. The acquisition of metamerically invaginated tracheae appears to have been one of the first specializations, and the locomotor limbs became efficient, secondarily uniramous walking legs.

Tentatively, the Arthropleurida (Figure 44) are considered as a sister group to the Atelocerata. These were rather large (over 150 cm long) animals bearing a long series of trunk segments, each of which bore a pair of uniramous limbs. The legs were robust, strengthened by longitudinal anterior and posterior grooves, and the podomeres each bore a pair of stout spines ven-

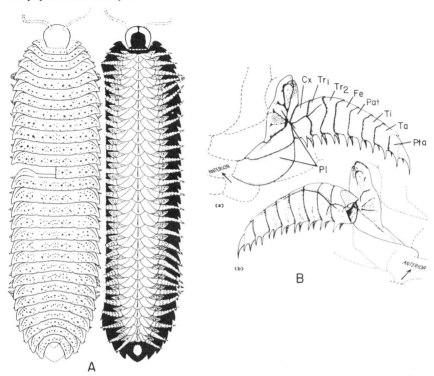

A

Figure 44 *Arthropleura armata* Jordan and Meyer, Arthropleurida. Reconstruction from Rolfe: *Arthropleurida*, 1967. A, Dorsal and ventral views of reconstructed whole animal; B, reconstruction of first leg, anterior view (a) and posterior view (b). Leg segmentation according to views in this book, Cx, coxa, Fe, femur; Pat, patella, Pl, pleurites; Pta, pretarsus; Ta, tarsus; Ti, tibia; Tr, trochanter. *From Treatise on Invertebrate Paleontology,* courtesy of the Geological Society of America and University of Kansas.

trally (Rolfe and Ingham, 1967). Each leg was articulated with a complicated pleural plate, and consisted of at least eight podomeres, indicating that a patella was still a component of the telopodite. The terga were distinctly expanded into prominent lateral paraterga on all the known segments. The diet included a great deal of land plant tissue, as is indicated by fragments of lycopod epidermis and tracheid cells found in the intestine. No obvious respiratory organs have been identified and only a poorly preserved head is known. Because of the lack of knowledge of the mouthparts, antennae, the telson, and the nature of the respiratory system, it is not possible to determine whether they were terrestrial specialized myriapods, trilobites, or crustaceans.

The superclass Atelocerata includes the myriapods and insects. The name of the group (from the Greek *atelos*, defective and *keras*, horn) emphasizes a specialization not found in any other arthropod group. The appendage of the first metamere, which became the second antennae of the Crustacea, appears only as an embryonic rudiment, and fails to develop. The paired ventral ganglia of the segment persist, and merge with the acronal brain to form the tritocerebrum. The tritocerebral ganglia retain their original postoral transverse commissure, looping behind the mouth. Elongation of the longitudinal nerve trunks between the tritocerebral ganglia and the next resulted in a pair of secondary circumpharyngeal connectives.

The acronal antennae served by the deutocerebrum have persisted, however. The name Atelocerata is not quite appropriate, because it refers to the absence of *second* antennae, rather than of first antennae. It is probably unlikely that this group ever possessed elaborate second antennae so typical of Crustacea. At any rate, the name is the oldest one for the group which describes a fundamental feature, the suppression of the first metameric appendage.

The reasons for discarding the hypothesis that atelocerate mandibles represent whole limbs is discussed in Chapter 4. The hypothesis that only the coxa remains with suppression of the telopodite is in harmony with other sources of evidence that speak for a monophyletic theory of arthropod origin.

Additional derived features of the Atelocerata are as follows: metamerically arranged tracheae; absence of the patella; absence of levator muscle of the pretarsus; head capsule of three or four metameres plus acron; development of excretory malpighian tubules from the hind intestine; suppression of digestive function of midintestinal diverticula; and acquisition of a pair of anterior tentorial apodemes. There appears to be a fundamental tendency for one or the other of the maxillae to unite the coxal portions into a single piece. Insects and Symphyla have fused second maxillary coxae (labium), millipedes and pauropods lack a second maxilla but have fused first maxillary coxae (gnathochilarium), and centipedes tend to have the coxae of both first and second maxillae essentially united.

If sequences of embryonic development and adult morphology can be trusted as indicators of past events, it can be inferred that the original atelocerate head capsule included only the segments of the first three metameres fused with the acron, and the fourth metamere (bearing the second maxillae) later joined the others in the head, perhaps independently in the insects and again in the ancestor of centipedes and symphylans. In the latter two groups, the embryonic fourth metamere is distinct for a while (Tiegs, 1940), but before hatching the unitary head capsule becomes fully formed. In the closely related pauropods and millipedes, the fourth metamere remains separate from the head as the legless collum segment. A presumably persistent intersegmental groove, the postoccipital sulcus, remains on the adult insect head (Snodgrass, 1960), between the labial segment and the rest of the head. Since insects and myriapods appear to be sister groups of Atelocerata, the four-segmented insect head probably originated in the insect line independently of the four-segmented chilopod and symphylan head.

The production of a spermatophore is typical in all groups of Atelocerata, and the spermatophore is primitively transferred indirectly by being deposited on a substrate, later to be picked up by the female. In a few millipedes the spermatophore is transferred directly to the female, and direct transfer of the spermatophore into the female genital tract is the rule in winged insects.

The class Myriapoda takes its name from an ancestral character, the presence of numerous limbs on a segmented trunk, different from the tagmosis of the trunk into a thorax and abdomen so characteristic of insects. There is an outstanding specialization in myriapods that is unique and more than any other suggests that myriapods form a natural assembly. The specialization is in the mechanism of biting with the mandible involving several derived characters. Instead of a dorsal monocondylic articulation taking advantage of the primitive promotor-remotor coxal rock (as in insects), the myriapod mandible either became broadly articulated with the head dorsally as in millipedes, or loosely articulated via a slender sclerite as in centipedes and pauropods. The originally antagonistic dorsal promotor and remotor muscles have come to operate in unison with the ventral muscles so that they contribute to a transverse biting action by the flexible gnathal endite lobe. The mechanism of abduction, or opening of the jaws, comes about through a specialization of the anterior tentorial apodemes. As in insects, these apodemes are invaginations from a premandibular region of the head. But they are so constructed that they are movable inside the head, and their action is a thrust outward against the mandibles, effectively opening the jaws (Manton, 1964). The endite lobe of pauropod jaws is not flexible on the mandible, which is usually reduced to a thin entognathous picklike structure. Whether pauropods ever had a flexible endite lobe is an unanswerable question, but in the primitive genus *Millitauropus* of Africa (Remy, 1950), the mandible is robust, bears a stong gnathal endite, and can bite its food. In each class of myriapods, there are further specializations of the basic transverse biting action of the jaws.

Additional derived characters common to the myriapods apparently include the articulation of the coxa ventrally with the sternum, rather than dorsally with a pleuron as in insects, and the general absence of paratergal lobes, except for some millipedes. In addition, myriapods molt by first separating the cuticle across the back of the head, pulling the head out of the head capsule, then more or less crawling out of the trunk cuticle, while insects split the head and trunk cuticle along the dorsal midline. Midintestinal diverticula were totally suppressed in the Myriapoda in contrast to the retention of caeca in the Insecta.

If the myriapods represent a monophyletic assembly by virtue of the characters discussed above, the grouping of the four types of myriapods on phylogenetic lines cannot be done with certainty. Not only are there specializations found only in each type, but one can group any type with either of three others on the basis of an apparent sharing of possibly homologous characters not found in the other two.

The Diplopoda and Pauropoda both bear a head of three metameres, with a single maxilla united into an underliplike structure, and the fourth metamere is separate from the head and is legless. The Symphyla and Pauropoda both develop embryonic ventral organs that contribute cells to the ventral ganglia and persist to transform into eversible vesicles. The Diplopoda and Chilopoda have flexible gnathal lobes and additional flexibility in the coxa or body of the mandible, and these two groups bear segmental tracheae. They both include species which have retained the ventral intersegmental tendons in the trunk. Symphyla and most Pauropoda lack tracheae on the trunk, although in Symphyla a pair of tracheal openings occur on the head, and in Pauropoda the genus *Millitauropus* bears tracheal openings on all its leg coxae.

The classification here proposed appears to call for the least amount of convergence. Collifera, including Pauropoda and Diplopoda, represents a sister group to the Atelopoda, including Symphyla and Chilopoda. The derived features in common in the Collifera are partly in the structure of the head. All indications suggest that only the acron and the first three metameres are included in the head. As usual, the tritocerebral metamere is vestigial. The appendages of the first maxillary segment are represented by a flat labiumlike structure consisting of the fused coxae of the first maxilla. Maxillary palps are either nonexistent, or may be short structures as are found in the pselaphognathid (palpjaw) millipedes. The metamere of the second maxilla forms the first trunk somite, and is entirely legless (the collum segment, which is the basis of the name for the group—Latin *collum*, (collar) plus *ferre*, (to bear)). The newly hatched larva bears the legless collum plus three leg-bearing segments, and rudiments of future segments just ahead of the zone of growth of the telson. The genital openings are on the second leg-bearing somite (metamere 6). The antennae of Collifera are short, consisting of eight or less segments, and the gonads are located ventrally below the intestine.

The derived characters of the Pauropoda include the branching of the antennae with muscles operating each branch, the fusion of a possibly originally segmented mandible into a single unit correlated with very small body size (less than 2 mm long) and reduction of the trunk to 11 or 12 segments (Figure 45). The primitive genus *Millitauropus* bears 11 pairs of legs, and each of the 12 trunk metameres bears a normal tergal plate. Other pauropods usually bear only nine pairs of legs on the second through tenth trunk metameres. In these, the terga of the somites bearing the third, fifth, seventh, and ninth legs are very narrow and unsclerotized, giving the impres-

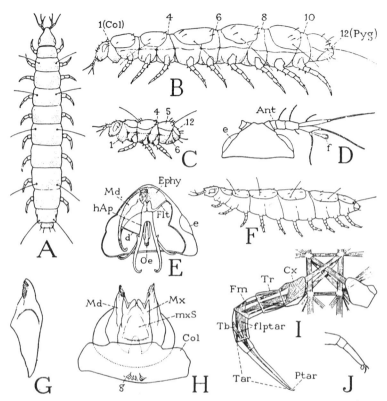

Figure 45. Pauropoda. A, unidentified pauropod with regular segmentation on dorsum; B, *Pauropus silvaticus* Tiegs, adult; C, same, first instar; D, *Stylopauropus pubescens* Hansen, head and right antenna, dorsal; E, *Pauropus silvaticus*, head from below showing right mandible and apodemes; F, *Decapauropus cuenoti* Remy; G, *Allopauropus brevisetis* Silv., mandible; H, same; head and collum segment, ventral; I, same, segmentation and musculature of a leg; J, a pauropod leg with accessory pretarsal claw and empodium, Ant, antenna; Col, collum; Cx, coxa; d, tendon from mandible to hypopharyngeal apodeme; e, pseudoculus; Ephy, epipharyngeal surface; f, globulus of antenna; flptar, flexor muscle of pretarsus; Flt, anterior tentorium (fultura); Fm, femur; g, ventral papillae of collum segment; hAp, hypopharyngeal apodeme; Md, mandible; Mx, gnathochilarium (maxilla); mxS, maxillary sternum; Oe, oesophagus; Ptar, pretarsus; Pyg, periproct; Tar, tarsus; Tb, tibia; Tr, trochanter. Reprinted from R.E. Snodgrass: *A Textbook of Arthropod Anatomy.* Copyright © 1952 by Cornell University. Used by permission of Cornell University Press.

sion of diplosegments. The eleventh somite just ahead of the telson is usually legless, except in males of some genera (Remy, 1950). Most pauropods lack tracheae altogether, but *Millitauropus* bears tracheal openings on the coxae of the legs.

The intersegmental trunk tendons in pauropods have been suppressed.

The head bears the only remaining vestige of intersegmental tendon in the form of tendinous tissue between the tentorial arms.

The Diplopoda, or millipedes (Figure 46), are a sister group to the Pauropoda, Most millipedes have specialized in burrowing by pushing into the substrate of loose soil. This is made possible by the combined effect of a large number of legs and the peculiar fusion of the somites in pairs behind the fourth trunk segment (seventh metamere), forming solid rings of sclerotized integument. The pselaphognathid millipedes, however, have not acquired a large number of body segments, and generally bear only five to seven diplosegments in addition to the anterior legless collum segment and three simple somites, each with one pair of legs. They do not burrow as do other millipedes, and have a soft integument on which there are tufts of strong featherlike setae (Figure 47). The burrowing habit is secondarily developed in the higher millipedes.

It is claimed (Manton, 1956) that the Pselaphognatha have *lost* the burrowing habit and are highly specialized for clinging to smooth rocky surfaces in crevices. Manton claimed that diplosegments originated through the need for strength and rigidity after millipedes had become burrowers. However, many pselaphognathids live in the open on tree trunks and on the ground, not in smooth crevices in rocky caves. Furthermore, their musculature is more suggestive of a primitive nature, in that they are the only millipedes whose legs bear dorsal muscles and which have well developed ventral intersegmental tendons. It is more likely that the burrowing forms are the specialists for that behavior, and that the Pselaphognatha are relics of the ancient ancestral millipedes that never acquired the burrowing habit.

The mandibles of most millipedes have become rather large, occupying the sides of the head (see Figure 24, p. 65). The coxal portion is divided into two parts, flexible on each other, reminiscent of the cardo and stipes of the insect maxilla. A flexible endite lobe forms the biting surface, and biting involves flexing the endites and the coxae through simultaneous contraction of the dorsal muscles originating on the cranium and the powerful ventral muscles originating on the transverse mandibular tendon and inserting into the body of the mandible.

Most millipedes have eliminated the transverse tendons of the trunk, and the muscles are attached to cuticular structures. The dorsal limb muscles are not developed, and all the coxal muscles are ventral, except that the pselaphognathids retained the primitive dorsal leg muscles and elaborated them. Pselaphognathids also have retained the ventral tendons and ventral extrinsic leg muscles and the dorsoventral muscles.

In contrast to the Collifera, the Atelopoda share fewer derived characters in common. If the latter is a natural group, it evolved soon after splitting away

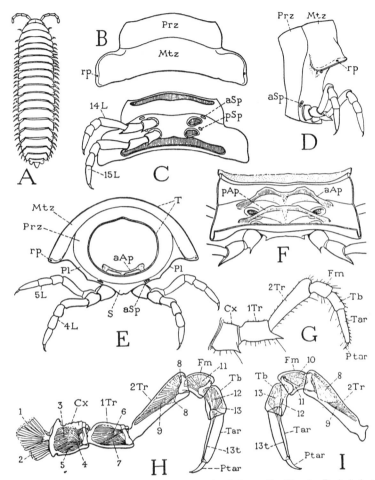

Figure 46 Diplopoda, Polydesmoidea. A, a typical polydesmoid millipede; B, *Apheloria coriacea* (Koch), tergum of tenth segment; C, same undersurface; D, same, lateral view of a diplosomite; E, same, fifth segment, anterior; F, same, inner surface of ventral wall of a diplosomite; G, same, a left leg, anterior; H, *Euryurus* sp., left leg and musculature, anterior; I, same, distal part of leg, posterior. aAp, anterior apodeme; aSp, anterior segmental spiracle; Cx, coxa; Fm, femur; L, leg; Mtz, metazonite; pAp, posterior apodeme; Pl, pleuron; Ptar, pretarsus; Prz, prozonite; pSp, posterior segmental spiracle; rp, pore of repugnatorial gland; Tar, tarsus; Tb, tibia; Tr, trochanter; 1, promotor, and 2, remotor muscle of coxa; 3, 6, levator muscles; 4, 5, 7, depressor muscles; 8, 10, extensor muscles; 9, 11, 12, 13, flexor muscles. Reprinted from R.E. Snodgrass: *A Textbook of Arthropod Anatomy.* Copyright © 1952 by Cornell University Press. Used by permission of Cornell University Press.

109

Figure 47 *Polyxenus* sp., a primitive pselaphognathid millipede. Scanning electron microscope photograph of nearly mature specimen. Original.

from its sister group into two specialized lines. The head capsule in the Atelopoda includes four metameres. It appears that in this line the fourth (second maxillary segment) may have joined the others after a three-segmented head capsule was established in the myriapods, because in Symphyla the second maxillary segment in the embryo remains separate from the head for a while before joining the embryonic head. If this is the case, a head consisting of the acron and four metameres may have originated three times during arthropod evolution: Crustacea, Atelopoda, and Insecta.

The Atelopoda share additional similar characters that could be homologous. Both centipedes and symphylans bear long filamentous antennae with numerous musculated antennomeres. Anamorphosis occurs in both groups, with the newly hatched larva bearing five to seven pairs of legs, in contrast to only three pairs in the colliferans. The name (Greek *atelos*, defective plus *pous*, leg) refers to the absence of legs on the last two trunk metameres. The absence of legs on the last metamere of pauropods is a specialization within the group, and the legless posterior rings of millipedes are found in those that have not completed their anamorphic development, or in a few species of specialized genera that are possibly neotenous in failing to generate limbs on the last few segments.

The Symphyla have derived characters of their own, and in some ways are quite specialized (Figure 48). The gonopore is on the venter of the fourth trunk somite (metamere 8). The anterior position of the gonopores of Collifera was the basis for grouping those into the Progoneata (Pocock, 1893). He grouped the Symphyla with centipedes and insects as Opisthogoneata because he believed symphilids possessed a posterior gonopore. Later, Symphyla was transferred to the Progoneata. However, there are other characters that seem to relate the Symphyla to the pauropods and diplopods. In the three groups, the gonads lie below the intestine, there are at least vestiges of eversible coxal vesicles, and the fat body develops from embryonic cells related to the midintestine. The coxal vesicles and type of fat body are ancestral features retained by these groups and therefore cannot indicate close relationship. The position of the gonads in the body is unusual, and could be either the result of convergent evolution, or the result of common ancestry, or an ancestral arthropod character. The gonads are dorsal to the intestine in Onychophora, Crustacea, centipedes, and insects, and are below the intestine in the Chelicerata.

The endodermal fat body has its counterpart in chelicerates and crustaceans, in which cells associated with the midintestine diverticula appear to act as storage sites for food reserves, especially lipid. In insects and centipedes, the fat body develops from "mesodermal" coelomere cells. Coxal eversible sacs are terrestrial organs of water absorption in the Atelocerata. That these are all homologous is uncertain, but similar organs also occur in Onychophora and the arachnid order Amblypygi.

The anterior position of the gonopores of millipedes and pauropods is only superficially similar to that of symphylans, since in the first two groups the gonopore is on the sixth postoral metamere (third trunk somite) rather than on the eighth (fourth trunk somite) as in Symphyla. Furthermore, Tiegs (1940) demonstrated that the anterior position of the symphylan gonopore is probably secondary as shown by embryonic studies. The symphylid gonopore is described as on the third trunk segment (Snodgrass, 1952), but this is not true. In some genera (e. g., *Symphylella*) the first trunk segment is

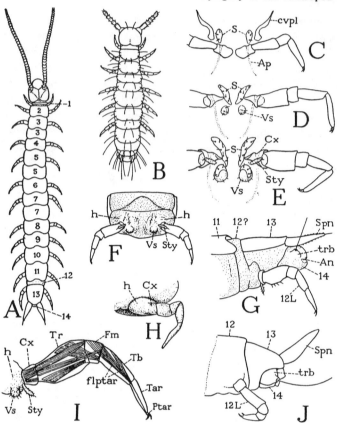

Figure 48. Symphyla, general structure of body and legs. A, *Scutigerella immaculata* (Newp.), adult; B, same, first instar; C, first legs and associated sclerites; D, same, second legs, with imperfect vesicles; E, same, third legs, with styli and fully developed vesicles; F, same, ventral segment from midbody; G, same posterior end; H. *Hanseniella agilis* Tiegs, embryonic leg; I, *Scutigerella immaculata*, leg of adult; J. *Symphylella* sp., posterior end with fully developed twelfth segment. An, anus; ap, apodeme; cvpl, cervical plate; Cx, definitive coxa (? first trochanter); flptar, flexor muscle of pretarsus; Fm. femur; h, lobe of body supporting leg (? true coxa); L, leg; Ptar, pretarsus; S, sternum; Spn, spinneret; Sty, stylus; Tar, tarsus; Tb, tibia; Tr, trochanter; trb, trichobothrium; Vs, eversible vesicle. Reprinted from R.E. Snodgrass: *A Textbook of Arthropod Anatomy.* Copyright © 1952 by Cornell University. Used by permission of Cornell University Press.

legless, and the gonopore appears to be on the third segment, since it is between the third pair of legs. In various symphylids, there are varying degrees of reduction of the first legs and of sclerotization of the first trunk segment.

Additional symphylan derived characters include a pair of first maxillae consisting only of the coxae (but bearing endites), a somewhat united pair of second maxillary coxae also without palps that provide a labiumlike structure, a tendency to reduce or eliminate the first pair of trunk legs, total suppression of eyes, the development of a line of flexibility on the terga of several trunk segments so as to stimulate duplicated terga, the suppression of all trunk intersegmental tendons and their replacement with sternal apodemes, the restriction of tracheae to the anterior end with a single pair of spiracles in the head, the standardization of the number of trunk somites to 14, and the presence of a pair of cercuslike spinning organs on the thirteenth (legless) trunk segment. There are 12 pairs of legs except in species whose first legs are suppressed. Near the coxa of each leg other than the first there is an eversible vesicle and a styluslike spur of unknown function.

Extreme lateral flexibility is possible in symphylids because several terga are in two pieces and loosely articulated by intersegmental and intratergal arthrodial membrances. The fewest double terga are found in *Scutigerella*, where they occur on the third, fifth, and seventh trunk segments. In *Symphylella* the terga are double on the third, fifth, seventh, and ninth trunk segments. In *Geophilella* the terga are double on the third through the twelfth trunk segments. In all symphylids the thirteenth trunk segment bearing the spinnerets bears a simple tergum, and the fourteenth is unsclerotized, legless, and is united with the telson. In *Scutigerella* the twelfth trunk segment (bearing the last legs) is narrow dorsally and has no sclerotized tergal plate. Thus the apparent number of trunk tergites is 15 in *Scutigerella*, 17 in *Symphylella*, and 23 in *Geophilella*.

The symphylan spinnerets (cerci) have been homologized with the trunk limbs, and therefore with insectan cerci. According to Ravoux (1962), the newly hatched larva of *Scutigerella* bears a pair of spinnerets on the legless ninth trunk segment. After three molts, the ninth segment develops a pair of legs, but retains a reduced pair of dorsal spinnerets and their glands along with legs. A new pair of spinnerets develop on the eleventh segment, which at this time is legless. After two more molts the original spinnerets vanish, those on the eleventh segment are reduced, and a new final pair form on the thirteenth trunk segment. The thirteenth segment remains legless, as does the fourteenth, but the eleventh developed a pair of legs while it still bore a reduced pair of spinnerets. Ultimately, no traces of the first two pairs of spinnerets remain. It appears that the spinnerets are not leg homologues, but are specialized tergal silk spinning organs peculiar to the Symphyla.

The Chilopoda (Figure 49) have several unique derived characters, the best known of which is the transfomation of the first trunk legs into a pair of poison fangs (toxicognaths, maxillipeds). Fully formed compound eyes were retained and even enlarged only in the scutigeromorph centipedes. In others the compound eyes are highly reduced. The sides of the body have developed various pleural sclerites often collectively called subcoxae. The number of trunk segments varies, from 15 to 170 or more, not including the

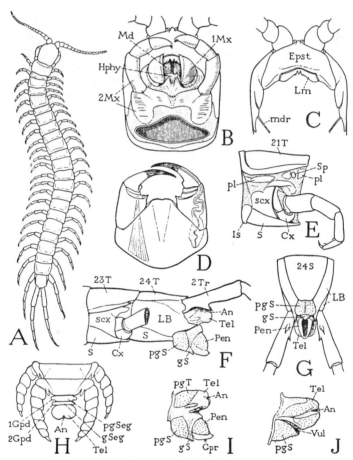

Figure 49. *Otocryptops sexspinosa* (Say), Chilopoda, Scolopendromorpha, except H. A, entire centipede; B, head and mouth parts, ventral; C, anterior part of head, ventral, mouth parts removed; D, maxillipeds (toxicognaths), ventral; E, left side view of middle segment of body and leg; F, terminal body segments, male, left side; G, same, ventral; H, terminal part of body of embryo of *Scolopendra*, dorsal; I, genitonal region of male, left; J, same, female. An, anus; Cx, coxa; Epst, epistome; Gpd, gonopod; Gpr, gonopore; gs, sternum of genital segment; gSeg, genital segment; Hphy, hypopharynx; Is, intersternite; LB, limb base resulting from union of first trochanter, coxa, and subcoxal sclerites; Lm, labrum; Md, mandible; mdr, articular rod of mandible; Mx, maxillae; Pen, penis; pgS, sternum of pregenital segment; pgSeg, pregenital segment; pgT, tergum of pregenital segment; S, sternum; T, tergum; Tel, telson; 2Tr, second trochanter of last leg; Vul, vulva. Reprinted from R. E. Snodgrass: *A Textbook of Arthropod Anatomy.* Copyright © 1952 by Cornell University. Used by permission of Cornell University Press.

114

fang segment and the two posterior legless genital segments and telson. The last pair of legs is always larger and longer than all others, and may be sensory (*Scutigera*) or grasping in function. The first and second maxillae retained the telopodite, but that of the first maxilla is reduced to two short flattened podomeres that are useful as an underlip behind the mandibles. The telopodite of the second maxilla is more elongate and somehwat leglike. In parallel with the insects, centipedes are provided with a storage organ, the fat body, which develops in the embryo from coelomere derived cells instead of from vitallophage cells as in other myriapods. The gonopore invaginates into the last trunk somite to join the ducts of the dorsally located gonds, another seeming convergence with insects.

The class Insecta represents the sister group to the Myriapods. Their peculiar derived characters are discussed in greater detail in the following chapters. This chapter concludes with a suggested formal classification of the arthropods based on a monophyletic concept. Each group is characterized with a summary of the principal derived characters that are presumed to have originated in each of the taxa and that in effect might characterize the ancestral progenitor of each taxon. The time of earliest appearance in the fossil record is given when known, but this is not to be interpreted as the time of origin of any taxon. The classification that follows is reprinted from Boudreaux (1979), by permission of Van Nostrand Reinhold Publishing Company, with minor changes.

A PHYLOGENETIC CLASSIFICATION OF THE ARTHROPODS

Superphylum Lobopodia Snodgrass, 1938

Hypothetical Prearthropod. Loss of coelom except in embryos or in gonads and coelomoducts; acquisition of hemocoel; dorsal tubular heart with metameric ostia; meroblastic yolky eggs, with ventral germ band developing from blastoderm; loboform paired metameric appendages moved by extrinsic muscles, and with intrinsic muscles operating a terminal sclerotized piece; cuticle with chitin, primitively unsclerotized; specialized epidermal cells forming typical sensory setae; body muscles consisting of paired longitudinal dorsal, lateral, and ventral muscles, plus a primitively circular muscle layer; septal muscles reduced to paired dorsoventral intersegmental muscles.

I.   Phylum Onychophora Grube, 1853. Early Cambrian to Recent.

Development of hemocoelic hydrostatic skeleton with additional oblique muscle layer of body wall; extensive collagenous tissue under epidermis; conversion of dorsoventral intersegmental muscles into continuous ostiate diaphragms; development of circular vascular channels (rings) under epidermis of body and legs; use of claw of leg 1 as jaw; leg 2 converted to slime papilla; originally eversible tubelike organs on limbs serving aquatic respiration (*Aysheaia pedunculata* Walcott), presently restricted to one eversible vesicle per leg in some, used as water absorbing organ. Recent species with aerial respiration through numerous fine tracheal openings in grooves between body rings; reduction of gonads to one pair, the gonoducts uniting to a median posterior gonopore; head of acron plus two somites. Relatively primitive: unjointed lobopod (= coxa plus simple telopodite or claws); unsclerotized cuticle; ciliated gonoducts and nephric tubes; all muscles smooth except those of jaw; eyes simple. [The Onychophora, Tardigrada, and Pentastomida were combined into the phylum Oncopoda by Weber (1954)].

II.   Phylum Arthropoda Siebold and Stannius, 1848

Sclerotized chitinous cuticle with periodic molts; ecdysial glands in cephalic region; limbs basically a coxopodite (limb base) formed from the main lobopod plus a telopodite consisting of several podomeres derived by segmentation of the terminal segment; eversible vesicle of limb base elaborated into respiratory gill (epipodite); endite lobes formed on medial part of coxopodite (= gnathobases); all muscles striated; longitudinal muscles became segmented, attaching to intersegmental tendons; somatic and extrinsic leg muscles primitively attached to intersegmental tendons; acronal palp formed into a short segmented antenna (= antennule); development of compound eyes; initital tagmosis of acron with metamere 1; lateral tergal folds above limb bases (= paranota); subepidermal collagen elaborated into an intersegmental tendinous endoskeleton system, variously modified; loss of somatic circular muscle.

A.   Subphylum Cheliceromorpha (Boudreaux, 1979).

Tagmosis of body into a trophic-locomotor prosoma of 6 or 7 somites, plus an abdominal opisthosoma of 12 or 11 somites; epipodites suppressed on prosomal limbs; opisthosomal limbs modified as gill-bearing structures or reduced; acronal antennules suppressed; limb of first metamere converted to a three-segmented pincer (chelicera); intracellular digestion in vacuoles of cells of midintestinal diverticula; meroblastic cleavage of eggs retained in the first several mitoses, secondarily becoming total in various ways, but always forming a blastoderm and embryonic ventral germ band plus extraembryonic dorsal blastoderm.

1.   Infraphylum Pycnogonida Latreille, 1810. Devonian to Recent. Sea spiders.

Opisthosoma reduced or suppressed without limbs; second legs palplike; third legs modified as ovigers; prosoma typically in two parts, the acron and first four somites fused into a cephalon, the next three somites typically not fused together; cephalon extended forward as stiff proboscis with terminal mouth;

compound eyes, paranota, coxal endites, intersegmental tendons all suppressed; precocious protonymphon larva with only three or four somites; tarsi subsegmented. A persistent relatively primitive condition is in the multiple paired gonads with gonopores on trochanters of locomotor legs.

2. Infraphylum Chelicerata Heymons, 1901.

Prosomal terga and paranota fused into a solid structure; opisthosomal appendages of ninth through thirteenth somite modified into respiratory organs consisting of epipodites constructed in the form of flat lamellae (book gills-book lungs); appendages of fourteenth through eighteenth somites suppressed; gonopores fixed on genital segment (eighth somite); ventral intersegmental prosomatic tendons united into a tendinous "endosternum"; ganglia of prosoma concentrated forward near pharynx; sulfur-bonding acquired as a cuticular tanning process; compound eyes without externally separate lens facets.

(1) Superclass Xiphosurida Latreille, 1802. Cambrian to Recent. Horseshoe crabs.

Opisthosomal appendages shortened, flattened, serving as gill covers, their motion as paddles ventilating the book gills, used in swimming or burrowing; seventh metamere (pregenital) united with prosoma; stomodaeum with a masticatory gastric mill. Evolution within the group resulted in the following: restriction of opisthosoma to six metameres; fusion of opisthosoma into a solid tagma; elongation of telson spine; development of chelae on prosomatic legs; reduction of trochanters to one podomere; enlargement of coxal endities of five prosomal legs into gnathobases; reduction of seventh leg to a lobelike structure (chilarium).

(2) Superclass Cryptopneustida Boudreaux, 1979.

Opisthosomal book gills or book lungs in covered chambers; seventh somite free of prosoma, becoming first opisthosomatic; coxal mobility restricted; paranota reduced or suppressed.

(a) Class Eurypterida (Burmeister, 1843). Ordovician to Permian (extinct). Sea scorpions.

Changed to freshwater life; posterior prosomal limbs usually elongated into swimming paddles; body scorpionlike, with prominent telson spine.

(b) Class Arachnida (Latreille, 1810). Silurian to Recent. Scorpions, spiders, mites, and so on.

Changed to terrestrial, air-breathing animals; book gills converted to sunken book lungs, and/or tracheae developed; second legs modified into sensory-feeding palps (pedipalps); typically four pairs of ambulatory legs (on third through sixth metameres); first opisthosomal metamere (seventh, pregenital) highly reduced or suppressed; compound eyes suppressed; ability to excrete nitrogen as guanine (insoluble) acquired; coxal mobility lost, with main leg-body joint between coxa and trochanter; tarsi subsegmented; food ingested suspended in intestinally produced liquid or ingested in liquid form; semen enclosed in spermatophores in most.

B. Subphylum Gnathomorpha Boudreaux, 1979.

Functional tagmosis into an anterior sensory-gnathal region including the acron and at least three (secondarily in various lines four, five, or seven) metameres, and a posterior trunk section with locomotor, respiratory, digestive functions and the like; acronal antennae developed into the primary tactile- chemosensory organs; coxal gnathobases always well developed on appendage of second somite, but primitively present on most coxae.

1. Infraphylum Trilobitomorpha Størmer, 1944.

Sensory-gnathal region fused into a solid head capsule including first three metameres; acron expanded laterally and posteriorly partly surrounding the head somites; gnathobases of head limbs larger than other coxal endites. This group apparently retained respiratory epipodites on all the limbs.

(a) Class Trilobita Walch, 1771. Lower Cambrian to Devonian (extinct). Trilobites.

Terga heavily sclerotized, with prominent paratergal lobes (pleura); terga firmly articulated permitting only dorsoventral bending. Early trilobites without tagmosis in trunk, later developing a pygidium consisting of unseparated variable numbers of posterior metameres. The three head limbs with telopodite fully developed, but with gnathal endites on coxae.

(b) Class Trilobitodea Størmer, 1959. Named emended from Trilobitoidea. Mid-Cambrian Burgess Shales (extinct).

Paraterga not markedly set off from body, joining more or less smoothly; head appendages variously enlarged and adapted to seizing food.

2. Infraphylum Mandibulata Snodgrass, 1938.

First metameric appendage modified by suppression of epipodite and gnathobase, and changed to function as accessory sensory structure (second antennae) or lost; telopodite of second appendage reduced to three podomeres (mandibular palp) or lost; epipodite of mandible lost; coxal endites suppressed on legs of trunk, but well developed at least on mandible; food digestion in lumen of midintestine through action of enzymes released by rupture of epithelial cells.

a. Subterphylum Crustaciformia Boudreaux, 1979.

Tagmosis initially included union of part of the first metamere with the acron, forming a protocephalon, and union of second and third metameres (mandibular and maxillulary) into a single part. The first appendages became primarily tactile-chemosensory second antennae.

(a)   Class Cheloniellida Broili, 1933. Devonian (extinct).

Strong gnathobases on coxae of second, third, fourth, and fifth gnathal legs, with reduction of one podomere (? trochanter); paranota widely extended laterally resulting in forming a broad, disc-shaped body form. A patella appears to have been retained in the trunk appendages.

(b)   Class Crustacea Pennant, 1777. Ordovician to Recent.

Gnathal segment tagmosis extended to include the fourth metamere (second maxillary); epipodites further suppressed on appendages behind the twelfth metamere (abdomen); the first trochanter (basipodite) of all appendages developed a prominent exite lobe, the exopodite; the patella became suppressed in all forms; antennules became reduced to eight or less musculated antennomeres; the mandible became the main biting organ; the first maxilla (maxillule) was highly reduced; posterior growth of the maxillary tergum produced a carapace in some; typical locomotor legs became reduced posteriorly on an abdominal tagma distinct from a postcephalic locomotor tagma (thorax); hatching became precocious with establishment of the typical nauplius larva with only three metameres. Modifications typical of the various subclasses and orders.

b.   Subterphylum Myriapodomorpha Boudreaux, 1979.

Derived characters related to emergence from aquatic to terrestrial life: Air-breathing developed by invaginated metameric tracheae; return to uniramous limbs by suppression of epipodites (? = coxal eversible vesicles ?); limbs adapted to terrestrial locomotion.

(1)   Superclass Arthropleurida Waterlot, 1934. Carboniferous (extinct).

Very large (up to 180 cm long) centipedelike plant eaters; paraterga prominent, demarked from body; legs with all podomeres bearing spiny endites; heavy pleural plates around legs. Head poorly preserved, telson unknown. Legs still retained patella. Relationship indicated provisional.

(2)   Superclass Atelocerata Heymons, 1901.

First metamere suppressed except for its persistent ganglia (tritocerebrum) and embryonic limb buds; head capsule of acron plus three or four somites; levator muscle of pretarsus eliminated; patella suppressed; malpighian tubules from proctodaeum; diverticula of midintestine, when present (caeca), devoid of digestive function; telopodite of mandible eliminated; anterior tentorium established; coxal eversible vesicles redeveloped as water absorbing organs, variously modified or secondarily repressed in each class; distinct storage organ, the fat body, established; metameric tracheae established (? also in Arthropleurida); spermatophore established, with indirect transfer to female (convergent with Onychophora, Arachnida, some Crustacea). Primitive features retained, but variously modified or later eliminated elements in each class include the following: individually musculated antennomeres; transverse intersegmental tendons; precocious hatching into a larva of nine or more metameres with anamorphic growth; continued molting beyond sexually adult stage.

(a) Class Myriapoda Leach, 1814.

Mandibles bearing movable, articulated endite (gnathal lobe, gnathobase), abducted through outward thrust of swinging anterior tentorial apodemes; paratergal lobes reduced or eliminated; coxae with single ventral articulation with sternum; digestive diverticula entirely suppressed; molting ecdysis through transverse split behind head followed by exit from rear of head capsule and from front of trunk exuviae. Primitive atelocerate features include body behind head not formed into locomotor and nonlocomotor tagmata and limbs retained on most postcephalic somites.

i.   Subclass Collifera Boudreaux, 1979.

Collum segment without legs, formed from fourth metamere, not joined with head as in other atelocerates; first (and only) maxillae consisting of a pair of united coxae whose telopodites (palps) became highly reduced; gonopore ventral between second legs (sixth metamere); antennules of eight or less musculated antennomeres; larvae at hatching with fully developed head and usually four postcephalic somites, with three pairs of legs, plus two or more incomplete somites and the growth zone; dorsal intersegmental tendons suppressed.

(i)   Infraclass Pauropoda Lubbock, 1866. Fossils unknown.

Trunk segment number reduced to 11 or 12; leg number reduced to 9, 10, or 11 pairs; antennae with four or six antennomeres, bearing two branches at tip on which a ringed flagellum forms; tracheal openings when present on coxae of legs, but tracheae usually entirely suppressed; tergal plates usually suppressed on collum and fourth, sixth, eighth, and tenth trunk segments (*Millitauropus* bears 12 trunk somites, each with a tergal plate); mandibles entognathous consisting of a solid unsegmented unit; eyes suppressed; eversible coxal vesicles restricted to one pair ventrally on legless collum segment; intersegmental tendons lost except between mandibles; trochanter single; tarsus subsegmented with two tarsomeres.

(ii) Infraclass Diplopoda Blainville-Gervais, 1844. Carboniferous to Recent. Millipedes.

Trunk somites beyond the fourth fused in pairs, forming diplosomites each bearing two pairs of legs, two pairs of tracheal openings, two pairs of ventral ganglia and two pairs of heart ostia; paraterga reduced or absent; mandible with coxa divided into two, with flexible gnathal lobe; tracheae in tufts on inner ends of hollow pleural apodemes; intersegmental ventral trunk tendons (Pselaphognatha) lie below the nerve cord (above, in all other arthropods), or are suppressed except between the mandibles; anamorphic growth from larva bearing head, four trunk segments, three pairs of legs, two diplosomites with limb buds and telson with zone of growth in most, but secondarily a trend toward epimorphosis in some (*Pachyiulus* hatches with 17 pairs of legs); compound eyes highly reduced to a few facets; lateral repugnatorial glands on sides of body. Burrowing established by pushing into substrate, correlated with loss of tergo-coxal muscles and segmental tendons, sclerotization of cuticle into solid segmental rings, and extension of anamorphic growth to provide up to 100 or more diplosomites. The Pselaphognatha retained ancestral characters such as remaining nonburrowing, the persistent ventral intersegmental tendons, tergo-coxal muscles and paranota, fewer trunk segments (collum, three single and five double somites), soft exoskeleton, and palps on the gnathochilarium.

ii.   Subclass Atelopoda Boudreaux, 1979.

Fourth metamere united with head to form a head capsule of four somites plus acron; typical legs suppressed on last two trunk somites; antennae elongated consisting of many musculated antennomeres; anamorphic larvae hatch with head and five to seven pairs of trunk appendages and two or more partly formed somites.

(i)   Infraclass Symphyla Ryder, 1880. Oligocene Baltic Amber to Recent.

Gonopore simple, anterior, on fourth trunk segment (eighth metamere); attain sexual maturity before completing anamorphosis; first maxillae palpless, with two endite lobes similar to galea and lacinia of insects, but without division into flexible cardo and stipes; coxae of second maxillae also palpless, and flexibly united into a labiumlike structure; eyes suppressed; first trunk legs generally reduced, sometimes totally suppressed; tergites doubled and jointed flexibly on three or more trunk segments; intersegmental trunk tendons suppressed, replaced by sternal and coxal apodemes (mandibular and maxillary transverse tendons retained); tracheae restricted to head, with openings on side of head; thirteenth trunk segment legless, bearing tergal spinning organs; fourteenth trunk segment small legless and undetached from telson; eversible vesicles and styluslike pegs borne on venter near coxae; legs with single trochanter; tarsus undivided.

(ii)   Infraclass Chilopoda Leach, 1814. Cretaceous to Recent. Centipedes.

Mandible retracted in head pouch, loosely articulated basally through sclerotic rod, protrusible and retractile, with several coxal sclerites; coxae of both maxillae generally united; first maxilla with short flat two-segmented telopodite serving as a labium; second maxillae with elongate palplike telopodite; first leg with robust coxae bearing strong endites ("can openers" used in opening arthropod prey); telopodite of first leg modified into poison fang, with poison gland; compound eyes generally reduced or suppressed, but well developed in scutigeromorph centipedes; pleural sclerites acquired around leg bases; minimum number of metameres, 22, up to 170 in some; last two metameres without legs (may be gonopods); gonopore on last metamere; anameric forms complete anamorphosis before sexual maturity; fat body (as in insects) from coelomere cells rather than from endoderm as in other arthropods; gonads dorsal to intestine rather than ventral as in other myriapods; silk glands found in genital chamber, except in scutigeromorphs; tarsus subsegmented into two or more tarsomeres.

(b) Class Insecta Linnaeus, 1758, s.s. (= Hexapoda Latreille, 1825). Devonian to Recent. Insects.

Head capsule consisting of four metameres fused with acron (convergent with pycnogonids, crustaceans, and atelopodans); initial mandibular specialization a dorsal articulation with promotor-remotor rolling motion; coxa of first maxilla (insect maxilla) subsegmented into cardo and stipes, gnathobases articulated (galea and lacinia), telopodite (palp) shortened; coxae of second maxillae subsegmented, joined medially into a flexible platelike labium, telopodite (palp) shortened; posterior tentorium established; first three postcephalic somites (fifth, sixth, and seventh metameres) specialized into a locomotor thoracic tagma; thoracic legs with single trochanter; abdominal tagma of 11 somites with limbs initially reduced to coxal plates bearing short segmented telopodite (abdominal styli) and eversible coxal vesicles; total number of metameres limited to 18 or less; eleventh abdominal styli modified into sensory cerci; tenth abdominal coxal plates and styli suppressed; pleural areas of thorax initially sclerotized forming anapleure; thoracic coxae subsegmented, the upper portion joining the anapleure as a catapleure (anapleure and catapleure became pleuron in pterygotes); insectan median and lateral ocelli established in addition to compound eyes; secondary, simple gonopore posterior on abdomen; ecdysis through median dorsal split from head through thorax; midgut diverticula lost the digestive function. Ancestral atelocerate characters in primitive insect, variously modified or suppressed in more advanced forms include the following: individually musculated antennomeres; transverse intersegmental tendons in entire body; paranotal expansions; unsegmented tarsus, simple pretarsus; embryo not originally developing an amnion; molting continued beyond sexual adulthood; no external genitalia; simple rolling mandible; compound eyes; spermatophore with indirect transfer. New features originating in various evolving lines are as follows: paranota becoming wings; return to aquatic life; social organization; external genitalia; internal fertilization; parasitism; pupation; extraembryonic membranes (amnion and serosa); dicondylic biting mandibles; sucking mouthparts; segmented tarsi; suppression of pretarsus; tarsal claws; secondary winglessness; filter chamber of intestine; and so on.

CHAPTER 6

The Insects: Monophyletic or Polyphyletic?

The Linnaean class Insecta originally included arthropods of all types, but the name is used presently to include only the hexapodous atelocerates. The name Insecta persists because it is well-known, and the other arthropod classes have been removed one by one, leaving only the hexapods. There is a general feeling that Linnaean names for orders and classes should be given priority, even though not strictly descriptive, when no confusion would result in their use. The name Hexapoda is strictly synonymous with Insecta as used here, and is preferred by those who want a more applicable, descriptive term for the class.

For most entomologists, the use of the name Insecta for this group is standard usage, and in most textbooks of entomology there is no question that the group includes all the hexapodous atelocerates. The question of whether to restrict further the use of the term to include only the winged insects and the Thysanura, excluding the entognathous orders Diplura, Protura, and Collembola, or even to abandon the use of the term altogether, is still being debated. Some recent advocates of the further restriction of the taxon are Sharov (1966a) and Mackerras (1970). Manton (1972, 1973b) took the hexapod state of insects to have arisen independently five times, and would erect five classes of hexapod insects, each having supposedly evolved from a multilegged, soft bodied stock having a sclerotized head capsule. These five hexapod classes are represented by the Collembola, Protura, Diplura, Thysanura, and Pterygota. In Manton's opinion hexapody represents a grade of evolution, and cannot be a monophyletic insect character. Anderson (1973) agreed with Manton in erecting the subphyla Hexapoda, Myriapoda, and Onychophora to be included in a new phylum Uniramia, but he thought

123

that the hexapods had more in common than just six legs. The Chelicerata and Crustacea would join the Uniramia as phyla of convergently derived arthropods.

The exclusion of the Diplura, Protura, and Collembola from the insect class has been urged by past workers either because they resemble the myriapods in retaining various ancestral characters, or because they possess derived characters not found in other insects. The antennae of Diplura and Collembola bear musculated antennomeres as in myriapods. Proturans have suppressed their antennae, retaining only small external vestiges, known as pseudoculi, and remnants of internal antennal head muscles and nerves (Tuxen, 1959). The simple unsegmented tarsus is an ancestral character, and so is the absence of typical insect external genitalia. Whereas these insects have remained primitive in many ways, their specializations such as entognathy, small size, lack of compound eyes and so on are derivable from the ancestral states that were the initial derived characters of the ancestral insect, as enumerated at the end of the previous chapter.

The conclusion of Manton that the hexapodous state evolved independently five times to produce five lines of insects is the result of an intensive prolonged study of arthropod locomotor mechanisms covering a period of over 25 years (Manton, 1950 to present). The muscular and skeletal structures of representatives of all arthropod classes were studied in detail for the first time, and at the same time Manton elucidated the walking and biting movements which arthropods perform, correlated with specialized structure. Part 10 of her series (1972) described the functional anatomy of the skeleto-muscular system of apterygote insects in detail. It was demonstrated that the details of body and leg muscles, leg movements, jumping activities, and body-leg articulations are widely different in various apterygote orders. Each order includes specializations not found in any other order. As a result, she properly concludes that the specializations are such that no specialization in one order could have been the precursor of the specializations in any other order. This is to be expected since the subjects of study were extant species, each of which is the end result of a long separate evolutionary history. Manton has concluded that each specialization type arose in parallel with others not only because there appear to be mutually exclusive specializations in habit, function, and structure, but also, simply because hexapod locomotion has arisen also in arachnids and crustaceans, "there is nothing improbable in the parallel evolution of a hexapodous thorax in the five pterygote and apterygote classes." However, six-legged walking, which is the rule in the arachnid orders Solfugida, Amblypygida, Uropygida, and Palpigradida, is accomplished by the last three prosomal pairs of legs, borne on the fourth, fifth, and sixth metameres. The legs of the third metamere are modified and serve as sensory structures. In crabs which utilize hexapodous walking, it is the legs of the ninth, tenth, and eleventh metameres that serve this function. The legs of the twelfth metamere are swimming paddles, and those anterior to the ninth metamere serve for feeding and grasping (maxillipeds and chelipeds). The six legs of insects are on a different set of metameres, the fifth, sixth, and seventh. Manton said that, "since functional continuity must have been maintained through all evolutionary stages, and since the leg mechanisms of the hexapod classes are mutually exclusive, these classes cannot have given rise one to another and must have evolved in parallel *from ancestors with lobopodial limbs and little trunk sclerotization*" (italics added). The supposition that these ancestors became hexapodous before the body and legs became sclerotized and segmented forces the conclusion that leg segmentation also is a five times convergent event in the insects in spite of the essential similarity of all insect legs. Any zoologist with a good knowledge of insects would immediately recognize an isolated insect leg as such, and never would confuse it with the leg of any other arthropod, even though some insect thoracic legs are greatly specialized. The probability of separate origins of hexapody in insects is lessened when it can be shown that insect hexapody can be derived from one common origin, as is suggested by the serial homology of the segmentation of the insect thorax, the relatively subordinate development of the prothorax compared with the other thoracic segments even in the apterygote groups, and the general similarity of all insect legs in the single short trochanter and relatively long femur and tibia. The unity of origin of the insects is further suggested by insectan-derived characters such as the basically similarly specialized maxilla and labium, the apodemal two pairs of tentorial arms, the head consisting of the acron and four metameres, the typically single gonopore situated subterminally on the abdomen, the retention of midintestinal diverticula, and the 11-segmented legless abdomen. It is not necessary to resort to assumptions of parallelism unless a great deal of evidence for nonhomology exists. The work of Manton is important in that it emphasizes the details of specialization that the process of evolution has accomplished. The relegation of similarities to the process of convergent evolution without considering seriously the principles of phylogenetic theory in terms of changes from deduced ancestral character states does little to provide answers to problems of phylogeny but does act as a challenge for further support or rejection of hypotheses through further study.

One of the axioms of phylogenetic systematics is that ancestors cannot be recognized or identified but can only be hypothesized (Cracraft, 1974). The hypothetical ancestor of any line can only be characterized in a general way by endowing it with the derived character states which distinguish the line, in that these are not found in any line not descended from the presumed ancestor. Such derived characters, and therefore the characterization of a hypothetical ancestor, may also include features that may be considered primitive for the group.

THE DERIVED CHARACTERS OF INSECTA

The derived characters (synapomorphies) which may be assumed to have originated as monophyletic events in the ancestral insect line may now be discussed in greater detail than in the outline at the end of Chapter 5. The early terrestrial mandibulates probably bore limbs consisting of at least eight more or less subequal podomeres, as is suggested by the extinct Arthropleurida. If that was the case, the original insect limb lost two podomeres. Judging from the fact that most myriapods bear two trochanters, and insects bear only one trochanter, it seems that a trochanter was one of the lost podomeres, but only if the trochanters in the ancestral atelocerate first were shortened relative to the other podomeres. This latter possibility is suggested by the presence of two short trochanters in most myriapods. The patella is generally assumed to be the other lost podomere. The postfemoral podomerelike ring in the legs of some millipedes (Figure 50, PoFe) lies between the femur and tibia, which is the position of the chelicerate patella. But this leg segment in some cases has a muscle to the next distal podomere, the tibia, as would be expected if the postfemur were equivalent to other podomeres. The musculature suggests that the postfemur is either a subsegment of the tibia, or a vestige of the patella often lacking patellar muscles. It is thus barely possible that the atelocerate tibia resulted from the lack of separation, during ontogensis, of the patella from the tibia, making the present atelocerate tibia a composite of two former podomeres. In the Protura, the broad membrane at the femur-tibia joint of the middle leg contains a narrow sclerite in the form of a partial ventral ring. It bears no muscles. Is this sclerite the result of partial expressions of persistent patellar genes?

In an earlier chapter, the functional workability of an acondylar coxa-body union was discussed. A freely swinging leg base (coxa), joined to the side of the body through flexible arthrodial membrane only, is here postulated for the earliest land precursor of the Atelocerata. It is further hypothesized that the dorsal and ventral extrinsic muscles originating on the primitive interseg-mental tendons (Figure 8) caused movement of the coxae in all directions, but mostly about a laterally tilted vertical axis. Hemocoelic pressure in general opposed the individual muscle tensions that caused specific movements such as forward and backward swing of the coxa. From this beginning, an explanation of the development of various insect pleurons can be developed.

Much has been written about the origin of the insect pleuron. Most of the discussion can be summarized into three principal theories about the nature of

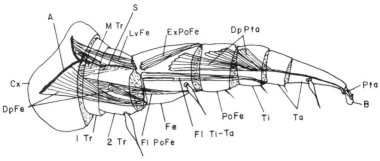

Figure 50   Leg of *Polyxenus* sp., bearing a postfemur, possibly a vestige of the patella. A, coxal apodeme; B, adhesive empodium; Cx, coxa; DpFe, depressor muscle of femur; DpPta, depressor muscle of pretarsus; ExPoFe, extensor muscle of postfemur; F1PoFe, flexor muscle of postfemur; F1 Ti-ta, flexor muscle of tibia and tarsus; LvFe, levator muscle of femur; M Tr, muscles of trochanter; Po Fe, Postfemur; Pta, pretarsus; S, trochanteral sulcus; Ta, tarsus; Ti, tibia; Tr, trochanter. Original.

the insect pleuron. One of these is that the various lateral sclerotizations about the arthropod coxae are nothing more than mechanically suitable sclerites arising from the sclerotization of the pleural membrane, many times and in response to the particular needs demanded by the specialized locomotor movements of the various arthropods (Manton, 1972). The oldest and most popular theory of the pleuron is one originated by Heymons (1899) and elaborated by Snodgrass (1927) and others. In this case it is supposed that originally there was a podomere lying between the body and the present coxa, the subcoxa. The subcoxa presumably joined the body, flattened out, became broken up into two concentric rings, and the coxa-subcoxa joint became the definitive coxa-body joint. According to this theory, the present insect coxa is homologous to the crustacean first trochanter and the true coxopodite became the insect pleuron in all its variations. Sharov (1966a), following Størmer (1939), assumed that a precoxa and the coxopodite both became implanted into the side of the insect body, producing a pleuron derived from two podomeres. Størmer had suggested that an obscure short supracoxal ring he found on trilobite legs, on which the epipodite emerged, was an additional ninth basic arthropod podomere, which he called the precoxa. The third theory, which is a compromise, assumes that the upper part of the coxa became detached from the coxa proper and merged with the pleural wall to become the lower part of the insect pleuron. The upper portion of the pleuron formed from direct sclerotization of the originally membranous pleuron. Matsuda (1970) discussed at length the various theories of the pleuron, and supported the third theory. He showed that the insect pleuron appears to be a structure that evolved in the insects independently of the pleural sclerotizations of other arthropods. There is evidence available that strengthens the interpretation of the insect pleuron as a composite structure. When examined phylogenetically the evidence provides for a monophyletic theory of insects as hexapods. It is assumed here that the original insect pleuron formed through the sclerotization of a membranous area between the lower edge of the tergum and the upper edge of the coxa, combined with a fragment of the upper part of the coxa which was first a subsegment of the coxa (coxomere), and which later became independent of the coxa. The strictly pleural sclerotization has been named the anapleurite, and the coxal fragment is called the catapleurite.

The interpretation of the anapleurite of insects as a sclerotized part of the originally membranous side of the body above the leg is suggested in several ways. In the extinct Arthropleurida a large "rosette plate" with sulci (Rolfe and Ingham, 1967) appears to provide a firm structure onto which the limb articulates. The coxa (Rolfe and Ingham's second podomere) bears a small sclerite dorsally. This sclerite was interpreted as the first podomere, but it resembles a coxal sclerite. Myriapods generally bear one or more pleurites, and Manton (1966) demonstrated that such pleurites in myriapods are not associated with condylar articulations with the coxa, but rather provide a means for the coxa to rock in a parasaggital plane at a membranous joint. The anapleurite of Protura is also fixed on the side above the catapleurite, and the catapleurite moves slightly forward and rearward against the anapleurite by virtue of a membranous joint between them (Manton, 1972). In all other insects except Collembola, parasaggital mobility of the catapleurite has been suppressed, and rocking of the leg is performed by various other means. In Collembola, the catapleurite is still a part (coxomere) of the meso- and metathoracic coxae and an anapleurite remains only on the prothorax.

Subsegmentation of the insect coxa resulting in the formation of the catapleurite and the definitive coxa is suggested by the following. The most obviously subsegmented insect coxa is the maxilla. A basal sclerite, the cardo, is flexibly articulated with a distal part, the stipes. The dorsal muscles originating on the cranium insert some on the edge of the cardo, some on the edge of the stipes and some on the flexible lacinial endite lobe. Muscles from the tentorium or transverse tendon insert into the body of the cardo and stipes. These muscles are all derivable from the hypothetical primitive coxal extrinsic muscles (Figure 8). Other muscles to the palp (telopodite), galea, and lacinia arise in the stipes (Figure 51). On the insect labium, the distal prementum is apparently the homolog of the stipes and bears the extrinsic labial muscles. The basal postmentum may represent a highly specialized fused pair of cardines. It is possible that the postmentum incorporates sternal elements of the labial segment because it bears no extrinsic muscles, but a pair of muscles originating on the postmentum insert into the prementum. These muscles may be all that is left of the ancestral ventral coxal labial muscles. The

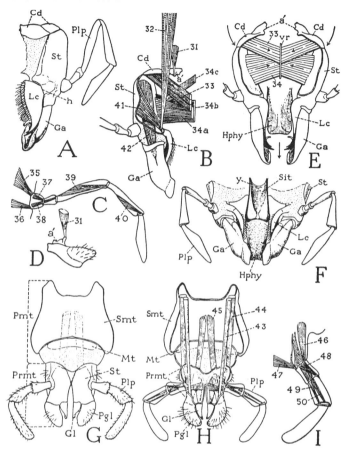

Figure 51   Maxilla and labium of *Periplaneta americana* (L), Insecta, Blattaria. A, right maxilla, posterior; B, right maxilla and muscles, anterior; C, maxillary palpus and muscles; D, maxillary cardo and muscle; E, diagram showing relation of maxillae to hypopharynx and protractor action of adductor muscles of maxillae anterior; F, maxillae in position retraction against the sides of hypopharynx, anterior; G, labium, posterior; H, labium and its muscles, anterior; I, labial palpus and muscles. a' cranial articulation of cardo; Cd, cardo; Ga, galea; Gl, glossa; h, hinge of lacinia on stripes; Hphy, hypopharynx; Lc, lacinia; Mt, mentum; Pgl, paraglossa; Plp, palpus; Pmt, postmentum; Prmt, prementum; Sit, sitophore; Smt, submentum; St, stipes; vr, ventral ridge of tentorium; y, oral arm of hypopharyngeal suspensorium. Reprinted from R.E. Snodgrass: *A Textbook of Arthropod Anatomy.* Copyright © 1952 by Cornell University. Used by permission of Cornell University Press.

collembolan catapleurites remain as a part of the meso- and metathoracic coxae (see Figure 57, p 153). The incipient catapleurite forms a complete ring on the upper end of the mesothoracic coxa, but the hind coxa bears a partly separated portion only, designated as part "a" of the coxa by Manton (1972). The position of the catapleurites in Collembola, unincorporated into the pleural area, is one of the neotenous character states of this group of insects. A condition somewhat similar to that of the Collembola may be seen in the extinct monuran, *Dasyleptus brongniarti*. Sharov (1966a) illustrated distinct subcoxal podomeres in his reconstruction of the animal, but the published photographs (Rohdendorf et al., 1961) do not permit precise identification of such podomeres. Sharov favored the subcoxal theory of the pleuron, and his reconstruction may have been influenced by his precon- ceptions. The "subcoxa" of *Dasyleptus* may not be true podomeres, but may rather be coxomeres of the same nature as those of the Collembola. The fossil imprints do not permit positive identification of supposed podomeres. An implanted coxomere (catapleurite) which has retained some mobility occurs in the Protura, a sister group of the Collembola. The motion of the catapleurite is of a different sort in the Lepismatidae. In this case, the elongated coxa lies in a horizontal position directed backward, and the coxal motion is in a horizontal plane. When the coxa swings out and forward, two sclerites of the coxa slide past each other in overlapping fashion, flexing at arthrodial membranes between them, the anapleurite, and the coxa proper. This situation is highly specialized in lepismatids, but the more primitive state is seen in *Tricholepidion gertschii*. Both the anapleurite and the catapleurite in *Tricholepidion* consist of sclerotized arcs extending from a low point ahead of the coxa, over the coxa rearward to a point behind the coxa. Dorsally the coxa articulates weakly with the catapleurite near the point of invagination of a small catapleural apodeme (Figure 52). The Machilidae bear only the catapleurite as an easily seen sclerite above the coxae of the meso- and metathorax. On these two somites the anapleurite has joined the underside of the paranotal lobes (Barlet, 1950). The pleural apodeme of the catapleurite is greatly elongated and serves as an anchor point for the powerful leaping muscles of the middle and hind legs (Barlet, 1950). The possibility of the dissociation of a coxomere from the coxa and its union with a pleural area is further suggested by the analogous separation of the meron from the coxa of the middle leg of higher Diptera, and its union with the lower part of the pleuron, where it is known as the hypopleure (Snodgrass, 1935). Among arachnids, the whole coxa has become hardly more than a fixed sclerite on the body in some mites. The plecopteran larval thorax exhibits the state of bearing an anapleurite and a catapleurite, and the adult prothorax of some species remains this way, while the larval state changes to a typical pterygote pleuron in the adult pterothorax. It was demonstrated above that there is a discernible common pleural structure in the insects. The original state of a sclerotized pleural anapleurite, and a detached coxal catapleurite as the beginning of the pleuron is retained but variously modified in the Protura, Diplura, Microcoryphia, Thysanura, and some pterygotes. The absence of a catapleurite in the collembolan prothorax, of anapleurites on the other

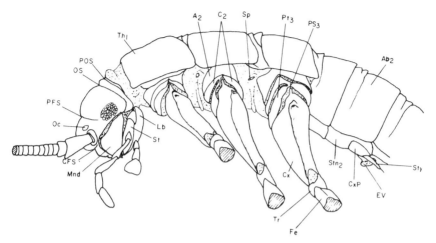

Figure 52   *Tricholepidion gerschii* Wygodzinsky, Thysanura, Lepidothricidae. Left side of anterior body. A_2, mesothoracic anapleurite; Ab_2, second abdominal tergum; C_2, mesothoracic catapleurite; CFS, clypeofrontal sulcus; Cx, coxa; CxP, left coxal plate of second abdominal segment; Ev, eversible vesicle; Fe, femur; Lb, labrum; Mnd, mandible; Oc, lateral ocellus; Os, occipital sulcus; PFS, postfrontal sulcus; POS, postoccipital sulcus; PS_3, pleural sulcus of metathorax; Pt_3, pseudotrochantin of metathorax; Sp, spiracle; St, maxillary stipes; Stn_2, second abdominal sternum; Sty, stylus; Th_1, prothoracic tergum; Tr, trochanter. Original.

thoracic segments, and only partial separation of the catapleurites of the middle and hind coxae are neotenous states and of secondary nature in the Collembola. The common basic pleural structure of insects is in harmony with a monophyletic concept of the insects.

The ancestral insect head is with hardly any doubt the result of the tagmosis of at least four metameres with the acron. There is no evidence that the tagmosis may have been progressive as has been suggested for the myriapods. It would seem that three gnathal metameres joined the protocephalon in a single step. In most insect embryos the head tagma is clearly set apart early as a unit along with the thorax and abdomen.

The original insect mandible was probably a simple rolling mandible bearing an enditic gnathobase, with which it was possible to rasp on small food particles such as algae, much in the fashion of present machilids, diplurans, and some collembolans. The early mandible was probably not firmly articulated with the head by any condyle, and was possibly slightly protrusible and retractile. A dorsal mandibular condyle was developed later in the line which led to machilids, but specializations in the ability to protrude and retract the mandible apparently occurred in the entognathous line which led to diplurans, proturans, and collembolans. In order to avoid repeated reference to the Diplura, Collembola, and Protura as a group in the following discussion, the group name Entognathata is used. This is the sister group to all other insects, here combined in the group named Ectognathata.

There is so much fundamental similarity in structure and musculature of insect maxillae that there is no doubt that the maxilla was acquired in the ancestral insect, but became variously modified in the various insect lines, along with other mouthparts. The insect maxilla bears a reduced telopodite in the form of a segmented palpus. The segments of the palpus are individually musculated by intrinsic muscles, just as in the usual telopodite. Judging from the large size of the six-segmented maxillary palpus (see Figure 60,p. 159) of the extinct *Dasyleptus brongniarti* (Monura), the original insect maxillary palpus was perhaps larger than is found in most insects and was more leglike. In one direction, machilids acquired additional subsegmentation in the palp femur and tarsus, and in the other direction, the maxillary palpus became much reduced in most insects. The coxal nature of the body of the maxilla has not been seriously questioned. The lacinia with its teeth is a gnathobasic movable endite, just as is the flexible gnathobase of myriapod jaws. In each case a sector of the anterior (secondarily dorsal) dorsal promotor muscle inserts on the base of the flexible lobe as a lacinial flexor (Figures 23, 24, and 51). Other maxillary muscles are basically similar to mandibular muscles. If the insect maxilla did not have a palpus, it would be comparable in structure to the basic myriapod mandible, although myriapods have developed highly specialized mandibular mechanisms of their own.

There is very slight evidence that may suggest that the original insect mandible may have been subsegmented in a fashion similar to the maxillary coxa. In the machilid mandible there is a sclerotic internal ridge (Figure 29, b) which apparently provides some basal rigidity near the articular condyle. A similar internal ridge can be seen basally on the mandible of *Heterojapyx* (see Figure 54, E). The endite of the machilid mandible consists of a strongly sclerotized molar area and an extended picklike structure (Figure 29) which are set off from the mandible by a similar internal sclerotic ridge, (Figure 29,b). It is just barely possible that these sclerotic areas, which now add strength to the mandible, may have been points of former flexibility. One might point also to the flexible lacinia mobilis or prostheca found on dipluran and larval ephemerid mandibles. The prostheca of *Campodea* is flexed by a dorsal muscle originating in the head and inserting on a long cuticular apodeme (Tuxen, 1959), exactly in the same manner of flexion of the maxillary lacinia in most insects, and of the gnathal lobe of myriapods.

The insect labium is fairly uniform in its basic structure. It arises in the embryo through union of the limb buds of the labial somite to become the functional underlip. The equivalents of the maxillary cardo and stipes are generally considered to be represented by the basal postmentum and the distal prementum, but the extrinsic labial musculature is much reduced and arises on the tentorium only. Endites known as glossa and paraglossa seem to be homologous with the maxillary lacinia and galea. The Symphyla bear maxillae and a labium remarkably similar to the insect counterparts, but in view of the many derived characters of Symphyla that they share with Myriapoda, it is most likely that the symphylan maxilla and labium represent a remarkable case of convergence with insects.

Another probable early ancestral insect character was the establishment of the posterior tentorial apodemes. Such apodemes are not present in the myriapods, but both insects and myriapods bear the anterior tentorial apodemes. Posterior tentorial apodemes are formed by invagination on each side at the presumed intersegmental line between the maxillary and the labial segments. The genus *Campodea* appears to have lost the anterior tentorium, but the posterior tentorial apodemes can be found in all insects except in those whose posterior tentorium is suppressed (Heteroptera).

It is difficult to determine the position and nature of the gonopore in the first insect. Judging from the gonopores of diplopods, pauropods, crustaceans, and chelicerates, the primitive gonopores opened on the leg bases as paired structures. The exceptional paired male gonopores of proturan and dermapteran insects, and of both sexes in mayflies, are the only instances among insects in which the gonopore is not a single median ventral opening. But in proturans the openings are retracted into a pocket of the eleventh abdominal segment, and in earwigs and mayflies they are on structures of the tenth segment in the male, and the dual mayfly gonopores of the female open just behind the seventh segment. The gonopore of both sexes is simple and on the ninth abdominal segment in Diplura, and on the hind edge of the fifth segment in Collembola. In all other insects, the male gonopore is simple and opens on phallic structures of the tenth segment, while in the female the exit pore of the eggs may be on the eighth, ninth, or tenth segment. The phylogenetic relationships of insects that can be inferred through study of derived insect characters (discussed in subsequent chapters) suggest that possibly the gonopores were originally double but unstable in position. The Protura seems to be the only group in which the double gonopore might have been retained. But if this is so, then the collembolan and the dipluran simple gonopores, as well as the simple gonopores of other insects, were each developed in parallel. If a simple gonopore was original in the Ectognathata, the double gonopore condition in mayflies and earwigs must be a secondary return to a mimic of the primitive arthropod state, because the closest phylogenetic relatives of these two orders bear simple gonopores. The explanation involving the least complications, considering all the advanced characters of all insects, is one supposing that the ancestral insect had already joined the two gonopores to an invaginated epidermal structure which became the simple median genital passage, as suggested by the ontogenesis of insects. A simple suppression of the median invagination would restore the original double exit of both sexes in mayflies. In proturans (Figure 59, C), the double exits of the vasa deferentia end on a pair of styletlike structures and the whole phallic complex is invaginated just behind the eleventh sternum, and the invagination forms in effect a simple genital passage into which the ends of the genital ducts protrude (Tuxen, 1970). Thus the exterior opening of the genital passage, or gonotreme, may be the equivalent of the ancestral insect's simple secondary gonopore. The ancestral position of the original secondary gonopore of insects remains unknown. Since it is in such a variety of positions among the insects, it probably became fixed through mutational changes in each of the various insect lines.

The insect abdomen has no counterpart in other arthropods. It is composed of the eighth through the eighteenth metameres, plus the nonmetameric telson segment. Abdominal limbs were retained only as vestiges in the ancestral insect. The vestiges are variously secondarily modified in various orders, but it would seem that the original condition in the ancestral insect was a reduced limb which consisted of the coxa and a very short telopodite. The remnant of the telopodite became the abdominal styli of the apterygote insects, as is suggested by the persistence of what appear to be coxal muscles that insert on the base of the styli. The coxa apparently remained a free podomere in the Protura, and it is movable by way of two extrinsic muscles. In some Protura the first abdominal coxa bears a short podomere that is flexed by muscles originating in the coxa (Denis, 1949). In other insects whose abdominal coxae are still evident, they are flat plates on each side of a small sternum as in the Microcoryphia, or on the hind border of a larger sternum as in some Thysanura (Lepidothricidae), or male mayflies and grylloblattids. The only remaining traces of abdominal limbs on the tenth abdominal segment are embryonic limb buds which in most insects are completely suppressed before hatching. External genitalia of some male insects form on the venter of the tenth segment during postembryonic stages, as discussed later.

The insectan cercus seems to have been established in the ancestral hexapod as a modification of the limb of the eleventh abdominal segment. The ontogenetic origin of cerci is from growth of the embryonic limb buds of the last metamere. When it is segmented, as in the lower insects, the segments are not individually musculated. It appears to be homologous with styli elsewhere on the abdomen. The uropod of malacostracan Crustacea is homologous with insect cerci only to the extent that they are independent modifications of the metameric limbs of the eighteenth metamere. The cerci, or spinnerets, of Symphyla are often made homologous with insect cerci, but in Symphyla these are on a somite ahead of the cercus-bearing somite of insects, and in fact they are tergal structures (p. 113) rather than leg homologues (Ravoux, 1962).

The manner of exit from the exuviae during ecdysis is typical in insects, so that in this respect the process probably originated with the ancestral insect, when compared with ecdysis of myriapods. In the latter, the old cuticle separates transversely behind the head, and the head is first pulled out of the head capsule. Then the body exits from the trunk exuviae through the anterior transverse crack. In all insects in which the process has been observed, there is first a dorsal longitudinal split over the thorax. The head may be then pulled out of the head capsule, as in Entognathata, or the top of the head splits along a middorsal line of weakness continued from the thoracic ecdysial line, as in other insects. In most cases the abdomen exits from the exuviae without splitting. The dorsal thoracic ecdysial line is evidence against any supposition that the earliest insects were unsclerotized.

Another new feature to characterize the ancestral insect is the near total loss of a digestive function in the midintestinal diverticula, known as caeca in insects. Caeca occur at the anterior end of the midintestine only in the orders Collembola, Microcoryphia, Thysanura, Orthoptera, Blattaria, Isoptera, Plecoptera, Coleoptera, Megaloptera, Anoplura, and Mallophaga, but are not necessarily present in all members of each order. The three pairs of caeca of the machilid *Dilta littoralis* were described as containing secretory cells which rupture in releasing the secretion, as is typical of mandibulate midintestine, but the nature of the secretion is not known (Denis, 1949). New lining cells are regenerated in the machilid caeca at molting time. The caeca of termites also contain nests of regenerative cells, and apparently the lining is also renewed. But digestive enzymes have not been identified as a product of insect caeca. Some beetles harbor symbionts in their caeca, and in other insects the caeca appear to be able to absorb products of digestion.

Finally, the insectan spermatozoa have been found to be remarkably consistent and differing from those of other arthropods in bearing an extra ring of nine doublets of axonemes accompanying the usual 9 + 2 arrangement of most animal spermatozoa, and their axoneme pattern can be characterized by the formula 9 + 9 + 2 (Baccetti, 1970, 1979). The insectan sperm also had acquired two mitochondrial derivatives that extend into the flagellum, as a ground plan condition. Further evolution of the spermatozoa has occurred in various lines, and is discussed when appropriate.

ANCESTRAL ARTHROPOD CHARACTERS OF THE FIRST INSECTS

There are additional features of the less evolved insects that can be assumed to have been retained by our hypothetical insect ancestor, inherited from ancient preinsect arthropodan progenitors. As evolution went on in the various insect lines, most of these primitive characters were variously modified, often in parallel in different lines. The relatively primitive arthropod characters of the ancestral insect are briefly mentioned, and the evolutionary changes are mentioned where appropriate.

Individually musculated antennomeres are typical of the Myriapoda. Only the Diplura and Collembola bear fully musculated antennae. Other insects bear muscles in only the basal segment, or scape, which presumably operate the second segment, the pedicel. The only antennomere remaining beyond the pedicel is the flagellum, a nonmusculated, subsegmented antennomere.

A sulcus extending over the back part of the head, from one posterior tentorial invagination to the other, is suspected to be the persistent intersegmental groove between the labial and the maxillary somites. The sulcus corresponds to an internal postoccipital ridge on which dorsal longitudinal muscles are attached. The groove can be seen in the heads of all insects except the Entognathata. The postocciput behind the groove is often very narrow or unsclerotized, but it is easily seen in the Microcoryphia and Thrysanura.

Intersegmental tendons have been retained but highly modified in the apterygote orders. Among these the transverse tendons between both the maxillae and mandibles were retained, except in most Thysanura. In this order, the ventral mandibular and maxillary muscles became attached to the anterior tentorium and hypopharyngeal apodemes, as in pterygotes. The intersegmental tendons of the trunk persisted ventrally in the thorax of Japygidae, Collembola, Microcorphia, and Thysanura. Dorsal thoracic tendons between the ends of the longitudinal dorsal and ventral muscles also persisted in these three groups. The Thysanura apparently have not retained the dorsal abdominal intersegmental tendons. Among the Pterygota, fragments of tendons remain above the thoracic spinae in larval *Corydalus cornutus* (Barlet, 1977).

The tergal paranota of ancestral arthropods were retained by the first insects, but became variously reduced, suppressed, or enlarged into wings. The Entognathata have greatly reduced the thoracic paranota, but slight expansions remained in the Diplura and some Collembola. The lateral edges of the abdominal terga, slightly overlapping the abdominal sterna, are all that remain of abdominal paranota in the Entognathata. The Monura, Microcoryphia, and Thysanura bear paranota on the trunk segments of the thorax and the first nine abdominal segments, and abdominal paranota persisted on the extinct Paleodictyoptera. Campodeid embryos develop distinct lateral tergal flaps above the limb buds of the abdomen, but these do not keep pace with growth of the rest of the embryo, so that the paranota are weak in postembryonic stages.

Since the Entognathata and Monura have an unsegmented tarsus, the subsegmented tarsus of higher insects must be a secondary derived character state (autapomorphy). These lower insects also bear a pretarsal podomere, which appears to be naked in Monura, but it bears tiny spines in many Protura (see Figure 55, A; p. 148), and in some Collembola one of the spines is spurlike and enlarged, even often resembling a claw (Figure 55, B). The claws of Diplura are articulated on the body of the pretarsus, suggesting that in this order the claws are enlarged pretarsal spurs (Figure 55, C, D, E, F). In the Ectognathata other than Monura, the two tarsal claws are articulated directly on the end of the tarsus, and the pretarsus is reduced mostly to a small sclerite (unguitractor plate) on which the pretarsal flexor apodeme invaginates, and the pretarsal sclerite is free of the claws (Figure 55, G, H). It seems that the tarsal claws of the higher insects are tarsal structures convergently evolved with the pretarsal claws of the Diplura and of arachnids.

It is very likely that the first insect did not have complex external genitalia, and the male probably produced its semen enclosed in a packet, or spermatophore, which was deposited so that the female could pick it up from the substrate. All the male apterygotes deposit a spermatophore that is later picked up by the females in various ways. The males of pterygote insects other than the Odonata generally deposit their spermatophores directly into the single genital opening of the females, but apparently the males of Mecoptera, Diptera, and perhaps Siphonaptera produce fluid semen not enclosed in a spermatophore. The evolution of external genitalia is discussed in Chapter 7.

As discussed in Chapter 4, it is highly probable that compound eyes were retained as well developed functional organs in the first insects, and have variously been modified by total or partial suppression, or by an increase in size and efficiency. Simple eyes, or ocelli, also may be presumed to have originated very early in arthropod evolution. Ocelli are commonplace in arthropods of all types except myriapods. Among insects, ocelli typically appear as a single median eye plus a pair of lateral eyes on the head, served by the protocerebrum as the visual center. Various insect evolutionary lines have suppressed the ocelli in various ways. Those of Collembola are almost entirely suppressed, in that the cuticular cornea is not differentiated, and the three ocelli lie under the epidermis as small groups of specialized visual cells generally seen only in histological sections. Median and lateral ocelli are entirely suppressed in Diplura, Protura, Dermaptera, Embiidina, Mallophaga, Anoplura, and Strepsiptera, but ocelli are present in some or all members of all the other orders of insects. In some beetles only the median ocellus remains (Dermestidae), and in some other Coleoptera (Staphylinidae), in Siphonaptera, Lepidoptera, and some Homoptera only the two lateral ocelli are present.

The tracheal openings of insects are mostly metamerically placed, but abdominal spiracles are always lacking on the ninth, tenth, and eleventh segments. Typically, there are mesothoracic and metathoracic spiracles. It seems that the first insect retained at least 10 pairs of tracheal openings. Various lines, especially the Entognathata, have suppressed spiracles in different metameres, some to the point of total suppression of the tracheal system, such as in the suborder Arthropleona of Collembola. The early insectan spiracles were probably simple and permanently open. A special closing apparatus is found only in the higher insects.

Since eversible coxal vesicles appear on many myriapods and lower insects, their presence in insects is probably a retention of a primitive atelocerate condition. These persist on some abdominal segments of Diplura, Microcoryphia, and Thysanura. In Collembola, the eversible collophore appears to have been modifed from the coxal sacs of the first abdominal segment.

Repeated molting cycles occuring after attainment of sexual adulthood is so commonplace among arthropods that the phenomenon can be regarded as an ancestral arthropod character. Adult xiphosurids continue molting several times a year. Theraphosid and filistatid spiders are very long lived and molt many times after sexual maturity. Postnupial molts are commonplace among crustaceans and are the general rule in apterygote insects. The more specialized arachnids, crustaceans, and winged insects never molt after becoming adult. The suppression of imaginal molting appears to have occurred independently in each of these groups of arthropods.

In their embryology, there are some features shared by some insects which are heritages from their atelocerate ancestors. One such is the absence of the amnion in the apterygote insects. The pterygote amnion forms from closure of the blastoderm around the embryo after it sinks into the yolk. In the apterygote insects, the embryo is at first superficial, but later a flexure forms so that the embryo bends ventrally. In Collembola and Diplura, the embryo folds over without extensive sinking into the yolk, similar to the condition in Symphyla, but the sinking is profound in the Microcoryphia and Thysanura. In the latter two orders, the sunken embryo becomes partially surrounded by the blastoderm, leaving a small opening, but without the fusion of blastoderm that produces the amnion of pterygote insects.

The final solution to the problem of whether insects are a monophyletic assembly is yet to be presented. The discussion in this chapter demonstrates that it is possible to assume that insects are monophyletic in their origin, following the principles of phylogenetic study.

CHAPTER 7

Evolution in the Class Insecta

PART 1
OVERVIEW

In Chapter 6 the discussion emphasized the hypothetical beginnings of insects by considering the ancestral states. This chapter attempts to follow some of the changes that may have taken place during the evolution of various insect lines, producing the great variety of orders. A phylogenetic classification is offered which can be depicted by a family tree (Figure 53) or presented in outline form. Each taxon represents a monophyletic line opposed to a sister group, each of which is presumed to have evolved with the appearance of derived characters in each line, independently of derived characters of other lines. The ranking is nonarbitrary, in that rank is determined by the presumed sequences of bifurcations. In order to accommodate the splits, it was necessary to employ 13 categorical ranks between that of class and order.

The class group and section group categories include taxa whose names were standardized with the ending "-ata", except for the group Pterygota, which was retained in its well-known form. The cohort group names were standardized with the ending "-odida" or "-ida." Superorders are made to end with "-odea" because the "-oidea" used extensively should be avoided. The International Code of Zoological Nomenclature (Stoll et al., 1964) has reserved the ending "-oidea" for superfamily names. New names in the outline are in italics. Emended names previously used, or sometimes at a different rank, are indicated by the authorship placed in parentheses. In using names for taxa the principle of priority has been followed in most cases.

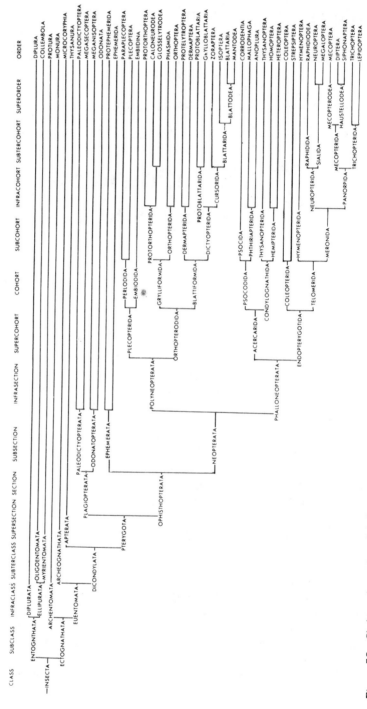

Figure 53 Cladographic outline of phylogenetic insect classification.

140

PHYLOGENETIC CLASSIFICATION OUTLINE OF CLASS INSECTA (LINNAEUS, 1758) S. S.

I. Subclass Entognathata (Hennig, 1953).
  A. Infraclass Diplurata (Hennig, 1969).
    Order Diplura Börner, 1904.
  B. Infraclass Ellipurata (Börner, 1910).
    (A). Subterclass Oligoentomata Berlese, 1909.
      Order Collembola Lubbock, 1873.
    (B). Subterclass Myrientomata Berlese, 1909.
      Order Protura Silvestri, 1907.
II. Subclass Ectognathata (Hennig, 1953).
  A. Infraclass *Archentomata*.
    Order Monura Sharov, 1957.
  B. Infraclass Euentomata Berlese, 1909.
    (A). Subterclass Archeognathata (Börner, 1904).
      Order Microcoryphia Verhoeff, 1904.
    (B). Subterclass Dicondylata (Hennig, 1969).
      A. Supersection Apterata (Linnaeus, 1758) s.s.
        Order Thysanura Latreille, 1796, s.s.
      B. Supersection Pterygota Brauer, 1885.
        (A). Section Plagiopterata (Lemche, 1940).
          1. Subsection *Paleodictyopterata*.
            i. Order Paleodictyoptera Goldenberg, 1854.
            ii. Order Megasecoptera Brongniart, 1893.
          2. Subsection Odonatopterata (Martynov, 1938).
            i. Order Meganisoptera Martynov, 1932.
            ii. Order Odonata Fabricius, 1792.
        (B). Section Opisthopterata (Lemche, 1940).
          1. Subsection Ephemerata (Handlirsch, 1908).
            i. Order Protephemerida Handlirsch, 1908.
            ii. Order Ephemerida Leach, 1817.
          2. Subsection Neopterata (Martynov, 1925).
            (1). Infrasection Polyneopterata (Martynov, 1925)
              1. Supercohort *Plecopterida.*
                (1). Cohort *Perlodida*
                  i. Order Paraplecoptera Martynov, 1938.
                  ii. Order Plecoptera Burmeister, 1839.
                (2). Cohort Embiodida (Handlirsch, 1908).
                  Order Embiidina Enderlein, 1903.
              2. Supercohort *Orthopterodida.*
                (1). Cohort *Grylliformida.*
                  a. Subcohort *Protorthopterida.*
                    i. Order Protorthoptera Handlirsch, 1908.

    ii. Order Caloneurodea Martynov, 1938.

    iii. Order Glosselytrodea Martynov, 1938.

  b. Subcohort Orthopterida (Handlirsch, 1908) s.s.

    i. Order Phasmida Leach, 1815.

    ii. Order Orthoptera Oliver, 1789. s.s.

(2). Cohort Blattiformida (Handlirsch, 1908).

  a. Subcohort Dermapterida (Martynov, 1925).

    i. Order Protelytroptera Tillyard, 1931.

    ii. Order Dermaptera Leach, 1815.

  b. Subcohort Dictyopterida (Leach, 1815).

    (a). Infracohort *Protoblattarida*.

      i. Order   Protoblattaria (Handlirsch, 1908).

      ii. Order Grylloblattaria (Brues and Melander, 1932).

    (b). Infracohort *Cursorida*.

      a. Subtercohort *Zorapterida*.

      Order Zoraptera Silvestri, 1913.

      b. Subtercohort *Blattarida*.

      (*a*). Superorder *Isopterodea*.

      Order Isoptera Brulle, 1832.

      (*b*). Superorder Blattodea Brunner, 1882.

        i. Order Mantodea Burmeister, 1838.

        ii. Order Blattaria Burmeister, 1829.

(2). Infrasection *Phalloneopterata*.

  1. Supercohort Acercarida (Börner, 1904).

  (*1*). Cohort Psocodida (Weber, 1933).

    a. Subcohort *Psocida*.

    Order Corrodentia Burmeister, 1839.

    b. Subcohort Phthiriapterida (Weber, 1939).

      i. Order Mallophaga Nitzch, 1818, s.s.

      ii. Order Anoplura Leach, 1815, s.l.

  (*2*). Cohort Condylognathida (Börner, 1904).

    a. Subcohort Thysanopterida (Weber, 1933).

Order   Thysanoptera   Haliday, 1836.
b.   Subcohort Hemipterida (Linnaeus, 1758) s.s.
   i.   Order   Homoptera   Leach, 1815.
   ii.   Order Heteroptera Latreille, 1810.
2.   Supercohort Endopterygotida (Sharp, 1893).
   (1).   Cohort Coleopterida (Handlirsch, 1908).
      i.   Order Coleoptera Linnaeus, 1758, s.s.
      ii.   Order Strepsiptera Kirby, 1813.

   (2).   Cohort Telomerida.
      a.   Subcohort   Hymenopterida   (Handlirsch, 1908).
         Order Hymenoptera Linnaeus, 1758.
      b.   Subcohort Meronida.
         (a).   Infracohort Neuropterida (Handlirsch, 1908).
            a.   Subtercohort   Rhaphidida (Handlirsch, 1908).
               Order   Raphidiodea   Burmeister, 1839.
            b.   Subtercohort Sialida (Handlirsch, 1908).
               i.   Order Neuroptera Linnaeus, 1758, s.s.
               ii.   Order Megaloptera Latreille, 1802, s.s.
         (b).   Infracohort   Panorpida   (Handlirsch, 1908).
            a.   Subtercohort Mecopterida.
               (a).   Superorder Mecopterodea.
                  Order   Mecoptera   Packard, 1886.
               (b).   Superorder   Hanstellodea (Clairville, 1798), s.s.
                  i.   Order   Diptera   Linnaeus, 1758.
                  ii.   Order   Siphonaptera Latreille, 1825.
            b.   Subtercohort Trichopterida.
               i.   Order   Trichoptera   Kirby, 1826.
               ii.   Order Lepidoptera Linnaeus, 1758.

PART 2
THE APTERYGOTE INSECTS

SUBCLASS ENTOGNATHATA

The three orders of this group, Diplura, Collembola, and Protura, together form a sister group to the rest of the insects, the Ectognathata, by virtue of their many specializations in common. In many ways they have also retained primitive features some of which can be considered to have originated in pre-insect arthropods, and others which appear to be early insectan characters.

Synapomorphies of the Entognathata

The insects in this group have evolved peculiar modifications of the mouthparts and head that are so fundamentally similar that there is hardly any doubt that the members of the group are descended from a common entognathous ancestor. During embryogenesis, lateral folds of the head (plica orales) form over the buds of the mouth parts, and grow downward to fuse with the base of the labium, or postmentum (Figure 54, F). The result is to enclose the elongated mandibles and maxillae in a gnathal pouch. The two pairs of jaws are turned forward to lie nearly horizontally, and their bases apparently are sunken somewhat in the posterior part of the gnathal pouch, since their proximal ends lie near the back part of the head. In insects in general, the maxillae can be protruded and retracted by virtue of the sharp bend between the cardo and stipes. In addition, the mandibles of the Entognathata are also protrusible and retractible by means of dorsal muscles having origins on the head capsule (Tuxen, 1959).

The tentorial apparatus of the Entognathata has become modified in its own peculiar fashion. Manton (1964) described the tentorial apodemes of a collembolan and a campodeid employing tissue sections and staining to differentiate the cuticular apodemes from the transverse gnathal tendons, for the first time. No such study of the proturan head is available, but the description by Tuxen (1959) can be reinterpreted from the descriptions of Manton. The ancestral state of the insect tentorium can be envisioned as two pairs of apodemes. The anterior pair form by invaginations just ahead of the mandibles. The posterior pair invaginate between the bases of the maxilla and labium. In all insects, the posterior tentorial apodemes become united forming an internal bridge.

With the development of the gnathal pouch, the posterior apodemal invaginations of Entognathata have shifted forward, lying on each side of the

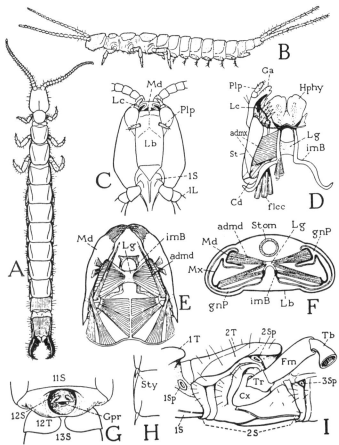

Figure 54. Diplura. A, *Heterojapyx gallardi* Tillyard; B, *Campodea* sp.; C, *Heterojapyx gallardi*, head and part of prosternum, ventral; D, same, hypopharynx, posterior tentorium, and right maxilla, ventral; E, same, mandibles and their muscles, dorsal; F, same, cross section of head, somewhat diagrammatic; G, same, ventral genital region of male; H, same, right halves of two abdominal sterna, showing styli; I, same, mesothorax and parts of adjoining segments, ventrolateral. admd, adductor muscle of mandible; admx, adductor muscle of maxilla; cd, cardo; Cx, coxa; flcc, cranial flexor muscle of lacinia; Fm, femur; Ga. galea, gnP, gnathal pouch; Gpr, gonopore; Hphy, hypopharynx; imB, posterior tentorium; L, leg; Lb, labium; Lc, lacinia; Lg, intergnathal transverse tendon; Md, mandible; Mx, maxilla; Plp, palpus; S, sternum; Sp, spiracle; St, stripes; Stom, stomodaeum; Sty, stylus; T, tergum; Tb, tibia; Tr, trochanter. Reprinted from R.E. Snodgrass; *A Textbook of Arthropod Anatomy.* Copyright © 1952 by Cornell University. Used by permission of Cornell University Press.

forwardly displaced hypopharynx, and then immediately joining as a short tentorial bridge (further modified in Protura). On each side, from the bridge, growth of the posterior apodemes into the head forms a pair of arms ("fultura") which reach backward as the skeletal support of the cardo of the maxillae (Figure 54, D), and between them stretches the remains of the transverse tendons of the maxillae and mandible (Figure 54, D, F). No other insects have developed the internal posteriorly directed tentorial apodemes. The anterior tentorial apodemes have remained distinct only in the Collembola. In these, two short anterior apodemes occur, each separate from the other, invaginated in the gnathal pouch above the mandibles on each side of the mouth. Some mandibular muscles originate on the transverse mandibular tendon, as in Machilidae. The anterior tentorium is suppressed in Diplura, and has not been positively identified in Protura.

The compound eyes are highly reduced, and have been retained as a small group of visual units only in some Collembola. Protura and Diplura have neither compound eyes nor ocelli, but the median and lateral ocelli in Collembola are still present as vestigial internal structures, often detached from the epidermis, without the differentiation of the cuticle into corneal facets (Denis, 1949).

Besides the specialization represented by the loss of eyes, the Entognathata have suppressed other ancestral insect features. In all of them, the malpighian tubules are barely perceptible. The anterior gastric caeca are practically gone, with some Collembola retaining small caeca. The ancestral paranota are highly reduced, with just traces of such remaining on the thoracic and abdominal terga of the adult, but there are distinct paranotal lobes above the limb buds of dipluran embryos that disappear before hatching.

It is not possible to determine with confidence if the absence of a distinct clypeus, marked off from the head by a clypeofrontal sulcus, is secondary in this group, or if a clypeus is an advanced feature of the higher insects, and never formed in the Entognathata. A "clypeofrontal sclerite" has been described in some Collembola; but it is very obscure and there is no suture with the frons. It appears to be a heavily sclerotized arc above the labrum, and it does not bear the dilators of the epipharynx. These muscles originate at the upper part of the labrum in *Tomocerus* (Manton, 1964). Thus it appears that the collembolan clypeofrontal sclerite is not the homologue of the clypeus of other insects.

Another specialization in this group is the great reduction of the palpi. The maxillary palpus is two-segmented in Protura and Diplura, and further reduced to a single segment in Collembola. In all, the labial palpi are short simple pegs. The $9 + 9 + 2$ pattern of spermatozoan axonemes and the two mitochondrial derivatives typical of insects was retained only in the Campodeidae. The pattern was reduced to the 9 + 2 state in Japygidae, Collembola, and Protura, and the sperm became nonmotile in the Collembola and Protura.

Persistent Atelocerate and Other Arthropod Features

The individually musculated antennomeres of ancient arthropod origin were retained in the Entognathata, in common with the Myriapoda and Crustacea. The original Protura probably bore musculated antennae that became entirely suppressed later. The neck of the Entognathata has not yet acquired the lateral cervical sclerites which are typical of the Ectognathata. In the latter, a sclerite apparently derived from the preepisternum on each side forms an articulating structure with the head forward, and the pleuron rearward. The simplest lateral cervical sclerites are found in the Microcoryphia (Matsuda, 1970), where they are part of the preepisternum and not yet detached. As in other arthropods and a few other insect orders, molting continues after sexual adulthood. In common with other ateloceratans, ecdysis does not involve a median dorsal split of the head cuticle, but the head is pulled out of the intact head capsule. Another primitive atelocerate character is the simple tarsus, further modified in Collembola. The pretarsus is still a distinct podomere attached to the ventral depressor tendon, but it bears variously modified setae or spines (Figure 55) that mimic tarsal claws. In the Devonian springtail, *Rhyniella praecursor*, the pretarsus was a bare clawlike podomere, very similar to the pretarsus of Protura and Monura. This further lends support to the idea that the original insect pretarsus was a simple pointed structure, and that tarsal claws evolved twice among insects, once in the Diplura and again in the ancestral Euentomata. Stepping is performed employing the primitive stance of contacting the ground with the tips of the pretarsi. Embryos in this group have not evolved either the extraembryonic amnion, which is a pterygote specialization, or the sinking of the embryo into the yolk as happens in Microcoryphia and Thysanura. Nothing in the characters of the Entognathata suggests that there were ever any external genitalia of the sort found in other insects. The primitive state of producing a spermatophore and depositing it on a substrate is retained, and internal copulation has not been acquired. Diplura and Collembola deposit a stalked spermatophore, which is later picked up by the female (Cassagnau, 1971; Paclt, 1956a; Schaller, 1971). The primitively rolling mandible with a single articulation, still found in the lower Crustacea and somewhat modified in Pauropoda and Symphyla, persisted in the Entognathata, least modified in Diplura and picklike and piercing, highly modified in Protura. The first entognathans apparently retained eversible coxal vesicles, which persisted as such in Diplura, but were

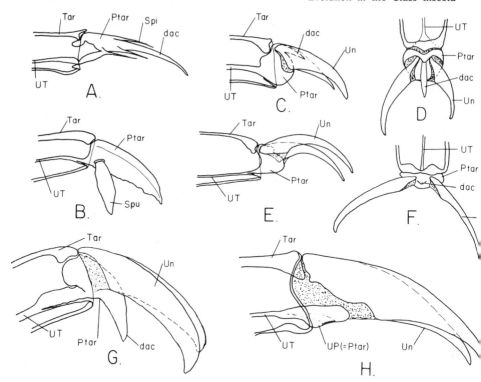

Figure 55. Variations of apterygote insect pretarsi. A, *Eosentomon* sp., lateral; B, a collembolan, lateral; C, a japygid (Diplura), lateral; D, same, dorsal; E, a campodeid (Diplura), lateral; F, same, dorsal; G, a lepismatid (Thysanura), lateral; H, a machilid (Microcoryphia), lateral. dac, dactylus of pretarsus; Ptar, body of pretarsus; Spi, spine; Spu, spur; Tar, tarsus; Un, unguis or claw; UP, unguitractor plate; UT, an unguitractor tendon. The unguis of entognath insects (C-F) articulates on the pretarsus apart from the tarsus. The unguis of ectognath insects articulates on the end of the tarsus, and only indirectly with the pretarsus or its remnant. Original.

mostly suppressed or modified in Collembola, and entirely lost in Protura. The transverse tendon system has been retained in the head and trunk, but in Protura there has been considerable loss of tendons except in the head.

Typical Ancestral Insect Characters Variously Changed

The simplest insectan pleuron was retained in this group in the form of two sclerites. An upper sclerite, the anapleurite, forms an arc over a ventral catapleurite. The catapleurite is slightly mobile in Protura, is still part of the upper coxa in Collembola, and is fixed in the Diplura. Abdominal limb vestiges remain as styli, with a set of up to seven pairs in Diplura, three pairs in Protura, and one remaining pair on the jumping organ of Collembola. The original three-lobed hypopharynx was retained in the Diplura and Collembola, and was lost in the Protura. Only the Diplura have retained cerci.

INFRACLASS DIPLURATA

The name Diplurata is used here to indicate a sister group relation with the Ellipurata. Hennig (1969) used the name Diplura in this fashion, as well as for the order Diplura, named by Börner (1904). The group was originally a suborder of the Thysanura, known as Entotrophi (Grassi, 1889). The latter name is older, but the order name Diplura is better known and most commonly used.

Synapomorphies of the Ancestral Diplura

The anterior tentorium was suppressed in this order. Since eyes are still present in some Collembola, their absence in Diplura is convergent with their loss in Protura. The embryo exhibits 11 abdominal segments, but on hatching only 10 appear. The tenth and eleventh abdominal metameres unite with a barely perceptible embryonic telson segment to form the definitive tenth segment, on which the cerci are borne. The pretarsal spines are enlarged and articulated laterally on the pretarsus, so that the end of the leg (Figure 55, C, E) superficially resembles the two-clawed foot of the Ectognathata. In the latter group the claws articulate directly on the end of the tarsus, and the pretarsus is highly reduced. The coxae are articulated ventrally with the sternum, rather than with the pleuron. The primitively rolling mandible can rotate, but it bears well-developed teeth terminally, suggesting a biting function. The abdominal coxal plates in the Diplura are completely fused with the sterna, but the styli are still present. Styli are fairly prominent in campodeids, and drag along supporting the abdomen as in machilids and thysanurans, but they are small and peglike in japygids. The gonopore of both sexes is on the venter of the ninth segment, a unique condition among insects. In the male of japygids, on either side of the gonopore, two small pegs of unknown homology appear. Their position on the hind border of the sternal plate suggests that they might be styli of the ninth segment. Eversible vesicles remain in most Diplura on the first eight abdominal segments, but in some they have been lost in the first segment.

Autapomorphies Among Modern Diplura

The ancestral cerci are generally believed to have been somewhat elongated and segmented. This condition still exists in the family Projapygidae, where the cerci have acquired ducts of silk glands. The cerci have become long and many-segmented in the Campodeidae, and have become short, robust, and opposable as grasping organs in the Japygidae. Tracheal evolution has proceeded in two directions in the Diplura. Most japygids have the usual meso- and metathoracic spiracles, plus two additional pairs of thoracic spiracles (Figure 54 I, 1sp, 3sp). The japygids bear seven pairs of abdominal spiracles. Japygids of the subfamily Parajapyginae bear only the two usual pairs of thoracic spiracles; thus these bear nine pairs of spiracles. In the opposite direction, abdominal spiracles are absent in the Campodeidae and Projapygidae, and there are three pairs of thoracic spiracles.

The dipluran head differs from that of the Ellipurata in that the oral folds join a rather broad labial base, resulting in the persistence of two lines of fusion some distance apart. The Ellipurata have advanced this feature in that the basal piece of the labium is almost or totally obliterated by encroachment of the oral folds ventrally, leaving the prementum protruding ahead of the oral folds. The ground plan spermatozoan of insects was retained by the Campodeidae, with the 9 + 9 + 2 pattern of axonemes in the sperm tail. This was secondarily reduced to the more primitive 9 + 2 state of animals in general in the Japygidae, in parallel with the same reversion in the Mecoptera, Siphonaptera, and some Diptera. The neck has remained membranous without cervical sclerites.

INFRACLASS ELLIPURATA

The name for this group appropriately refers to the absence of cerci (Greek *elleipsis*, defect plus *oura*, tail). There are no abdominal spiracles, and the antennae are reduced to four segments or less. The postmentum of the labium is very narrow between the oral folds, or the oral folds join ventrally. The result is that only the prementum seems to remain ahead of the ventrally apposed oral folds of the head. The Ellipurata have reduced their body size to 8 mm or less in length. The gonopore became fixed on the last abdominal metamere, the eleventh in the Protura, and the fifth in the neotenous Collembola. The spermatozoa of the Ellipurata have become coiled and immotile, together with a loss of one of the outer ring of axonemes, returning to the primitive 9 + 2 axoneme pattern. No cervical sclerites have been developed in this group. The orders Collembola and Protura represent sister groups of the Ellipurata.

The subterclass Oligoentomata was so named because the abdomen consists of only five segments. The group is represented by a single order, Collembola. The oldest known insects are represented by the fossil collembolan *Rhyniella praecursor* (Hirst and Maulik, 1926) (Figure 56, D), described from

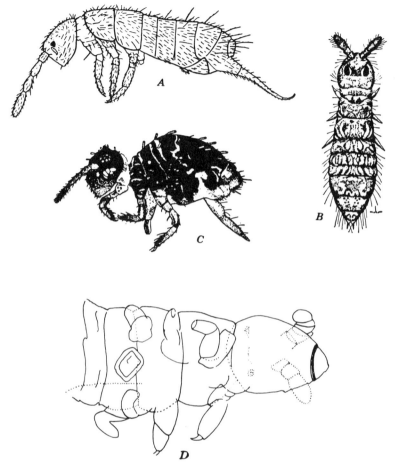

Figure 56 Representative Collembola; A, *Isotoma andrei*, and B, *Achorutes armatus*, suborder Arthropleona; C, *Neosminthurus clavatus*, suborder Symphypleona; D, *Rhyniella praecursor*, a Devonian fossil collembolan, the oldest known insect. Reproduced from Ross: A *Textbook of Entomology*, 3rd. ed., 1965, with permission of John Wiley and Sons, Inc.

the middle Devonian (Scourfield, 1940). Such a highly specialized insect type occurring in the Devonian strongly suggests that the first insect originated in the Silurian or earlier.

Synapomorphies of the Collembola

The springtails are the most highly specialized of the Entognathata. The order name refers to the highly developed eversible vesicles (collophores) of the first abdominal segment. These are highly extensible by blood pressure, and are useful as grooming organs. They were originally believed to produce a sticky secretion (Greek *kolla*, glue plus *embolon*, peg), which is now believed to be saliva.

On the head, the eyes are highly reduced, and may be absent in some species. The compound eyes consist of eight or less facets, and the ocelli are vestigial internal structures without corneal areas. The antennae are primitive in bearing muscles, but the segments number only four (neoteny?). The maxillary palpi appear as a small single segment, and there are no stipitopalpal muscles (also neoteny?).

The thorax of springtails appears to be simple, but it is highly specialized in some ways. Most springtails are very small, extending from 0.25 mm up to 8 mm in length. Their small size is correlated with lack of extensive sclerotization, especially of the sternum, and with a generally neotenous appearance. The anterior part of the prothorax is often indistinguishable from the neck, and the major part of the pronotum is insensibly joined with the mesonotum. Dorsal prothoracic muscles operating the forelegs originate on the apparent mesonotum in *Orchesella* (Manton, 1972). Only the prothoracic pair of spiracles were retained by some Collembola, and these are in the unsclerotized pleuron of the prothorax, so that the spiracles appear to lie in the neck. "Neck" spiracles are found in the suborder Symphypleona, and in the genus *Actaletes* of the suborder Arthropleona. In the latter, tracheae reach only into the head, but extend into the trunk in the Symphypleona. Tracheae are mostly lacking in the Arthropleona. The pleuron of springtails is the simplest found among insects. The simple condition is probably the result of their neoteny. The anapleurites are very poorly sclerotized, and appear at their best in the metathorax (Manton, 1972). The catapleurites are still part of the meso- and metacoxae, and are not at all incorporated in the pleuron. The legs are joined to the body through arthrodial membrane without condyles. However, the legs and membranous pleura are firmly anchored by strands of specially developed intersegmental tendons that prevent ballooning of the soft cuticle and dislocation of the coxa-trochanter joint when blood pressure rises in jumping. Another neotenous state is seen in the lack of separation of the tibia from the tarsus (Figure 7). The leg segment past the knee is a tibiotarsus because the depressor muscle of the pretarsus originates in part on the tibial portion near the knee, and partly more proximally in the femur, as in other insects. A very short part at the distal end of the tibiotarsus is discernible in some species, where it is marked off by an inflexible cuticular groove which may be all that remains of a former tibiotarsal joint. Although the pretarsus is a prominent elongate structure, it may appear "two-clawed" because a spur near the base may be enlarged into an accessory empodial appendage (Figure 55, B).

Neoteny is well shown in the abdomen. The embryonic abdominal segments are not generated beyond the fifth, and the telson surrounding the anus may be indistinct. A parallel condition occurs in pycnogonids, whose opisthosoma develops into a single structure. The ancestral insects may have been anamorphic, and hatched before completing the generation of the abdominal segments. The Collembola are epimorphic in the sense that a mutation in this line may have eliminated further development in an originally anamorphic line. In addition to the development of the embryonic limb buds of the first abdominal segment into the collophore, the embryonic limb buds of the third segment end up as a functional latch (retinaculum) that retains the jumping organs when it is not in use (Figure 57, C). The jumping organ (furcula) grows from the embryonic limb buds of the fourth abdominal segment. The basal fusion of the two coxae is similar to the fusion of the labial coxae, and forms the part called manubrium. Beyond the manubrium the two telopodites, which are homologous with the styli of other insects, are separate and can be clamped over the retinaculum by adductor muscles from the manubrium (Figure 57, D). The furcular telopodites are the only surviving segmented abdominal styli of Recent insects excepting the genital styli of mayflies. The terminal piece (mucron) often is constructed much like the pretarsus of the springtail, sometimes bearing a seta similar to the setalike accessory empodium. The longitudinal and some dorsoventral abdominal

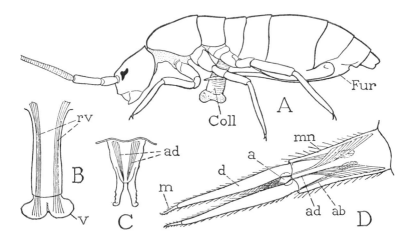

Figure 57   Abdominal appendages of a collembolan, *Tomocerus vulgaris*. A, entire insect with furcula in flexed position; B, collophore; C, retinaculum; D, furcula; a, part that closes over retinaculum in repose, ab, abductor muscle; ad, adductor muscle; d. dens of furcula; Fur, furcula; m, mucron (? pretarsus); mn, manubrium of furcula; rv, retractor muscle of eversible vesicle; v, eversible vesicle. From R.E. Snodgrass: *Principles of Insect Morphology*. Copyright 1935 by McGraw-Hill, Inc. Used with permission of McGraw-Hill Book Company.

muscles are powerful in springtails. These muscles build up the blood pressure, which extends the furcula suddenly after its release from the retinaculum. The sudden extension of the furcula can hurl the animals high in the air during a leap (Manton, 1972).

The fact that cleavage in springtail eggs is total has been offered to indicate a primitive state in springtails. But yolky eggs undergoing meroblastic cleavage were presumably the standard state for ancestral lobopods, and were retained in most arthropods, insects, and many myriapods. It makes better sense to regard the totally cleaving yolk-poor egg of springtails as a secondary specialization. Ultimately, the springtail embryonic blastomeres undergo tangential cleavage, and a central yolk mass surrounded by yolk-free blastoderm forms (Anderson, 1973). The Collembola have retained the ellipuran ground plan of a coiled immotile spermatozoan.

There seem to be only a few primitive characters in springtails. The musculated antennomeres have been mentioned. The mandibles generally are primitive in being hollow, and bear a molar area distally, beyond which there is a picklike extension, as in machilid jaws. The action of the mandibles consists of a primitive rolling motion during which food particles are crushed by the molar areas, but there is an additional action typical of the Entognathata, in which the mandibles are protruded and retracted. The picklike tips then are employed in scraping up food particles which are suspended in saliva.

The subterclass Myrientomata was given the name because the members of the only included order Protura appear to have 12 distinct abdominal segments, one more than the usual number in insect abdomens (Figure 58), and were thought to be myriapodlike in having abdominal limbs. The name Protura (primitive tail) suggests that full segmentation of the abdomen is an ancestral character.

Synapomorphies of Ancestral Protura

The Protura lack segmented antennae. A pair of small structures known as pseudoculi appear to be vestiges of the antennae. Tuxen (1959) described an antennal nerve and a pair of muscles attached to the pseudoculi, and suggested that these might be slightly movable. The embryos of Protura are unknown. Studies of proturan embryos should settle the question of the antennal nature of pseudoculi, if it can be shown that antennal buds give rise to the pseudoculi. Four-legged walking is the rule, and the antennal functions are performed by the forelegs. The mandibles have become solid styletlike slender rods, which with the styletlike laciniae perform piercing actions in feeding. The primitive rolling motion of the mandibles was eliminated, and the muscles were changed to perform the protrusion and retraction functions

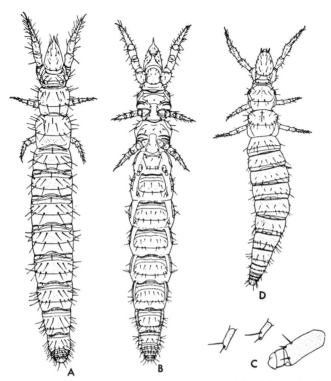

Figure 58. Proturans. A, dorsal, and B, ventral aspects, and C, abdominal limb vestiges of *Acerentomon doderoi* Silvestri; D, dorsal aspect of *Eosentomon ribagai* Berlese. From Essig: *College Entomology.* Copyright 1942 by the MacMillan Company. Used with permission of the MacMillan Company.

(Figure 59, B). Food particles loosened by the slender pointed mouthparts are sucked up suspended in saliva. The posterior extensions of the tentorium are fused along the midline before diverging and forming a support for the maxillary cardo. The anterior tentorial arms seem to be separately fused with the posterior tentorium near where the latter is invaginated into the head. Both anterior and posterior tentorial invaginations are in a forward position (Figure 59, B). The eyes and the hypopharynx have been suppressed.

An innovation on the thorax can be seen in the ventral coxal articulation with the sternum (Figure 59, A). In some species the coxa articulates with the catapleurite also, and the catapleurite can slide back and forth below the anapleurite, providing a parasaggital rock to the coxa. When tracheae are present, there are only two pairs of spiracles on the thorax.

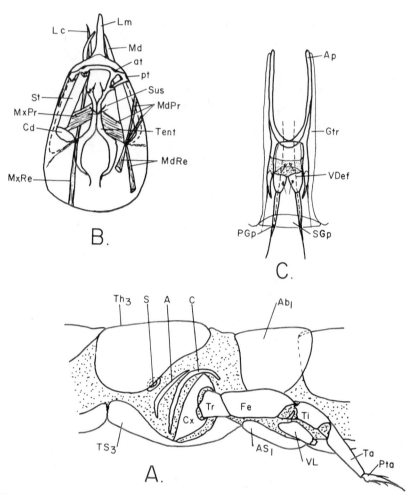

Figure 59   Some details of proturan anatomy. A. *Eosentomon pallidum* Ewing, left side view of metathorax and first abdominal segments; B. *Acerentomon* sp., head partly dissected, dorsal, left maxilla, right mandible, tentorium and muscles; C. *Eosentomon* sp.,male genitalia, dorsal. A, anapleurite; Ab₁, first abdominal tergum; Ap apodeme of genitalia; AS₁, first abdominal sternum; at, anterior tentorial pit; C, catapleurite; Cd, cardo; Cx, coxa; Fe, femur; Gtr, gonotreme; Lc, lacinia; Lm, labrum; Md, mandible; MdPr, mandibular protractor muscles; MdRe, mandibular retractor muscles; MxPr, maxillary protractor muscles; MxRe, maxillary retractor muscles; pt, posterior tentorial pit; Pta, pretarsus; PGp, primary gonopore; S, spiracle; SGp, secondary gonopore; St, stipes; Sus, dorsal suspensory muscle of tentorium; Ta, tarsus; Tent, posterior extension of tentorium; Th₃, metathoracic tergum; Ti, tibia; Tr, trochanter; TS₃, metathoracic sternum; VDef, vas deferens; VL, vestigal leg. Part A, is an original; Parts B and C, were redrawn slightly modified and relabeled from Grasse: *Traite De Zoologie*, Vol. 9 Masson, Paris, 1949, with permission of Masson et Cie.

The proturan abdomen has retained vestiges of the first three abdominal legs, but these vestiges are highly reduced. The second and third abdominal legs apparently consist of highly reduced coxae, without styli. In the Eosentomidae the first abdominal legs also are reduced as above, but in the Acerentomidae the first legs consist of a coxa with extrinsic muscles, and with intrinsic muscles that insert on a small podomere that appears to be all that is left of the stylus (telopodite). The tip of the stylus is membranous, and mimics the coxal eversible sacs of other insects, since it is inflatable and retractile. The proturan genitalia have apparently evolved independently of all other insects, except that perhaps the terminal position of the proturan gonopore can be related to the terminal position of the collembolan gonopore. However, the Collembola are neotenous in the development of the abdomen, and fail to develop the usual 11 abdominal segments. The vasa deferentia of the proturan male open separately into a genital chamber invaginated ventrally into the eleventh abdominal segment. Each vas deferens is extended into a stylet-like prolongation (Figure 59, C). The double phallic complex can be extruded out of the genital chamber by evagination of the chamber. When the genitalia are retracted at rest, the external opening of the genital chamber forms a single secondary gonopore.

The alimentary canal is very simple in the Protura. Malpighian tubules are represented only by enlarged cells at the upper end of the hind intestine, and there are no caeca. The fore and midintestines lack muscles and the suspended food particles are digested intracellularly in midintestinal cells that do not disrupt, as they do in arachnids (Denis, 1949). Proturans have become some of the smallest insects, with a maximum length of about 2 mm. The proturan spermatozoa are coiled and immotile, as in the Collembola, but the 9 + 2 pattern was further reduced to 9 + 0, while in some cases the sperm cells are of the 12 + 0 or 14 + 0 pattern, with disc-shaped sperm cells.

Plesiomorphies of the Protura

The hatching of proturans into a form whose abdomen has eight metameres and a zone of growth at the telson is apparently a retention of the ancestral arthropod feature of anamorphosis, which is commonplace in the Myriapoda and Crustacea. A new abdominal segment is added with each of three successive molts, producing the final 11 segments plus the separate telson. Two more molts may occur, without the addition of segments, before sexual maturity occurs. Proturans are the only insect group in which the telson segment remains distinct from the other segments.

The proturan pretarsus has remained a single clawlike podomere, but it often bears tiny spines, none of which are jointed at the base (Figure 55, A). The spines on other insect pretarsi are enlarged to resemble claws, and usually the pretarsus is highly reduced.

SUBCLASS ECTOGNATHATA

The name Ectognathata descibes an ancestral character, the generally exposed mouthparts not surrounded by facial overgrowth. Some of the Ectognathata bear seemingly enclosed mouthparts evolved independently, such as the suctorial mouthparts of Anoplura, Condylognathida, and Siphonaptera.

The head in this group usually has evolved sulci that strengthen the head capsule. A clypeofrontal (epistomal) sulcus forms the boundary between the clypeus, a new structure, and the frons. The clypeus bears muscles to the preoral cavity (cibarium) that pull on the front wall of the cibarium, enlarging it so as to provide a sucking function. Extrinsic muscles of the labrum originate on the frons and insert on the base of the labrum, and the frons usually bears the median ocellus. The external pits of the anterior tentorial apodemes generally lie in the lateral boundary of the clypeofrontal sulcus. The posterior tentorial pits are in the lower ends of the postoccipital sulcus. This latter sulcus is believed to represent a persistent intersegmental line between the maxillary and labial somites, but it may be a derived character of the Ectognathata, since there is no such sulcus in the Entognathata. Dorsal thoracic muscles are attached internally on the ridge of the sulcus. The postoccipital sulcus and internal ridge functions as a muscle attachment point and as a posterior strengthening structure of the head. The posterior tentorial arms are joined together to form a sclerotic bridge (corporotentorium) across the lower part of the cranium and serve for muscle attachments.

The antennae of the Ectognathata became reduced to a maximum of three antennomeres. The basal segment, known as the scape, is the only segment that is musculated. Extrinsic head muscles insert on its base, causing most of the motion of the antennae. In the scape, intrinsic muscles insert on the base of the next segment, the pedicel, and flex the pedicel slightly against the scape. The third segment is the flagellum, which has become subsegmented into flagellomeres. There are no muscles from the pedicel to the flagellum.

A derived character of the thorax is seen in the suppression of the prothoracic spiracles. The two pairs of thoracic spiracles are on the mesothorax and metathorax. Another derived thoracic character is the acquisition of a pleural apodeme in the catapleurite. It is a large apodeme in machilids and is much smaller in the Thysanura. In the pterygotes the pleural apodeme invaginates along a more or less vertical sulcus. The sulcus provides rigidity in the well developed pterygote pleuron.

The insect ovipositor originated in the early Ectognathata. The indication of the beginning of an ovipositor consists of the enlarged coxal plates on the eighth and ninth abdominal segments of the Permian *Dasyleptus brongniarti* (Figure 60). Further elaboration of the ovipositor occurred in the Euentomata

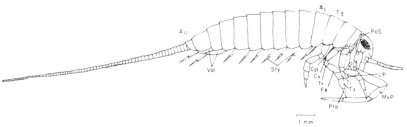

Figure 60 *Dasyleptus brongiarti* Sharov, Monura, Lower Permian. Reconstruction based on illustrations in Sharov, 1966. A, abdominal segments; Cpl, catapleurite resembling a coxomere; Cx, coxa; Fe, femur; LP, labial palpus; MxP, maxillary palpus; PoS, postoccipital sulcus; Pta, pretarsus; Sty, stylus; T₃, metathoracic tergum; Ta, tarsus; Ti, tibia; Tr, trochanter; Val, valvifers of primitive ovipositor.

with the elongation of the eighth and ninth abdominal coxal plates, and the conversion of the eighth and ninth coxal vesicles into elongate structures forming a bundle of four annulated "valves" (Figure 61 K, 1 Gon, 2 Gon), such as are found in Microcoryphia and Thysanura (Bitsch, 1974a). The valves of the pterygote ovipositor became stiff and articulated one on the other through a tongue-and-groove mechanism, which permits a drilling process into a substrate and makes openings into which eggs can be laid. Apparently the early pterygote ovipositor was toothed, as seen in the Permian cockroach *Kunguroblattina microdictya* Bekker-Magdisova and Vishnjakova (Figure 62), and in the paleodictyopteran *Monsteropterum moravicum* Kukalova-Peck (Figure 63). Teeth on the ovipositor blades (valves 1 and 2) are characteristic of the lower Hymenoptera, and a suggestion of teeth occurs on the ovipositor of the primitive thysanuran *Tricholepidion* (Figure 67). Generally, the coxal styli of the genital segments were retained on the eighth and ninth coxae in machilids and thysanurans, and in nymphal female cockroaches. In other insects, coxal styli remain only on the ninth coxae, at best. Nymphal grylloblattids and adult female raphidians have coxal styli on the elongated ninth coxae. These latter are often elongated and modified to serve as a sheath enclosing the ovipositor valves at rest, and are generally known as the third valvulae (Figure 64). The further evolution of the ovipositor is treated in the discussions on various insect groups.

The primitive female gonopore of the Ectognathata appears to have been medial just behind the seventh sternum, formed by an epidermal invagination into which the lateral oviducts emptied close together. With the acquisition of an ovipositor, the gonopore shifted to the posterior border of the eighth segment (Figure 64), by closure of a median longitudian groove on the eighth sternum, forming an extension known as the oviductus communis or vagina.

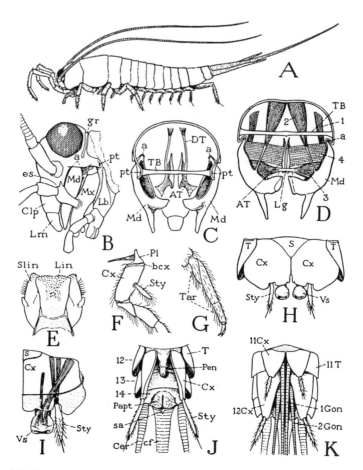

Figure 61 Microcoryphia, Machilidae. A, *Machilis* sp.; B, *Nesomachilis maoricus* Tillyard; C, same, posterior view of head, diagrammatic, with mandibles in place, showing tentorium; D, same, mandibles and their muscles, posterior; E, same, hypopharynx, posterior; F, *Machilis* sp., base of right metathoracic leg, lateral; G, same, tarsus and pretarsal claws; H, *Nesomachilis maoricus*, ventral surface of a pregenital abdominal segment, I, same, right half of ventral plates of an abdominal segment, with stylus and vesicle, dorsal; J, same, terminal part of male abdomen, ventral; K, *Machilis* sp., end of female abdomen, with ovipositor, ventral. a, primary articulation of mandible; AT, anterior tentorial arm; bcx, basicoxite; Cer, cercus; cf, caudal filament; Clp, clypeus; Cx, coxa; DT, dorsal tentorial arm; es, epistomal (clypeofrontal) sulcus; Gon, gonapophysis, or valvula, of ovipositor; gr, postocular groove; Lb, labium; Lg, transverse tendon; Lin, lingua; Lm, labrum; Md, mandible; Mx, maxilla; Papt, paraproct; Pen, penis; Pl, catapleurite with pleural apodeme; pt, posterior tentorial pit; S, sternum; sa, supraanal plate; Slin, superlingua; Sty, stylus; T, tergum; TB posterior tentorial bridge; Vs, eversible vesicle. Reprinted from R.E. Snodgrass; *A Textbook of Arthropod Anatomy*. Copyright © 1952 by Cornell University. Used by permission of Cornell University Press.

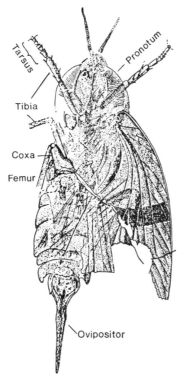

Figure 62 *Kunguroblattina microdictya* Becker-Migdisova and Wischnjakova, a Lower Permian cockroach with a well developed ovipositor. From Hennig, W: *Die Stammesgeschichten Der Insekten*. Verlag Waldemar Kramer, Frankfurt am Main, 1969, p. 179.

Above the ninth sternum and the ovipositor bases there generally forms a pocket, the genital chamber. In the Panorpida, with the suppression of the ovipositor, the genital chamber has closed into a tubular extension of the oviduct, and the definitive new female gonopore is often behind the ninth sternum.

The male gonopore of the earliest ectognath insect appears to have been formed by the union of a pair of internal vasa deferentia with an invagination from the anterior median border of the tenth abdominal segment. The ejaculatory duct thus formed from the ventral epidermis provided the primary median gonopore in this group, which was at the intersegment ventrally between the ninth and tenth abdominal segments. The external genitalia of male insects originate ontogenetically from several different organs, and are not universally homologous, although Smith (1969) was among the more re-

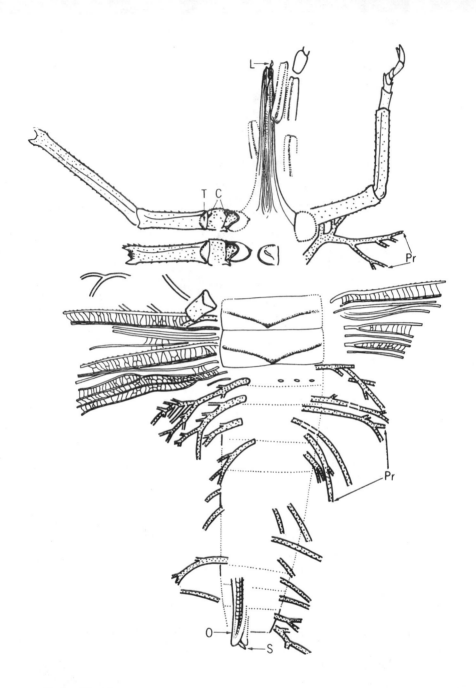

Figure 63 *Monsteropterum moravicum* Kukalova-Peck, Paleodictyoptera, Lower Permian. Note the sucking rostrum, ovipositor, sclerotized tergal projections, and the still separate catapleurite and anapleurite above the mesothoracic coxa. C, coxa; L, lacinia; O, ovipositor; Pr, setiferous projections; S, stylus; T, trochanter. Reproduced from Kukalova-Peck, 1972, with permission of *Psyche: Journal of Entomology.*

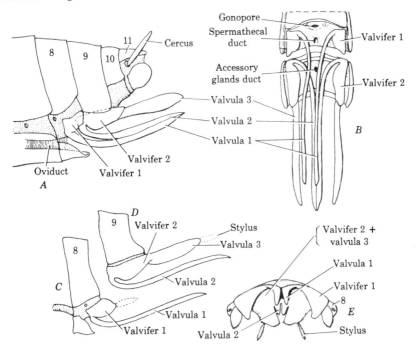

Figure 64   Structure of the ovipositor of pterygote insects (A-D, diagrammatic). A, B, showing segmental relations of the parts of the ovipositor: C, D, lateral view of the genital segments and parts of the ovipositor dissociated; E, nymph of *Blatta orientalis*, ventral view of genital segments with lobes of ovipositor. Reproduced from Ross: *A Textbook of Entomology*, 3rd ed., 1965, with permission of John Wiley and Sons, Inc.

cent of several authors who regarded all male external genitalia as modified appendages of the ninth abdominal segment, used as clasping organs during mating. The only insects which undoubtedly clasp using modified coxae and styli of the ninth abdominal segment are the mayflies and grylloblattids. A few orthopteroid insects (embiids and some grasshoppers) employ modified eleventh abdominal cerci as claspers. The clasping genital organs of the Phalloneopterata appear to be new structures elaborated from a pair of phallic lobes (Figure 65 A) that develop from the fore edge of the tenth sternum (Snodgrass, 1957) during postembryonic growth, and may be analogous to the genital papillae (penis lobes) of xiphosurids, crustaceans, and millipedes. The genital papillae of insects were called phallic lobes or phallomeres by Snodgrass. In thysanurans and machilids, the phallomeres became excavated medially and fuse dorsally and ventrally into a short "penis." In mayflies the phallomeres remain separate, each bearing the opening from a vas deferens.

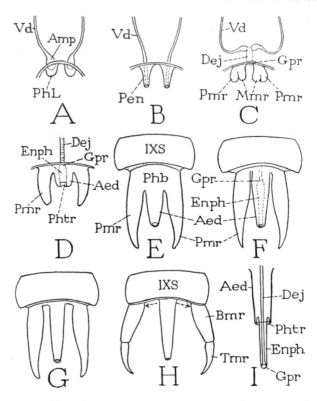

Figure 65   Diagrams illustrating the common origin and various types of development of male phallic organs of insects. A, the primary phallic lobes (PhL) invaded by the ampullas (Amp) of the vasa deferentia (Vd); B, penes of Ephemerida, giving exit to the genital ducts; C, the usual division of the phallic lobes into mesomeres (Mmr) and parameres (Pmr), ejaculatory duct (Dej) developed between the mesomeres, opening through the gonopore (Gpr); D, mesomeres united to form the aedeagus (Aed), aedeagal lumen, or endophallus (Enph), opening through the phallotreme (Phtr); E, aedeagus and parameres not completely separated from common phallobase (Pfb); F, aedeagus and parameres removed from the aedeagus; H, parameres more widely displaced from aedeagus on margin of ninth abdominal sternum (IXS), divided into the basimeres (Bmr) and telomeres (Tmr); I, aedeagus with endophallus everted, bringing the gonopore to its tip. From Snodgrass: *A Revised Interpretation of the Reproductive Organs of Male Insects*, 1957, Smithsonian Miscellaneous Collections.

In the orthopteroid orders the phallomeres frequently divide into median mesomeres and lateral parameres (Figure 65, C), but remain as simple lobes with hardly any clasping function. The members of the Phalloneopterata bear an intromittent organ, the aedeagus, formed by fusion of the mesomeres so as to provide a passage similar to the one in the thysanuran penis, the endophallus (Figure 65 F). The lateral parameres become the functional clasping organs. The parameres bear a terminal movable section operated by muscles in the Telomerida, in which case the basal piece is called basimere, and the distal piece the telomere (Figure 65 H). The basimere is commonly homologized with the ninth abdominal coxal plate and the telomere with its stylus, but in some insects the coxal plates and styli of the ninth segment of males occur together with phallomeres on the tenth segment (grylloblattids). The male genitalia of the Neopterata must represent a totally new derived set of characters. Some further evolutionary aspects of male genitalia are discussed later where appropriate.

The Ectognathata have evolved a segmented median caudal filament by a prolongation of the tergum of the eleventh abdominal segment. This filament appears to extend from the extreme end of the abdomen, but it is associated with the eleventh segment along with the cerci. The caudal filament is also known as the paracercus, or cercoid. It is well developed only in the Monura, Microcoryphia, Thysanura, and Ephemerida. In embryos of the Ectognathata the periproct appears as a distinct postsomatic structure, but only traces remain in postembryonic states, such as a tiny three-part structure in machilids. In these insects the dorsal lobe may be called the supraanal lobe. The supraanal lobe appears also in the anisopterous Odonata, under the shortened paracercus. The short paracercus of Odonata, and similar structures in Plecoptera and Dermaptera have been called epiproct (Snodgrass, 1935). Generally, the eleventh abdominal sternum consists of two sclerotized plates called paraprocts. The epiproct is most nearly homologous with the telson of Crustacea, which is a posterior prolongation of the dorsal periproct. The paracercus, or terminal filament, of insects is a new structure developed in parallel with similar structures in trilobitoids (*Emereldella, Burgessia*), some trilobites and some arachnids. The paracercus appears to have been suppressed independently in two lines, in the Plagiopterata and again in the stem line of the Neopterata. Along with the terminal filament, the cerci of the parent ectognath insect may have also become elongated and subsegmented, later to become each a single unit or suppressed. The anus-bearing telson segment failed to separate from the eleventh abdominal somite in all the Ectognathata.

The molting process of the Ectognathata became more specialized in that the median dorsal ecdysial line on the thorax was extended into the head as the epicranial suture, forking forward into two frontal sutures. During ecdysis, the old cuticle splits along the top of the head forward into the frontal sutures, as well as over the thorax.

The spermatozoan ground plan of two mitochondial derivatives and the $9 + 9 + 2$ axoneme pattern is typical of most insects. This pattern was variously modified in various lines of evolution, particularly in the panorpoid complex and in the Acercarida (hemipteroid insects).

INFRACLASS ARCHENTOMATA

This new taxon is erected to include the single order Monura, as a sister group to the Euentomata of Berlese. The Monura includes the fossil species *Dasyleptus lucasi* Brongniart, 1885, from the Upper Carboniferous of France, *D. brongniarti* Sharov, 1957, from the Lower Permian of southwest Siberia, and some recently uncovered fossil species of *Dasyleptus* from the Upper Carboniferous in Mazon Creek, Illinois and New Mexico, plus a Lower Permian species from Moravia (Carpenter, personal communication). *Dasyleptus brongniarti* is represented by about 90 imprints, from which a reasonable reconstruction can be offered (Figure 60), based on Sharov (1957, 1966a). Except for size and age, the newly discovered specimens agree in general with Sharov's Monura.

If fossil specimens of *Dasyleptus* had been found restricted to the Silurian or earlier the genus might well serve as a possible actual ancestor of the Euentomata, because it is difficult to recognize derived characters in the Monura other than those discussed above for the Ectognathata. *Dasyleptus* is contemporaneous with most orders of winged insects, and therefore cannot be the ancestral genus of the Euentomata. Furthermore, the tenth abdominal tergum is enlarged, the prothoracic tergum is reduced, and the caudal filament of the abdomen is as long as the body. No trace of cerci were found on the imprints, so it may be that the absence of cerci is a fourth derived character state of the Monura. Abdominal styli are visible only in one fossil specimen, on the first four abdominal segments of *D. brongniarti* (Sharov, 1966a). Judging from the presence of styli on the second through ninth abdominal segments in the Machilidae, it might be presumed that *D. brongniarti* also bore styli on the fifth through ninth segments. It is also barely possible that the cerci were no larger than the styli, and that these were not preserved either, but Carpenter (1977) suggested that the absence of cerci is a monuran character, since recently discovered additional ones also are without cerci. Because no cerci are known in the Monura, Carpenter (1977) thought that cerci first appeared later among the Microcoryphia. But cerci appeared much earlier in the insect stem line, to be suppressed in the Ellipurata, and again in the Monura.

Most of the other features of the Monura are rather primitive. The antennal flagellum is short, made up of 12-13 flagellomeres. The original description of *D. brongniarti* suggested that the lateral edges of the mandibular, maxillary, and labial terga were still evident. Riek (personal communication) has examined fossils of Monura in Moscow, and suggested that these so-called separate tergal edges may not be real, but rather appear so because of the possibility that the presumed intersegmental lines of the head are perhaps cracks in the cranium caused by pressure during fossilization. Carpenter (personal communication) is inclined to agree with Riek. The postoccipital sulcus is distinct over the back of the head, however, as in most insects. Compound eyes are well developed, and not close together as they are in Microcoryphia. Ocelli are not perceptible on the imprints, and were not mentioned in descriptions. The possession of ocelli by the Monura must remain an open question.

The maxillary palpus is leglike, with apparently six podomeres. It is not possible to be certain whether the distal palpal segment is the pretarsus or merely a subsegment of the palp tarsus, but the thoracic tarsi are simple. The pretarsi of the walking legs are still primitive as distinct elongated pointed podomeres. Since the pretarsus is well developed in the legs, it is possible that the terminal segment of the maxillary palp is a pretarsus. Two small basal segments on the palp might be persistent trochanters. The upper part of the thoracic coxae appears to be marked off as a short segment, which Sharov took to be a subcoxa. It is here regarded as a coxal subsegment not yet joined to the pleuron as the catapleurite (Figure 60, cpl.). If there are any anapleurites, they are hidden by the overhanging thoracic paranota.

The thorax of the Monura is recognizable by the presence of legs on the first three trunk segments. Otherwise, the trunk seems to consist of a more or less uniform series of 14 metameres, the last of which apparently has not separated from the telson, similar to most insects except Protura. Presumably there were abdominal coxal plates, but only the enlarged plates of the simple ovipositor are perceptible. The abdominal styli were two-segmented in contrast to the simple styli of other insects. If the simple ovipositor bore the equivalent of valvulae, they are not visible in the imprints. The paranota are simple on all the trunk segments, and hang downward over the sides as they do in Microcoryphia and Thysanura.

INFRACLASS EUENTOMATA

The synapomorphies of the typical insects as a group are relatively few. Most of the advances occurred in various evolving lines of pterygote insects. The ancestral typical insect may be characterized by the following derived characters contrasted with those of the sister group, the Archentomata. The head acquired a well defined clypeus generally marked off from the frons by a clypeofrontal sulcus, in or near the lateral ends of which the anterior tentorial invaginations are situated. The dorsal remotor muscle of the mandible became enlarged (Figure 29, DRM), later to become the chief adductor to cause biting when the axis of roll of the mandible tilted forward and the mandible acquired a secondary ventral anterior condyle. The enlargement of the mandibular remotor muscle is in contrast to its suppression in the myriapods. In the latter case, the former promotor muscle became elaborated into flexors of the gnathal lobe (Figures 23, 24, 25, DPM). The ventral mandibular muscles that insert into the body of the mandible partly shifted some insertions to the anterior border (Figure 29, VPM) and to the posterior border (Figure 29, VRM), and some of their origins shifted to the anterior tentorium, becoming accessory promotors and remotors of the mandible. The other ventral muscles (Figure 29) remaining still originated on a remnant of the transverse mandibular tendon (Figure 29, TT). The thorax became distinctly specialized as a locomotor tagma. The catapleurites became part of the lower pleuron by detachment from the proximal ends of the coxae, becoming incorporated in the pleuron in association with anapleurites, and a pleural apodeme formed in the catapleurite. The tarsi became subsegmented into three or more nonmusculated tarsomeres. The tarsi of Microcoryphia are three-segmented, but Tricholepidion in the Thysanura has pentamerous tarsi. Five-segmented tarsi appear to be typical of most insects. The reduction of tarsal segment number seems to have occurred repeatedly in various insect lines. The pretarsi became highly reduced, at best as a small median pointed structure such as in Thysanura, and the tarsal claws became structures separate from the pretarsus. The claws of the Euentomata articulate on the sclerotized upper tip of the last tarsomere, and may be movable through the pull of the unguitractor plate (vestigial pretarsus) on arthrodial membrane stretching from below the claws to the unguitractor plate, which in turn bears the cuticular apodeme of the pretarsal flexor muscle (Figure 55, H). The tarsal claws do not seem to be homologous with the clawlike structures of the Diplura. The dipluran claws are clearly appendages of the pretarsus, apparently homologous with the bristles on proturan pretarsi, and they do not articulate on the last tarsomere. Thoracic spiracles became fixed as two pairs, situated at the front borders of the mesothorax and metathorax, and abdominal spiracles became fixed as eight pairs on the sides of the first eight abdominal segments. Abdominal styli of Recent species became simple unsegmented peglike structures, and the cerci became multisegmented, in parallel with cerci of Campodeidae. Segmented cerci are found in many fossil insects, and they are still segmented in the Microcoryphia, Thysanura, Plecoptera, Ephemerida, Grylloblattaria, Blattaria, Mantodea, and some termites. The abdominal styli on the first abdominal segment became suppressed in addition to the earlier suppression of the tenth abdominal styli. No special male genitalia with a copulatory function were developed in the early Euentomata, but the paired genital papillae (phallomeres) of the tenth segment in males initiated a primitive penis by fusion, a condition retained in the Microcoryphia and Thysanura (Figure 61, J). The coxal plates of the eighth and ninth abdominal segments of females became enlarged and their coxal sacs became sclerotized, elongated and segmented, forming the initial ovipositor blades, or first and second valvulae. The abdominal coxal vestiges in the Machilidae appear to have separated into small "subcoxae" equivalent to the thoracic catapleurites, and those of the eighth and ninth abdominal coxae of the females became the gonangula, or basal valvifers, of the ovipositor (Bitsch, 1974a). The female primary gonopores joined a median invagination of the eighth abdominal sternum, forming the definitive gonopore opening on the hind border of the segment.

The Euentomata are represented by two sister groups, the Archeognathata represented by the single order Microcoryphia and the Dicondylata represented by the Thysanura and the winged insects.

Subterclass Archeognathata

This group was named Archeognatha by Börner (1904), and it is here emended to indicate its class group rank. This is the sister group to the remaining insects, the Dicondylata. The original rank was as a suborder of the Thysanura. Two old names are available for the single included order: Machiloidea Handlirsch, 1903, and Microcoryphia Verhoeff, 1904. The name Machiloidea was used as a thysanuran superfamily name by Remington (1954), who used Microcoryphia as a suborder or order. The latter name was again raised to order rank (Ross, 1955). Usage is tending to settle on Microcoryphia as the order name.

The microcoryphian head is usually retracted under the pronotum, exposing only a part, hence the name "small head." The compound eyes and antennae approach each other near the midline. The maxillary palpus has become elongated, and seems to consist of eight segments. The first segment (trochanter) appears to be two-segmented because of a sclerotic internal ridge near its middle, but the muscles are not separated in two segmental groups. The next segment (femur) also appears to be segmented at its middle, also because of an apparent joint, but the long muscles which operate the segment (tibia) continue the length of the two subsegments without attaching at the joint. The palptarsus consists of three tarsomeres, as in the leg tarsi. The mandibular muscles are modified somewhat, in a fashion that later in the Dicondylata made possible a strong biting action. The dorsal remotor (Figure 29, DRM) is much larger than the dorsal promotor muscle. In the Dicondylata the remotor muscle further enlarged, and with a shift in the axis of rotation of the mandible the muscle has become the chief adductor of the mandible in its biting action (Figure 30, DRM). The mandible of Microcoryphia is primitive in its vertical axis, single dorsal condyle, and rolling motion, but it has elaborated a grinding surface, or molar area, and a picklike slender lobe used in scratching food particles such as algae from a substrate.

The bristletails are noted for their leaping ability. This function is performed in a unique fashion, involving features evolved by the line. When in a hurry short hops are performed by operating the legs in unison. The middle legs, and especially the hind legs, are brought downward and rearward, the legs extended, in making these short hops. High leaps are performed by pushing the abdomen vigorously and quickly against the ground, along with the legs pushing in unison (Manton, 1972).

The anatomical modifications of the thorax that aid in leaping are typical of the Microcoryphia. The anapleurites of the middle and hind thorax have shifted to lie under the paratergal folds (Barlet, 1950). The catapleurites of these two thoracic segments bear strong pleural apodemes, on which the main trochanteral depressors provide a strong pushing action of the telopodite in hopping or high leaping. The coxal origin of the catapleurite is borne out by the origin of trochanteral muscles on the pleural apodeme. The dorsal and oblique longitudinal abdominal muscles are in ropelike twisted bundles, providing a means of pushing powerfully with the abdominal venter during high leaping. Various thoracic muscles originate ventrally on an elaborate complicated thoracic transverse tendon system. The abdominal transverse tendons are arranged as a dorsal and a ventral pair in each segment, on which the longitudinal, dorsoventral, and coxal muscles arise (Bitsch, 1973). Bitsch considered these tendons to be remnants of intersegmental phragmata, but they appear to be homologous to the intersegmental tendons of arthropods in general. The pretarsus has become reduced to a small unguitractor plate below the tarsal claws, without any extended dactylus (Figure 55, H).

The coxal styli of machilids have been the subject of much discussion. These are peglike structures of the second and third thoracic coxae, bearing no muscles. They have been regarded as persistent vestiges of exopodites, of epipodites, as homologues of the ventral styli of symphylids, or as homologues of the abdominal styli. Since such thoracic coxal styli are not found on any other insects, and since machilids are not really ancestral insects, the thoracic styli are best interpreted as new formations peculiar to this order. They are not yet developed at hatching time, and grow later after a few molts, but abdominal styli develop before hatching. These thoracic styli serve as lateral sensors for orientation in crevices (Smith, 1970b). Some machilids lack mesothoracic coxal styli, and all Meinertillidae lack thoracic coxal styli.

There are only a few derived characters of the abdomen. The powerful twisted leaping muscles were mentioned above. The abdominal spiracles are on the apices of the second through eighth abdominal paranota. Since the pleural anapleurites move into the lower surface of the thoracic paranota, and the abdominal ventral plates join laterally to what appears to be the base of the paranota, it appears that the abdominal pleural areas are represented by the membranous inner faces of the paranota (Bitsch, 1973), and the position of the spiracles on the apices can be explained by a shift of the pleural areas of the abdomen to lie on the inner faces of the paranota. Spiracles are usually pleural structures rather than tergal.

The Microcoryphia otherwise have remained somewhat primitive. Ocelli are generally present on the head. The hypopharynx is three-lobed, as in the Entognathata. The labium is a typical insect structure. The mandible is a primitive rolling mandible with a single dorsal condyle. There is no distinct boundary between the frons and clypeus in the form of a clypeofrontal sulcus, such as occurs in the Dicondylata (Bitsch, 1963) and in the Monura. The lack of distinct clypeus therefore might have resulted from a secondary suppression in this order. The intersegmental tendon system is well developed, including separate intermandibular and intermaxillary transverse tendons, ventral thoracic, and both dorsal and ventral abdominal transverse tendons. The abdominal tendons are separated medially into paired structures. The anterior tentorial arms are separate from each other and are not joined to the posterior tentorium. The anterior arms apparently invaded the region of the mandibular transverse tendon, and are connected together by means of very fine fibrillae (Figure 29). The main ventral mandibular muscles originate on what is left of the transverse mandibular tendon. The tarsi have evolved only to the point of developing three tarsomeres, rather than five as in the higher insects. A dicondylic femur-tibia joint is just barely formed in this order. The condyles consist essentially of the sclerotized ends of a dorsal hinge joint, so the axis is near the dorsal part of the leg, and the femoral extensor muscle of the tibia thus works at a poor mechanical advantage. The femur-tibia joint of machilids has been described as a single condyle, but the hinge extends across the top of the joint, and there are really two condyles. In the Entognathata and in myriapods, the femur-tibia joint consists of a single dorsal condyle. Abdominal styli are simple, on segments 2 through 9, but some apparently became secondarily subsegmented in a few modern species, and in the extinct *Triassomachilis* (Janetschek, 1957). The abdominal coxal plates are still separate from the triangular sterna, and eversible vesicles occur on the first seven abdominal segments. The genitalia are still simple, as were described for the ancestral typical insect, and the ovipositor merely serves as an egg guide. There is no internal copulation. Delany (1959) described the development of the penis of *Petrobius brevistylis*. At the fourth instar, two phallomeres on a common base have formed behind the coxal plates of the ninth abdominal segment, on the tenth sternum. At the next instar, the parameres, which are converted coxal sacs of the ninth segment, begin to form from the ninth coxal plates, and the developing penis is larger. At the seventh instar, the two phallomeres have fused to form the external passage of the penis, and the parameres are better developed. The eighth instar bears the fully developed penis. Thus, the penis is a tenth segment structure, and has no serial homology with vestiges of limbs, since tenth abdominal coxal plates, styli, and coxal sacs never have been positively identified in any adult insect, and probably were suppressed in the ancestral insect. The female gonopore opens behind the seventh sternum. The epidermal invagination that forms a vagina in the higher insects is, in this order, merely a groove or gutter on the eighth sternum, the inner sides of which bear a pair of seminal receptacles (Bitsch, 1974a). Traces of the ancestral periproct remain as a small supraanal lobe under the caudal filament and a pair of lobes beneath the anus. Molting continues after adulthood, and the embryo does not yet sink so far into the yolky egg so as to form an amnion. The cerci have not become very long, and remain much shorter than the segmented median caudal filament.

The lateral cervical sclerites that permit wide movements of the head have not yet developed in the Microcoryphia. But apparently a beginning state occurs, in that the anterior parts of the episterna [preepisternum (Matsuda, 1970)] are prolonged forward slightly and articulate on each side with occipital condyles. It appears that it was such a process of the episternum that evolved into the lateral cervical sclerites of the pterygote insects.

Subterclass Dicondylata

The name for this group, which includes all the remaining insects not yet discussed, is emended from Dicondylia (Hennig, 1966) so as to form a sister class-group with the Archeognathata. Dicondylata as a name refers to the occurrence of a secondary articulation of the mandible in addition to the primary originally dorsal mandibular condyle. It could just as well refer to the typically dicondylar femur-tibia joint.

The forward tilt of the mandible and the secondary articulation with the head occurred in parallel with a roughly similar modification in the malacostracan Crustacea. As the mandibular movement changed to a horizontal axis on the head, the former dorsal promotor muscles became abductor in action (Figure 30, DPM), while the former dorsal remotor and the ventral muscles (Figure 30, DRM, VRAM) became powerful adductors performing the biting action. The biting or molar lobe of the mandible became toothed, and shifted to a morphologically posterior, but actually ventral, position just as in the higher Crustacea. The transverse tendons of the head were suppressed (except in *Tricholepidion*), and the ventral mandibular and maxillary muscles shifted their origins to the anterior tentorium, and partly to hypopharyngeal sclerites. The anterior tentorial arms became fused to each other, forming an anterior tentorial bridge. The fused tentorial arms became important, along with the clypeofrontal sulcus, in resisting the strong pull of the powerful mandibular muscles. The maxillary palpi became reduced to structures always smaller than the legs, in contrast to the huge palpi of Monura and Microcoryphia. The hypopharynx became consolidated into a single unit from the primitive three-lobed condition.

With the new mandibular condyle near the anterior tentorial pit on each side, the clypeofrontal sulcus became a strengthening device across the front of the head, allowing strong transverse biting.

The first Dicondylata retained the original anapleurite and catapleurite of the pleuron, but in a modified condition. These otherwise became the highly modified pleuron of the winged insects. The femur-tibia joint has become distinctly dicondylic by the shift of the condyles from a dorsal position to a more nearly equatorial position at the joint, giving the femoral extensor muscle of the tibia a better mechanical advantage. The tarsi of the Dicondylata originally were pentamerous, but became secondarily of fewer tarsomeres in various orders by convergent evolution. With the appearance of pentamerous tarsi, the tips of the tibiae became the principal weight-bearing parts of the legs. The tarsi generally lie flat on the ground, in contrast to the condition in the previous groups in which the weight is borne on the claws. The transverse tendon system became highly suppressed, with a retention of thoracic and abdominal ventral tendons only in the Thysanura (Rousset, 1973). The dorsal tendons are mostly suppressed. Fragments of the ventral tendons remain in larval Megaloptera (Barlet, 1977). The muscles formerly originating on the transverse tendons became attached directly to the cuticle by way of transepidermal tonofibrillae. The tonofibrillae seem to be remnants of the tendon system.

The thoracic paranotal lobes of the Dicondylata did not incorporate the anapleurites as in Microcoryphia, but rather, especially on the middle and hind thoracic segments, the paranota remained somewhat elevated laterally exposing the pleuron. This is a more primitive condition, but it led eventually to the development of wings.

The ovipositor of Dicondylata became more efficient in that a gonangulum sclerite became articulated with the modified coxal plates, now called valvifers, so as to provide a means of moving the valvulae past each other on the tongue and groove union of the first and second valvulae. This movement led to an efficient means of inserting the ovipositor into a substrate while laying eggs.

The tracheal system became more complex in this group, with longitudinal tracheal trunks forming along with transverse trunks joining the segmental system into an intercommunicating tracheal system. The cerci became elongated, so that these at first were as long as the terminal filament, but were later reduced in most winged insects. The Dicondylata separated into two sister groups, the Apterata and the Pterygota.

Supersection Apterata

Linnaeus erected the insect order Aptera to include such wingless arthropods as silverfish, springtails, termites, lice, fleas, mites, phalangids, spiders, scorpions, crabs, copepods, millipedes, and centipedes. Over the years, each of these groups has been transferred into various arthropod and insect groups, and the name Aptera in the sense of Linnaeus has been abandoned. It is here resurrected, restricted, and emended for the only primitively wingless insects remaining in the Dicondylata. It happens to be the oldest available name also. The only included order, the Thysanura, is also restricted. The order formerly included the Diplura, Collembola, Microcoryphia, and lepismatids. As here restricted, only the lepismatid families are retained: Lepidothricidae, Nicoletiidae, Ateluridae, Lepismatidae, and Maindroniidae. Since the members of this order share unique specializations, no known thysanuran could have been ancestral to the Pterygota. They have evolved alongside the Pterygota from an unknown dicondylate ancestor.

Apomorphies. The Thysanura have greatly reduced compound eyes whose ommatidia are somewhat simplified, or the eyes are totally suppressed in some species. The ocelli generally are also suppressed, but secondarily within the order, since *Tricholepidion* still has ocelli. The body has become flattened dorsoventrally (Figure 66).

The pleuron of Thysanura (Lepismatidae) has been discussed widely as a prototype from which the pleuron of winged insects has evolved. Actually, the lepismatid pleuron is highly specialized. While it may have parts that could be homologous with some pterygote pleura, it is so modified that it is unlikely that the pterygote pleuron originated from the usual thysanuran type. The elongate coxae are flattened, with the outer border forming a sharp ridge. Behind the ridge, there is a concavity into which the femur fits when the trochanter is fully levated. But the elongate coxae do not hang vertically. The coxae extend rearward and swing forward in a horizontal plane around a vertical coxa-body axis. Therefore, the levated femur is actually brought horizontally forward. Lepismids walk by swinging the legs horizontally, performing in effect oar-like movements. The coxae of *Tricholepidion* (Figure 52) have not quite assumed a horizontal position. The pleuron of *Tricholepidion* is less modified than that of lepismatids, and this makes it possible to interpret the lepismatid pleural sclerites anew. The anapleurites of *Tricholepidion* are, as in the Entognathata, superior arcs which extend from ahead of the coxae to well behind each one. An unsclerotized membrane separates the anapleurite from the catapleurite. Each catapleurite also forms an arc over the head of the coxa. The single dorsal coxal condyle articulates at the middle of the catapleurite, at which point the external pit of the pleural apodeme is visible as a tiny pleural sulcus. The anterior arc of the catapleurite is split into two.

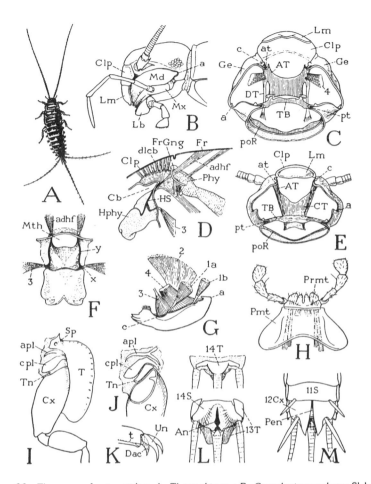

Figure 66 Thysanura, Lepismatidae. A, *Thermobia* sp.; B, *Ctenolepisma urbana* Slabaugh; C, same, section of head below level of tentorium; D, same, hypopharynx, cibarium, and pharynx, lateral; E, *Lepisma* sp., section of head below level of tentorium; F, *Thermobia* sp., hypopharynx, anterior; G, *Ctenolepisma urbana*, left mandible and muscles, dorsolateral; H, same, labium, ventral; I, same, base of left mesothoracic leg and adjoining parts, ventral; J, same, base of right mesothoracic coxa and pleuron, mesal; K, same, end of tarsus and pretarsus; L, same, eleventh abdominal segment, dorsal and ventral; M, same, end of male abdomen, ventral. a, primary articulation of mandible; adhf, adductor muscle of hypopharynx; An, anus; apl, anapleurite; at, anterior tentorial pit; AT, anterior tentorium; c, secondary articulation of mandible; Cb, cibarium; Clp, clypeus; cpl, catapleurite; CT, fused part of anterior tentorial arms; Cx, coxa; Dac, dactylopodite (pretarsus); dlcb, dilator muscles of cibarium; DT, dorsal end of tentorium; Fr, frons; FrGng, frontal ganglion; Ge, gena; Hphy, hypopharynx; HS, hypopharyngeal suspensorium; Lb, labium; Lm, labrum; Md, mandible; Mth, mouth; Mx, maxilla; Pen, penis; Pmt, postmentum; poR, postoccipital ridge; Prmt, prementum; pt, posterior tentorial pit; S, sternum; Sp, spiracle; T, tergum; Tn, pseudotrochantin; Un, unguis; x, loral arm, y, oral arm of hypoharyngeal suspensorium. Reprinted from R.E.Snodgrass: *A Textbook of Arthropod Anatomy.* Copyright © 1952 by Cornell University. Used by permission of Cornell University Press.

175

The lower portion is broader than the upper part, and this lower part corresponds to the so-called trochantin of lepismids. It is plainly a part of the catapleurite in *Tricholepidion* (Figure 52). The rear elements of the anapleurites and catapleurites are unsclerotized in lepismids, and the small pleural apodemes are in effect near the rear border of the catapleurites. The thysanuran trochantin is not the homologue of the pterygote trochantin. In the pterygotes it is a new structure formed in the same position as in Thysanura, from the lower anterior border of the episternum, but only in the Neopterata. The Odonata, Ephemerida, and extinct primitive flying insects have yet no trace of a trochantin. The thysanuran trochantin is a thysanuran development in parallel with the neopterous trochantin, and would better be named pseudotrochantin. In *Tricholepidion* as well as in lepismatids, the forward swing of the coxa brings its anterior border under the pseudotrochantin, which in turn folds under the remaining part of the catapleurite, and the latter underlaps the adjoining border of the anapleurite. The movements of catapleurites in this fashion are unique among insects, and represents a special feature of Thysanura.

The Thysanura are primitive insects in that some of their characters are not highly advanced. Since ocelli are still present in *Tricholepidion*, their absence is a derived feature in other species. There are yet no lateral cervical sclerites. The pentamerous tarsi of the ancestral Dicondylata were retained in *Tricholepidion*, but became secondarily reduced to two or three tarsomeres in other Thysanura. Abdominal styli are almost always present on the eighth and ninth abdominal somites, but on other segments they may be suppressed. Thus, *Tricholepidion* and nicoletiids bear styli on abdominal segments 2 through 9 (eight pairs), and most other species bear styli only on the seventh, eighth, and ninth segments. A few species bear styli on segments 3 to 9, on 4 to 9, on 6 to 9, or only on the ninth. Eversible coxal vesicles occur on segments 2 through 7 in *Tricholepidion* and *Nicoletia*, and on segments 6 and sometimes 7 in *Gastrotheus* (Nicoletiidae) (Paclt, 1956b). There are no coxal sacs in other Thysanura. The suppression of styli and coxal sacs within the order appears to have occurred in parallel with their loss in the Pterygota. The same appears to be the case in the merging of abdominal coxal plates with sterna. Independent coxal plates are situated on the posterior lateral borders of the broad sterna of segments 2 through 9 in *Tricholepidion* (Figure 67) and *Nicoletia*. All other Thysanura have independent coxites only on the eighth and ninth abdominal segments. In the Pterygota, coxites of the eighth and ninth abdominal segments were modified in females to form part of the ovipositor, are independent of the ninth sterna in male grylloblattids and some male ephemerids, and are imperceptible on other abdominal segments. The Thysanura are the only dicondylate insects in which there is a well developed transverse tendon system. It is represented by

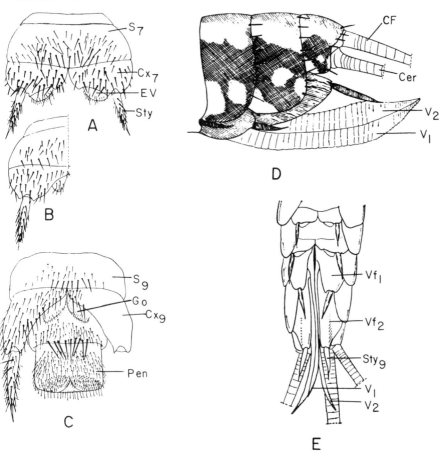

Figure 67 *Tricholepidion gertschii* Wygodzinski, Thysanura, Lepidothricidae. A, ventral view of seventh sternum and appendages; B, right half of eighth abdominal sternum and appendages, male; C, genital area of male, ventral; D, genital area of female, left side; E, genital area of female, ventral. Cer, cercus; CF, caudal filament; Cx, coxal plate; EV, eversible vesicle; Go, male gonopophysis (homologue of eversible vesicle); Pen, penis; S, sternum; Sty, stylus; V, valvulae of ovipositor (homologue of eversible vesicle); Vf, valvifers (homologues of coxal plates). Reprinted from Annals of the Entomological Society of America by permission of the Entomological Society of America, copyright owner.

a complex of tendons (endosternum) in the ventral regions of the thoracic and abdominal segments. Tendons have persisted in the head of *Tricholepidion*, but are lacking in other species. The genitalia of both sexes are essentially similar to that of Microcoryphia, with only traces of parameres left in males of *Tricholepidion* (Wygodzinsky, 1961). Mating is performed in the primitive fashion of indirect sperm transfer, by the deposition of a spermatophore to be picked up by the female. The Thysanuran spermatozoa typically unite in pairs, but retain the ground plan of 9 + 9 + 2 axonemes per sperm cell. Molting of sexually mature adults is repeated through life, as in the lower insects.

The name Apterygota can be discarded as a taxon. The union of the primitively wingless insects into a common taxon is artificial, and produces a taxon distinguished only by the primitive lack of development of wings. Such a paraphyletic group therefore merely represents a grade of evolution. The only reason for using the name is to emphasize a persistent primitive character in a group of taxa that represent a variety of evolutionary lines. The word Apterygota has practially no phylogenetic significance.

PART 3
SUPERSECTION PTERYGOTA

While the term Apterygota is phylogenetically meaningless, the Pterygota is a group especially distinguished by the development of winged flight, apparently as a monophyletic evolutionary process. Insect wings are fundamentally similar not only in their venation, but in their origin from tergal paranota of the mesothorax and metathorax. The first winged insects have not yet been discovered as fossils, and one can only speculate on the true nature of the ancestral wings. The oldest fossil insect wings known already had undergone some specialization.

WING ORIGIN AND STRUCTURE

For a few years, excitement and speculation were generated by the description of a presumed Devonian insect with short leathery wings, (*Eopterum devonicum* Rohdendorf, 1961), but with further study, Rohdendorf (1972) concluded that his species and another subsequently discovered (*Eopteridium striatum* Rohdendorf, 1970) were remnants of the tail fan of malocostracan Crustacea. The oldest fossil insect wings from the Upper Carboniferous include representatives of the paleopterous nonwing-folding Paleodictyoptera and of neopterous wing-folding orthopteroid insects. Since each of these types of wings is specialized in its own fashion, neither could have been the ancestral form of insect wing.

Attempts to suggest a hypothetical ancestral insect wing had their best start when Comstock (1918) utilizied the tracheation of the wing pads of immature insects as an indicator of vein homology in the various orders. Previously Redtenbacher (1886) had provided names for veins that were adopted with few changes by Comstock. Redtenbacher had emphasized the convex and concave alternation of the principal veins by naming the veins occurring at the tops of ridges (convex, or +) as the costa (C), radius (R), media (M), cubitus (Cu), and anal (A) veins. Other veins lying at the bottom of furrows (concave, or −) were given numbers, except that he named the subcosta (Sc) as the first concave (−) vein. It bore number II, the costa being number I. Comstock did not accept Redtenbacher's system of alternating high and low veins, but retained Redtenbacher's name system. The use of immature tracheation for establishing vein homologies became widely used for many years, in spite of the many exceptions warned against by Comstock. There are numerous cases (Fraser, 1938; Whitten, 1962; Carpenter, 1966) in which the tracheation does not follow the venation, and in which the veins appear before the tracheae invade the developing wing. Snodgrass (1909) introduced a method of tracing wing veins of neopterous wings by their basal attachments to the basal sclerites. Lameere (1922) and Martynov (1925) independently returned to the use of the alternating convex and concave veins, and suggested that the media was similar to the radius in having a convex (+) anterior branch (MA) and a concave (−) posterior branch (MP), deduced from studies of fossil insects and Recent Paleoptera. Lameere considered the concave veins as subbranches of the primary convex veins, and his influence in that regard has persisted with the notion that the base of the media was convex, continuing as a convex MA beyond the point of departure of the concave MP branch. Reexamination of published illustrations of well preserved Carboniferous and Permian fossil wings suggest that this notion is ill founded, since the stem of the media before its forking into MA and MP (see Figure 69, C; Figure 70, A) is concave. It appears that the anterior media branch may be a convex branch of the originally concave media. The vein commonly known as cubitus (Cu_1, or CuA) is often joined basally to a posterior branch (Cu_2, or CuP), but in many cases the two veins are independent at the base. This led Forbes (1943) to name CuP the plical (P) vein, and Snodgrass (1952) supported this interpretation by showing that this vein is usually independently joined to a basal wing sclerite, and called it the postcubitus (PCu). Hamilton (1972a) returned to the use of the name plical for CuP. Ross (1965) proposed that all the veins may have been basally independent originally, with branching in the radial sector (Rs), MA, MP and CuA. Riek (1970a) suggested the same, except that the original cubitoanal field consisted of a single vein, and that the original veins were unbranched except for Rs. (Table 1).

Sharov (1966a) elaborated a hypothesis of wing vein evolution based on

Table 1. Comparison of Various Schemes of Prototype Insect Wing Venation

Comstock (1918)	Lameere (1922)	Martynov (1925)	Forbes (1943)	Snodgrass (1952)	Ross (1965)	Sharov (1966)	Riek (1970a)	Hamilton (1972a)	Boudreaux (This Book)
C	C(+)	C(+)	C(+)	C(+)	C(+)	C(+)	C(+)	C(+)	C(+)
Sc	SC(−)	Sc(−)	Sc(−)	Sc(−)	Sc(−)	Sc(−)	Sc(−)	Sc(−)	Sc(−)
⌐R1	⌐R(+)	⌐R1(+)	R1(+)	⌐R1(+)		⌐R1(+)			⌐R(+)
R	R	R	R2-3(−)	R(+)	R(+)	R(+)	R(+)	R(+)	R(+)
⌐Rs (4br)	⌐Sr(−)	⌐Rs(−)	R4-5(+)	⌐Rs(−)	RS(−)	⌐Rs(o)	Rs(−)	S(o)	⌐Rs(−)
	⌐M(+)	⌐MA(−)		⌐MA(−)	MA(+)	⌐MA(o)	MA(+)	⌐MA(o)	⌐MA(+)
M (4br)	M	M(+)	M1-4(−)	M(+)		M(o)		M(o)	M(−)
	⌐SM(−)	⌐MP(−)		⌐MP(−)	MP(−)	⌐MP(o)	MP(−)	⌐MP(o)	⌐MP(−)
	⌐Cu(+)	⌐CuA(+)			Cu1(+)	⌐CuA(+)			⌐CuA(+)
Cu (2br)	Cu	Cu(+)	Cu(+)	Cu(+)		Cu(+)	CU+A(+)	Cu(+)	Cu(+)
	⌐Scu(−)	⌐CuP(−)	P(−)	Pcu(−)	Cu2(−)	⌐CuP(−)		P(−)	⌐CuP(−)
1A	⌐Pu(+)	1A	1Ax(+)	1V	1A(+)	1A(o)		E(+)	1A(o)
	Pu					A(o)		1A(+)	A—2A(o)
2A	⌐Spu(−)	2A	2Ax(+)	2V	2A(+)	2A(o)		2A(+)	3A(o)
3A	⌐U(+)	3A	3Ax(+)	3V	3A(+)			3A(+)	
	U			nV	4A(+)			Jb	
	⌐Su(−)			J	J(+)				

A, Stem of anal veins; Ax, axillary vein; C, costa; Cu, stem of Cubitus; CuA, cubitus anterior; CuP, cubitus posterior; E, empusial vein; J, jugal vein; Jb, jugal bar; M, stem of media; MA, media anterior; MP, media posterior; P, plical vein; Pcu, postcubitus; Pu, stem of Penultimate vein; R, stem of radius; Rs, radial sector; Sc, subcosta; Scu, subcubitus; Sm, submedia; Spu, subpenultimate vein; Sr, subradius; Su, subultimate vein; U, stem of ultimate vein; V, vannal vein. The symbol (+) indicates a convex vein, (−) indicates a concave vein and (o) a neutral vein.

the idea that the ancestral flying insect could partially fold its wings. He assumed that the oldest known fossil insect wings represent the original ancestral type. These are Upper Carboniferous wings of *Sustaia impar* Kukalova, represented by a pair of wings apparently partly folded backward, and a single wing of *Zdenekia grandis* Kukalova, both of which he assigned to a new order, Protoptera. He further strengthened his hypothesis by referring to the Devonian *Eopterum devonicum* Rohdendorf, which is a fragment of a malacostracan tail fan. In Sharov's opinion, the earliest modification of the ancestral venation was the fusion of the posterior media with the anterior cubitus, and that the present media vein includes only the persistent anterior median branch. He considered the cross vein m-cu so often found in modern insect wings to be the only vestige of the posterior media, but this cross vein is a convex vein, while the persistent posterior media of fossil Paleodictyoptera, and of Ephemerida and Neopterata, is a concave vein. The well preserved wing of *Zdenekia grandis* is more suggestive of the orthopteroid type of wing, and cannot represent an ancestral type for the Pterygota. The wing in question only attests to the notion that neopterous insects capable of folding their wings existed in the Carboniferous period contemporaneously with well-evolved insects that never acquired the ability to fold wings (Paleodictyoptera). The ancestral flying insect is yet unknown.

Hamilton (1971, 1972a, 1972b, 1972c) revived the name "plical vein" as a substitute for CuP. He renamed a vein, ordinarily known as first anal, as the "empusal vein" in reference to its distinctive features having been overlooked for so long. This vein in the Neopterata is usually separate from the other anal veins that are ordinarily joined together basally, and are independent of the basal wing sclerites. The empusual vein also lies before the vannal fold of the neopterous hind wing, while the other anal veins are behind the fold. The separation of the empusal vein from the other anal veins is a neopteran feature, correlated with the appearance of the vannal fold.

Hamilton (1972a) returned to the use of tracheae in immature wings as indicators of wing vein homology, in spite of the fact that the tracheae in immature wing buds enter the wing buds along already determined incipient vein channels. It appears that the growth pattern of tracheae is not controlled by the same genetic code as that of the venational pattern, and that each system has evolved in its own way. Sclerotization of cross veins and intercalary veins occurs in the absence of wing tracheae along the future courses of these veins.

Mayflies are generally regarded as "the most primitive winged insects living today" (Hamilton, 1971), and therefore they are cited as representing ancestral types. But it may be improper to adopt this thesis, because mayflies in many ways are highly specialized, although ancient in origin. Modern mayfly wings may have retained a few primitive features, such as the inability to fold over the abdomen, the persistence of both branches of the media vein, and the regular alternation of the veins forming a strong fluted pattern. But the retention of a few primitive features does not necessarily indicate overall primitiveness. Hamilton interpreted the possession of a single basal plate on mayfly wings as the primitive state. The basal plate in his opinion broke up into two plates in Odonata, and into three plates in the Neopterata.

Kukalova-Peck (1974) described previously unknown basal wing plates in Paleodictyoptera and Megasecoptera. From all indications the primitive condition consisted of individual basal platelike enlargements at the roots of veins Sc, R, M, Cu, and A (Figure 68, A). Various combinations of fusion occurred in various orders, such as a single subcostoanal plate in more recent Paleodictyoptera (Figure 68, C), a single mediocubital plate in some Megasecoptera (Diaphanopterodea) (Figure 68, E), a subcostocubital plate in Odonata (Figure 68, B), a mediocubital plate in Ephemerida (Figure 68, F), and retention of only small subcostal and anal plates plus a fused radiocubital plate in the narrow-winged Megasecoptera (Figure 68, D).

The basal sclerites in the Neopterata are difficult to interpret in the light of these new findings. The plates are highly modified and articulate with each other and with the vein bases in various ways that are associated with neopterous wing folding. It would appear that the first axillary sclerite of neopterous wings is the modified subcostal plate, and the second axillary sclerite is the radial plate. The median sclerites resulted from the fusion of the median and cubital plates, as in Ephemerida, but secondarily acquired a flexibility across the fused plate, forming the anterior part of the jugal fold (Figure 68, G). The third axillary sclerite of neopterous wings is the modified anal plate. The basal fold (Figure 68, E, bf) of the Diaphanopterodea appears to be a different method of wing folding developed convergently with neopterous insects. The fused mediocubital and anal plates seem to be able to crumple along three folding lines.

Considering that mayflies are highly specialized, the simple basal plates of their wings is probably the result of a secondary fusion of an original set of basal sclerites, correlated with their weak flight ability, short imaginal life, and lack of imaginal feeding. Agility in flight and wing folding are possible in part through the variety of movements permitted by basal flexibility between articulated basal sclerites, such as is found among the Neopterata.

Since there are no known fossil transitional protowings, it almost appears futile to try to reconstruct the hypothetical not-yet-functional prewing. Hamilton (1971) regarded the paranota found on the prothroax of Paleodictyoptera as an indicator of the protowing venation. But such paranota are on the bodies of insects bearing fully developed functional wings. It is to be expected that the venational pattern suggested by the ridges on paranota should be similar to that of the wings. The genetic code for fully developed wing

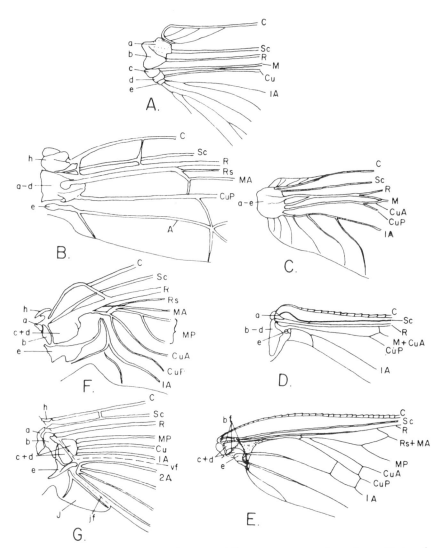

Figure 68   Pteralia (axillary plates) of extinct and Recent insects suggesting patterns of fusion of pteralia. A, base of fore wing of *Boltopruvostia nigra* (Paleodictyoptera), Namurian, Upper Carboniferous, with pteralia mostly separate; B, base of fore wing of *Anax junius* (Odonata), Recent, with fused subcostocubital plate; C, *Dictyoptilus sepultus* (Paleodictyoptera), Upper Carboniferous, with all pteralia fused into a single plate; D, base of fore wing of *Protohymen permianus* (Megasecoptera), Lower Permian, with fused radiocubital plate; E, fore wing of *Martynovia protohymenoides* (Diaphanopterodea), Lower Permian, with only a mediocubital and an anal plate remaining, correlated with wing folding; F, fore wing of *Hexagenia* sp. (Ephemerida), Recent, with a fused mediocubital plate; G, diagrammatic base of hind wing of generalized Neopterata, recent, with mediocubital plate broken along medial fold. a, subcoastal plate; b, radial plate; c, medial plate; d, cubital plate; e, anal plate; h, humeral plate; A, anal veins; C, costa; Cu, stem of cubitus; CuA, anterior cubitus; CuP, posterior cubitus; J, jugum; jf; jugal fold; M, stem of media; MA, anterior media; MP, posterior media; R, radius; Rs, radial sector; vf, vannal fold. A, C, D, and E, from Kukalova-Peck, 1974. Used by permission of *Psyche: Journal of Entomology*. B, F, and G, are original.

venation can be only partly expressed in the short stubby paranota of such insects. These paranota cannot be used as a model of the original presumably nonfunctional prewings.

By following the principles of phylogenetic study, the primitive state of a functional wing can be hypothesized from the characters shared by most of the descendants (Figure 69, A). One of the most regularly recurring characters is the fluting of the wings. The costa is nearly always a convex vein. The subcosta is concave. The stem of the radius is always convex. In fossil wings which bear both the anterior and posterior media, and in Recent mayflies, the stem of the media is concave. Cubitus is a convex vein, as are usually the anal veins. It may be that both the radius and the media were simple in the early evolving wing, but each is branched to some degree in nearly all known wings. The radius acquired a concave radial sector (Rs) branch extending behind the main vein. The oldest known radial sectors are also branched into concave veins. In some insects, additional "intercalary" veins were convergently acquired between the concave branches, so as to extend the fluting to the wing border (mayflies, Figure 69, C, D; and the Odonata, Figure 70, C, D). The stem of the media has been regarded as equivalent to the radius (Lameere, 1922), and therefore should be convex. The anterior media would have a posterior concave companion similar to the radial sector. But excellent photographs of well preserved fossil wings [for example, those of *Lithoneura mirifica* Carpenter, 1943a, and of *Dunbaria fasciipennis* Tillyard, reproduced in Ross (1965)] indicate that the stem of the media was a concave vein, which continues as the posterior media, while the convex anterior media is an anterior convex branch of the media. The stem of cubitus is always convex, and it continues into an anterior convex branch (CuA) that frequently forks at the tip. The posterior branch (CuP) is concave and usually simple (Hamilton's plical vein). The anal vein system consisted of at least three basally united convex or perhaps neutral veins, as in seen in many fossils of the late Paleozoic. The detached first anal vein (Hamilton's empusal vein) of the Neopterata is a derived feature of this group, correlated with the folding of the hind wing along the vannal fold. The prototype wing suggested in Figure 69(A) represents a possible functional wing such as might have appeared in the first flying insect. It probably was incapable of being folded over the body. Such a wing probably bore basal articulating sclerites permitting slight changes in rotation along an axis near the radial vein system. All the departures from this hypothetical prototype wing may be traced to venational shifts, fusions, separations at the bases, suppressions, or additions. The anterior media was retained by the plagiopterous insects and by mayflies. The posterior media apparently was suppressed in the odonate line, while the anterior media was suppressed in the Neopterata.

The acquisition of flight must have been accompanied with a mechanism

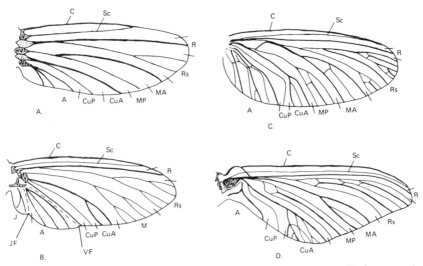

Figure 69 Hypothetical prototype wing (A), and wings of Opisthopterata. B, diagrammatic representation of hind wing of Neopterata, with suppressed MA and detached 1A; C, hind wing of *Lithoneura mirifica* (Ephemerida), with MA partly fused with Rs, and with intercalary veins; D, fore wing of *Hexagenia* sp. (Ephemerida) with R, M and CU bases nearly fused. The convex veins are drawn with a heavier line than are the concave veins. C from Carpenter, *Carboniferous Insects from the Vicinity of Mazon Creek, Illinois,* 1943. Scientific Papers of the Illinois State Museum, 3(1):9-20. A, B, and D are original.

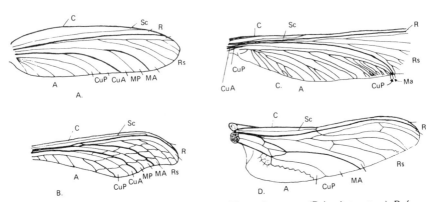

Figure 70 Wings of Plagiopterata. A, fore wing of *Stenodictya* spp. (Paleodictyoptera); B, fore wing of *Mischoptera nigra* (Megasecoptera); C, fore wing of *Triadotypus guillaumei* (Meganisoptera); D, fore wing of *Anax junius* (Odonata). The convex veins are drawn with a heavier line than are the concave veins. A, original, composite from various sources; B, from Carpenter and Richardson, 1968. Used by permission of *Psyche: Journal of Entomology*; C, from Laurentiaux, 1953, in J. Piveteau, Ed., *Traite De Paleontologie,* Vol. 3, by permission of Masson, Paris; D, original.

185

for flying. Tergosternal, tergocoxal, tergopleural, and pleurocoxal muscles already acquired in the apterygote insects (Matsuda, 1970) acquired the ability to move the wings in addition to the service of primitive functions of leg motion by the coxal muscles. A segment of tergosternal muscle and of tergocoxal muscle shifted to a tergal sclerite, the subalare, so that it became possible to depress the posterior wing base. The subalar sclerite is connected to the hind edge of the wing base through a tough axillary cord. A set of pleurocoxal muscle fibers stretched from an upper sclerite near the anterior base of the wing, the basalare, to the anterior catapleural part of the pleuron (the trochantin of Neopterata). This set of muscles also caused depression of the anterior wing base distal to the fulcrum. Tergosternal, tergopleural, and tergocoxal muscles caused depression of the wing base proximal to the fulcrum, resulting in the elevation of the wing tip as the tergal plate was pulled down. Pronation of the wings during downstroke and supination during the upstroke were produced by the differential actions of the basalar (pronator) and subalar (supinator) muscles, and by air pressure on the somewhat flexible posterior borders of the wings. This primitive wing motor mechanism was further developed into the direct wing motor system of the Odonata (Figure 71 A), and was the probable mechanism of all the Plagiopterata, because their thoracic terga were not divided into alinotum and postnotum.

OTHER PTERYGOTE FEATURES

The pterygote pleuron evolved into a strong sidewall of the thorax (Figure 72). The anapleurite and catapleurite became joined together and enlarged. The union between the anapleurite and catapleurite remained in many insects as the paracoxal suture (Matsuda, 1970). A strong pleural sulcus extended from the pleural wing process to the coxal articulation. The pleural sulcus is the external mark of the internal pleural ridge from which the pleural apodeme extends toward the furca (or sternal apodeme) on the sternum. The pleural ridge helps strengthen the pleuron, resisting the strong pull of the dorsoventral flight and leg muscles. The pleural region ahead of the pleural sulcus usually forms a section known as the episternum. Thus the episternum consists of the portions of the anapleurite and catapleurite that lie anterior to the pleural sulcus, known, respectively, as the anepisternum and catepisternum. The region posterior to the pleural sulcus forms the epimeron, consisting of the portions of the anapleurite and catapleurite that are posterior to the pleural sulcus, and are, respectively, the anepimeron and the catepimeron. All the variations of the modern pleuron may be regarded as evolutionary departures from this ground plan (Matsuda, 1970).

Much has been written concerning homologization of pleural areas in in-

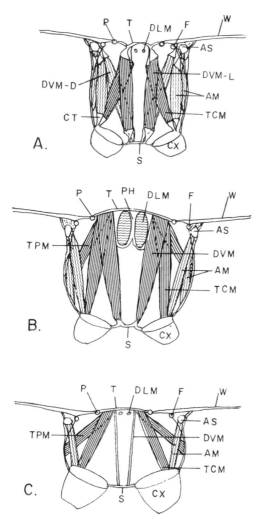

Figure 71 Illustration of the three main types of wing flapping mechanisms. A, odonate type, direct wing muscles. Elevation of wing tip through pull of one sector of dorsoventral muscle (DVM-L) and a tergocoxal muscle. Depression of wing tip through pull of another sector of dorsoventral muscle (DVM-D) on wing base lateral to fulcrum, and alary sclerite muscles. B, typical neopterous indirect flight muscles. Dorsoventral, tergocoxal, and tergopleural muscles lower the tergum, causing elevation of wing tips. Dorsal longitudinal muscles cause buckling of tergum, elevating wing base mesal to the fulcrum, while alary sclerite muscles pull on wing base lateral to fulcrum, lowering wing tip. C, blattiform modification of neopterous type. Tergopleural and tergocoxal muscles replace tergosternals to lower the wing base, causing elevation of wing tip. lowering of wing tip through pull of alary muscles. Dorsal longitudinal muscles reduced or lost. AM, alary sclerite muscles; AS, alary sclerite (basalare or subalare); CT, cap tendon; Cx, coxa; DLM, dorsal longitudinal muscle; DVM, dorsoventral (tergosternal) muscle; F, wing fulcrum; P, wing articulation with tergum; PH, phragma; S, sternum; T, tergum; TCM, tergocoxal muscle; TPM, tergopleural muscle. Original.

187

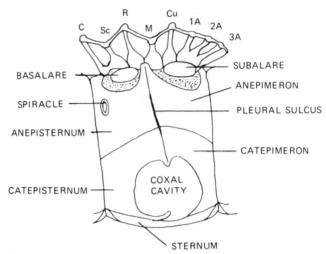

Figure 72   Reconstruction of hypothetical primitive pterygote pleuron, with wing base and un-fused pteralia. A, anal veins; C, costa; Cu, stem of cubitus; m, stem of media; R, stem of radius Sc, subcosta. Original

sects with the pleurons in other arthropods. It is not proper to compare too closely mechanisms that most probably have evolved independently in rather distantly related evolutionary lines. The pleurons of various arthropods appear in highly evolved lines, each in its own fashion. The only meaningful comparisons are the ones that can be made among members of more closely related evolutionary lines. Unfortunately, the only sources for direct comparison are living species, whose evolutionary histories are unknown and which have their own specializations. The pterygote pleuron appears to be a product of insect evolution, and should not be regarded as the homologue of the centipede or crustacean pleuron. The insect pleuron can only be properly regarded as the product of modification from simple beginnings in the earliest insects, as was suggested in Chapter 6.

The pterygotes have acquired additional derived characteristics. The anterior tentorial arms became fused with the posterior bridge, forming a solid unit inside the head. Various gnathal muscles shifted their origins to the tentorium as the intergnathal transverse tendons were suppressed. In all the pterygotes the intersegmental tendon system was highly suppressed, and body muscles formerly originating on the tendon system are attached to the cuticle by way of transepidermal tonofibrillae. The tonofibrillae are all that is left of the tendon system in pterygotes, other than traces at the intersternites of some insects, such as larval *Corydalus cornutus* (Barlet, 1977). The abdominal eversible sacs and pregenital abdominal styli and coxal plates were suppressed. Postnuptial molting was not retained in the pterygotes. Except for ephemerids, molting never occurs after the growth of functional wings, and in no winged insect after sexual maturity. The embryos of pterygotes sink into the yolk in various ways, so that the blastoderm serves as an amnion for a while during embryogenesis.

It is highly probable that a functional ovipositor capable of penetrating into oviposition substrates was developed in all early pterygotes, later to be partially or completely suppressed in various evolving lines. Numerous traces of well developed ovipositors have been found in fossils of ancient insects of all sorts (Figures 62 and 63).

Among insects, only the pterygotes bear free lateral cervical sclerites. The original ground plan state appears to have been in the form of a single lateral cervical sclerite on each side, articulating in front with the lateral posterior borders of the head. The posterior end possibly was detached from the preepisternum, since it is associated with the episternum and it bears prothoracic muscle insertions (Matsuda, 1970). The lateral cervical sclerites are single pieces in most insects, except in the Odonata, the Orthopterodida, and some beetles, in which the sclerite consists of two or three articulated pieces.

Two main evolutionary lines became established very early in the history of flying insects. The Plagiopterata represent one of the sister groups. The other is represented by the Opisthopterata. In the early members of these two groups, the wings were incapable of being folded at rest over the abdomen. Independently in each line, some members acquired the ability to fold the wings.

SECTION PLAGIOPTERATA

The insect taxon Paleoptera is generally used by most entomologists to include those insects that apparently never acquired the ability to fold their wings at rest. This grouping is phylogenetically undesirable. Odonata and Ephemerida are the only surviving paleopterous insects. The extinct Paleodictyoptera, Megasecoptera, Meganisoptera, and Protephemerida are generally included in the Paleoptera. The Megasecoptera are close relatives of the Paleodictyoptera, but are now known (Diaphanopterodea) to have acquired the ability to fold their wings (Carpenter, 1963b). Therefore some of the Megasecoptera do not fit well into the definition of "paleoptera," when the group is defined as primitively unable to fold their wings. The Ephemerida are allied with the neopterous wing folders (Lemche, 1940; Schwanwitsch, 1956; Matsuda, 1970), rather than with the Odonata. The taxon Paleoptera therefore is a paraphyletic taxon and should not be used in a phylogenetic classification.

The derived characters shared by the Plagiopterata do not concern only the wings. The group as used here includes the orders Paleodictyoptera, Megasecoptera, Meganisoptera, and Odonata. The median caudal filament was greatly shortened or suppressed. Cuticular outgrowths in the form of filamentlike structures formed on the dorsum of some adult representatives of each of the orders of the group (Kukalova-Peck, 1972). In some species these even became branched. In others they were spiny. The only surviving form of these outgrowths occurs as carinal denticles across the abdominal terga of Odonata. These projections have been interpreted as vestiges of nymphal tracheal gills, but no nymphal fossil in this group shows any evidence of ever having lateral tracheal gills. In each order additional derived characters appeared, as discussed below.

The Plagiopterata became specialized in the use of dorsoventral thoracic muscles as the main wing motor mechanism. The large dorsal longitudinal muscles that cause upward deformation of the thoracic terga of opisthopterous insects were not important in the Plagiopterata, and are highly reduced in the Odonata. Flight in the Odonata (and probably in other related extinct orders) employs the direct method, in which the vertical muscles pull on or near the wing bases on either side of the fulcrum (Figure 71, A). The thoracic terga of plagiopterous fossils are not adequately preserved to show much detail. It appears that these terga probably were not flexible as they are in other insects which use the indirect method of flapping their wings.

The nymphal wing pads extended sideways free from the body, although they were curved somewhat to the rear distally (Carpenter and Richardson, 1968). Odonate nymphal wing pads begin their growth as lateral expansions of the terga, but as they enlarge, the pads are turned upward and backward as the pleura enlarge (Matsuda, 1970). In older odonate nymphs the wing pads become turned so that the costal borders are medial and they lie along the abdominal surface. The primarily lateral growth of the nymphal wing pads is the basis for the name Plagiopterata.

The early Plagiopterata generally retained a few ancestral pterygote features, such as the sideways growth of the nymphal wings, the inability to fold the wings, the well developed meshwork of the archedictyon, the retention of the anterior media vein, unspecialized legs with pentamerous tarsi, and essentially similar fore and hind wings. In some lines changes occurred, such as the reduced archedictyon of Megasecoptera and of zygopterous dragonflies, loss of the posterior media, and reduction of tarsal segments in Odonata.

The genitalia of the Plagiopterata appear to have consisted of a well developed sclerotized ovipositor somewhat similar to that of Odonata, and of male structures comparable to those of the Microcoryphia. The male of

Stenodictya lobata Brongniart appears to have the well developed coxal plates of the ninth abdominal sternum, but without styli. The paired parameres described by Kukalova (1970) in this species, and in *Dunbaria* (Kukalova-Peck, 1971) consist of slender segmented structures similar to the structures of machilids that Bitsch (1974b) showed were modified coxal sacs of the ninth abdominal segment. The term paramere should be reserved for the genital claspers that are typical of the neopterous Phalloneopterata and that are new structures developed on the tenth abdominal segment. The Plagiopterata probably did not acquire the ability to transfer semen directly into the female gonopore, judging from the absence of true clasping organs. The two sister groups of the section are the Paleodictyopterata and the Odonatopterata.

Subsection Paleodictyopterata

An outstanding specialization was the development of a sucking type of proboscis (Figure 63), acquired convergently with the sucking apparati of various neopterous orders. The peculiar proboscis apparently consisted of stylets and an elongate labium, but differed from the hemipteroid proboscis in bearing a pair of long segmented palpi (Laurentiaux, 1952). Prothoracic paranota became well developed, and the wing venation pattern was often discernible on these. As mentioned earlier, the venational pattern of the prothoracic paranota is not necessarily indicative of the original pterygote venation. The genetic code for the fully veined functional wing is merely partly expressed in the prothoracic paranota of these insects. The abdominal paranota became prominent features of the early Paleodictyopterata, but became secondarily suppressed in the Megasecoptera. In all members of the group, there is no median caudal filament. The nature of the lateral cervical sclerites is unknown. Two orders are included, the Paleodictyoptera and the Megasecoptera.

The members of the order Paleodictyoptera lived during the Carboniferous and Permian times. It is in this group that the thoracic and abdominal paranota were especially well developed. They seemed to have been inhabitants of luxurious humid forests. There is no evidence that the nymphs were aquatic, with respiration carried on in the same manner as in the Odonata. But internal rectal gills would not be expected to be visible in nymphal fossils.

The basal wing sclerites of the Paleodictyoptera fused so as to produce a single subcosto-anal plate. The wing venation included all the main veins with separate stems. Veins R, M, Cu, and A were basally separate, and the archedictyon was well developed in the early species. In the Permian, the archedictyon became obscure or lost.

The members of the order Megasecoptera acquired several features as the group split off from the paleodictyopterous stem. The paranota became reduced to spinelike structures or were completely suppressed. The wings became narrowed, or even petiolate. The chief basal plate was the fused radiocubital plate in the narrow-winged species, but the wing folders had a fused mediocubital plate plus small radial and anal plates. The archedictyon was largely eliminated and the branching of the wing veins was reduced. Vein Sc ends in R_1, and does not reach the wing border. Veins M and Cu shifted forward basally so as to become closer to R, in some cases fusing into a complex of R + M + Cu before branching. Although most fossil representatives are preserved with the wings extended, many have the wings folded over the abdomen as in the Neopterata. The wing pads of the nymphs were laterally extended, however (Carpenter and Richardson, 1968). It appears that wing folding in this order arose as a new feature, independently of the wing-folding ability of the Neopterata (Carpenter, 1953).

The wing-folding Megasecoptera have been classified as order Diaphanopterodea (Carpenter, 1963b). If such a status becomes acceptable, the Diaphanopterodea and the narrow-winged Megasecoptera should be sister groups. Both differ from the Paleodictyoptera in the coalescence of the bases of M and Cu, nearly fusing with R, and in the absence of lobelike thoracic and abdominal paranota.

It appears that a nymphal ability to fly was acquired by some of the Paleodictyopterata. Kukalova-Peck (1976) found a new species of Megasecoptera in which the last two subimaginal instars had apparently functional wings, and Kukalova-Peck and Peck (1976) described some Calvertiellidae (Paleodictyoptera) whose subimaginal wings were fully formed and perhaps functional. Since these were upper Paleozoic (Permian) insects, it may be that nymphal flight was acquired by these highly advanced paleodictyopterous insects independently in each order. This is also convergent with the peculiar nymphal (subimaginal) flight of the Ephemerida.

Subsection Odonatopterata

This group is represented by the extinct Meganisoptera and the Odonata. The name at first glance appears to be nonsensical, but it means "odonatelike winged insects." Their derived characters in common include shortening of the cerci; great enlargement of the compound eyes; suppression of the paranota; conversion of the archedictyon to numerous cross veins; reduction of the media vein to a single vein, MA; reduction or loss of vein CuA. Basally, veins MA and Cu became crowded forward so that MA lies close to R. The axillary sclerites fused into a large radioanal plate plus the small subcostal plate. The precostal basal part of the wing became expanded so that there appears to be a sclerotized precostal area ahead of the costa near the wing base. It appears that the vein Cu became separated basally into independent CuA and CuP veins, but the fossils suggest that CuA became very short, then fused with CuP (Figure 70, C), later to become a single vein, CuP (Figure 70, D), in the Odonata (Laurentiaux, 1953b). The legs in the Odonatopterata acquired a series of spines and an elongate structure, permitting seizure of prey in flight. The adult mouthparts were adapted for preying. The pleura became tilted backward, so that the legs appear to have moved forward, as in present Odonata. Fossil meganeurids are known in which this specialization is strongly suggested (Carpenter, 1943b).

The order name Meganisoptera was erected in 1932 by Martynov for the genus *Meganeura* Brongniart, 1893, leaving *Protagrion* in Protodonata. Brongniart had included his *Meganeura monyi* and *Protagrion audouini* in his order Protodonata. *Protagrion* is a member of Paleodictyoptera Goldenberg, 1854 (Carpenter, 1943b), so Protodonata becomes a synonym of Paleodictyoptera. Carpenter, however, returned to using Protodonata for *Meganeura* and allied genera. He placed *Protagrion* in Paleodictyoptera and made Meganisoptera a synonym of Protodonata. Martynov was the first revisor, and so his name Meganisoptera is used here for the sister group of the Odonata. This usage follows the principle of priority, and does not cause confusion because the name was never used outside its present context.

The advanced features of the Meganisoptera are such that the members of the group must have evolved contemporaneously with the Odonata. Fossil representatives are known from the Upper Carboniferous to the Lower Triassic. Fossils of Odonata date from the early Permian. The Meganisoptera were large insects with a wing spread up to about 65 cm in *Meganeura monyi*, the largest known insect, down to about the expanse of the larger existing dragonflies. The fore wing was narrow, elongate, with a long narrow anal field bearing many anal veinlets. The meshworklike archedictyon became a regular series of cross veins between the various vein branches. Vein CuA (+) was separated basally from CuP (−), but soon rejoined CuP and lost its identity. Veins MA (+) and CuP (−) were unbranched, and basally were close to R, and these did not form the typical arculus and triangle such as in Odonata. The precostal field was well developed in most species.

The prothoracic paranota appear to have been highly reduced (*Meganeurula*). The tarsi became four-segmented and the cerci and median caudal filament were reduced to short two-segmented structures. Otherwise, the Meganisoptera retained the ancestral characters that include lack of pterostigma and nodus, fairly large unreduced thoracic terga, a short broad abdomen, and the tergal filaments similar to those in Paleodictyopterata.

Members of the order Odonata are the only surviving plagiopterous insects. Most of the characteristics of the Odonata are highly specialized. The young are aquatic, and bear their own specializations for aquatic life. The nymphal labium is highly modified into a prehensile organ. The tracheal gills are posterior on the abdomen, but evolved in two directions. The Anisoptera and Anisozygoptera bear internal rectal folds richly supplied with air-filled tracheae. Breathing is accomplished by the inhalation and discharge of water through the anus. In the Zygoptera, the nymphal paraprocts and the median caudal filament are expanded to serve as tracheal gills. Abdominal gill filaments are borne ventrolaterally on a few species of damselflies inhabiting slow shallow streams (Corbet, 1962). These lateral gills appear to be new structures adaptable to conditions of low oxygen tension. Some damselfly nymphs inhale and expel water through the anus, suggesting that rectal respiration was the original respiratory adaptation of Odonata, secondarily lost and replaced by caudal gills in the Zygoptera.

Although the nymphal wing pads lie along the abdomen, the first appearing wing buds grow laterally, as is typical of the Plagiopterata. Later, as the wing buds elongate and the pleuron grows upward, the immature wings turn upward and backward so that the hind wings lie over the fore wings, with the costal border medial (Matsuda, 1970).

In the adults, the antennal flagellum is reduced to a small bristle, as it is also in the nymphs. The compound eyes became much enlarged, so that the insects have an extremely broad field of vision. The lateral cervical sclerites consist of plates divided into three sclerites. Prey is captured in flight with the spiny-armed legs. Walking locomotion had been eliminated in the adult. The prothorax is highly reduced, and paranota are entirely suppressed. The tergal filaments have become reduced to rows of spines on the posterior borders of the abdominal terga. The mesonotum and metanotum have become reduced, bringing the bases of the wings closer together than in other pterygotes. The pteropleura are enlarged, and strongly tilted backward. The mesepimeron invaded the mesotergum, separating the prescutum from the scutum and almost meeting dorsally. In the wings, a pterostigma appeared distally between the R and C veins. The Sc vein ends abruptly at a notch near midwing, where there is a nodal cross vein. Vein MA became closely united basally with R, and with the stem of Rs forming a cross vein-like "arculus." Rs and MA extend from the arculus. The basal wing sclerites formed a fused radioanal plate separate from the subcostal plate. In the leg, the trochanter bears an encircling sulcus that gives the appearance of segmentation. The tarsomeres became reduced to three.

The Odonata probably never flew using the indirect muscles such as are found in the Opisthopterata. The dorsal thoracic phragmata did not develop and the dorsal longitudinal muscles, which are so prominent in the Opisthopterata, are highly reduced. The mesonotum and metanotum do not flex. Instead, the downstroke is accomplished by powerful dorsoventral muscles attached dorsally to cap tendons that pull on the wing bases slightly lateral to the fulcrum. The tergosternal muscles that elevate the wings also attach via cap tendons at the lateral edges of the reduced nota, very close to the wing base mediad to the fulcrum (Figure 71, A).

The abdomen of Odonata became very slender, in contrast to the robust abdomen of *Meganeurula*. Abdominal sterna are very narrow, somewhat hidden by the abdominal terga that almost completely encircle the abdomen. The cerci and epiproct became unsegmented. A highly unusual method of copulation was acquired. The male gonopore shifted to the front part of the ninth abdominal sternum, and phallic lobes were suppressed. Spermatophores are transferred to an accessory copulatory apparatus on the second abdominal sternum of the male before coupling with the female. In mating, the male grasps the neck or prothorax of the female using the cerci as clasping organs, and tandem flight may occur while the two sexes are coupled together. The female bends her abdomen down and forward, so that her genitalia receive the spermatophore from the accessory copulatory organs of the male. The ovipositor was retained in reduced form, still bearing a stylus on the "outer valve" (gonocoxa) of the ninth abdominal segment. The libelluloid families bear a highly reduced nonfunctional ovipositor. Odonate spermatozoa are typical insect sperm cells with the 9 + 9 + 2 axoneme pattern and two mitochondrial derivatives.

SECTION OPISTHOPTERATA

This section is the sister group to the Plagiopterata. The name refers to the growth of the nymphal wing pads as posteriolateral expansions of the thoracic paranota. The nymphal wing buds lie posteriorly against the abdomen so that the costal border is lateral, and the hind wing is partly covered by the fore wing. This is the adult position of the wings when folded.

With the exception of members of the cohort Blattiformida, flight is through the action of large indirect thoracic muscles, aided by the somewhat direct alary muscles which also perform some leg motion. At the dorsal intersegmental lines between the pronotum-mesonotum, mesonotum-metanotum, and metanotum-abdomen, strong flat transverse apodemes known as phragmata serve for attachments of the large dorsal longitudinal muscles (DLM) (Figure 71, B). Contraction of the DLM causes upward buckling of the notum, raising the points on the side of the notum on which the wing base articulates, mesad to the fulcrum on the pleural wing process. Raising the wing base by the notum causes the wing to stroke downward and forward with the costal border tilted down. The upstroke occurs when powerful dorsoventral muscles (DVM) lateral to the DLM depress the buckled notum, lowering the wing base at its notal articulation. During upstroke, the wings tend first to move slightly forward, then rearward during most of the upstroke, so that the twisted trailing edge provides a forward and downward push. Lift and forward motion generally are performed during both the upstroke and the downstoke. The indirect DLM and DVM are variously aided by lateral oblique tergopleural muscles and by muscles from the basalar and subalar sclerites to the pleuron and coxae. The latter muscles provide a means of controlling the angle of attack of the wings during flight and also serve as direct wing motors.

The Opisthopterata have acquired the method of direct spermatophore transfer from the male gonopore to the female gonopore without any intervening mechanism. The primitive atelocerate method of sperm transfer was through the deposition of a spermatophore by the male onto a substrate, later to be picked up by the female. This primitive mating process still occurs in apterygote insects. The Odonata differ from the Opisthopterata in having an indirect mating method in which the spermatophore is transferred twice. The sister groups of the Opisthopterata are represented by the Ephemerata and the Neopterata.

Subsection Ephemerata

The subsection consists of two orders, the Protephemerida and the Ephemerida. In both orders, the advanced characters in common include a slender abdomen lacking paranota, and in general a mayflylike appearance. The only known member of the Protephemerida is *Triplosoba pulchella* Brongniart, from the Upper Carboniferous, represented by a single poorly preserved specimen. Imaginative restorations picture long antennae, pentamerous tarsi, and male genitalia. The fossil exhibits none of these characters (Carpenter, 1963a). *Triplosoba* differs from the mayflies largely in having a more primitive wing venation, especially in that CuA is simple rather than branched, MA is unbranched and it does not join Rs briefly before diverging posteriorly. Both wings are of the same size, shape, and venation.

Triplosoba was contemporaneous with the Upper Carboniferous early mayflies (*Lithoneura*), so it is not ancestral to the mayflies. Whether the Protephemerida are ranked as a suborder of Ephemerida (Tschernova, 1970) or retained as a separate order is a matter of opinion.

The insects of the order Ephemerida (suborder Plectoptera of various authors) have acquired an impressive list of derived features since their origin. The prothoracic and abdominal paranota were eliminated. While the nymphs retained fairly long antennal flagella, the adults acquired a tiny bristlelike flagellum since the Permian. The Permian mayfly *Protereisma* had an obvious 11-segmented antennal flagellum. The hind wings became variously reduced in different lines. Veins MA and MP became branched, and both veins persisted in most species. The anterior media at first arched forward toward Rs, joining it briefly before turning posteriorly. CuA became branched and the base of Cu also moved forward toward the stem of M. Intercalary veins between the branches of Rs, MA, MP, and CuA extend the regular fluting to the wing margin. The wing shape ultimately became somewhat triangular. The basal wing sclerites fused into a large subcostocubital plate apart from the anal plate.

Five outstanding sets of characters shared by the mayflies and the Neopterata testify to their common descent from an opisthopterous ancestor. One is the growth of nymphal wing buds from the posterio-lateral border of the thoracic nota, with the growing wings in the typical folded position. Another is the employment of greatly enlarged dorsal longitudinal muscles attached to phragmata as the main mechanism for wing depression in flight. A third is the large phragmata to which the indirect flight muscles attach. The fourth is the mechanism of copulation involving the introduction of the spermatophore directly from the primary male system into the female gonopore. The fifth is the suppression of the archedictyon.

Mayflies are the only modern pterygotes that undergo molting after acquiring functional wings. Much has been written about whether the subimago, the first winged instar, represents an adult imago that undergoes an imaginal molt comparable to the adult molting of the apterygote insects and many other arthropods. Therefore, adult molting could be regarded as a persistent primitive character. Another view considerd the subimago as merely a young imago whose outer cuticular layers are delaminated (Lameere, 1917). But all the physiological phenomena accompanying a true molt occur at the subimaginal molt (Taylor and Richards, 1963). The subimago actually undergoes considerable change in becoming the imago. At the subimaginal molt the mouthparts are further reduced, the male genitalia become more slender and the basal parts better articulated, and the antennae are shortened and become bristlelike. The male fore legs greatly elongate, and microtrichiae on the subimaginal wings are eliminated (Ide, 1937). The male subimago with few exceptions does not mate in that state. Mating occurs normally only after the male and female subimaginal molt, except in a few genera in which the female remains somewhat neotenous by suppression of the subimaginal molt, and in one genus in which both sexes mate without a subimaginal molt. The subimago, therefore, cannot be a true imago which molts a second time. Flight during the subimaginal instar does not necessarily make this stage imaginal, or truly adult. There is fossil evidence that in the Permian, mayflies had not yet acquired the ability to fly as subimagoes, but rather during the last nymphal stadium the wing buds became quite large and extended posteriolaterally from the sides of the thorax, just as they do in the imago. But these nymphs appeared to be aquatic, bearing well developed abdominal gills (Kukalova, 1968). Younger stages of Permian nymphs still held the developing wings directed backward. The slightly spread out relatively large nymphal wings clearly exhibited a full venation, abundant and reticulate cross veins, and the adult type of fluting. The modern mayfly subimago is one evolutionary solution to the problem of emerging safely from an aquatic habitat full of predators at a critical stage of the life cycle. The last aquatic nymph remains active until the time of molting. About an hour before ecdysis, gas appears between the nymphal cuticle and the pharate subimago, but it continues activity until it floats or climbs to the surface. A sudden ecdysis, lasting as little as 15 seconds in some species, occurs and the subimago flies off. The mayfly subimago then becomes quiet and imaginal organs complete their development. Thus this stage more nearly resembles the pupal stage of the higher insects. Pupation has occured convergently in several other groups of insects: Thysanoptera, the scale insects and whiteflies in Homoptera, and in the Endopterygotida. The mayfly type of quasi-pupation is best regarded as a neotenous state of the last nymphal instar, in which certain adult features develop precociously. These include the subimaginal body form with functional wings, the premature suppression of nymphal gills and the extreme neoteny in these species which do not undergo the imaginal molt, but remain as subimagoes until death. A convergent acquisition of flight by subimagoes occurred in the Permian among some advanced Paleodictyoptera and Megasecoptera (Kukalova-Peck, 1976; Kukalova-Peck and Peck, 1976). These events are not related to emergence from an aquatic habitat, however. Perhaps precocious flight was an adaptation to escape terrestrial predators.

In addition to the advanced features above, mayflies have acquired several more derived characters that eliminate them as primitive insects. The juvenile stages are aquatic but secondarily so, as is shown by the fact that the nymphs bear air-filled tracheae. Their lateral paired abdominal tracheal gills seem to be morphologically unique. Their structure suggests they may have originated from abdominal paranotal winglets (Kukalova-Peck, 1968). Although the nymphs bear mandibulate mouthparts, the maxillae do not grow fully. The maxillary laciniae and galeae are mere flat lobes. In adults, the mouthparts are highly reduced and nonfunctional. The adult alimentary canal became nonfunctional also, and is reduced to a simple epithelial tube filled with air. The gastric caeca of all stages were eliminated. Adult male compound eyes have become enlarged and close together, and in some species the compound eyes are divided into two groups of facets. The male fore legs became elongated and adapted as claspers for holding the female during mating by bending the fore tarsi *upward* in holding on to the wings of the female. The prothorax of both sexes is highly reduced and is poorly sclerotized. The hind wings and metathorax have assumed a minor role in flight, and have become reduced in size. The anal field of both wings is highly reduced, and a costal brace was developed in the fore wing. The gonads undergo most of their development during late juvenile stages, so that the imagos are ready to mate and reproduce right after the imaginal molt. The imaginal life is greatly shortened to a few hours, sufficient only to reproduce without feeding. Mayflies produce a greatly elevated number of eggs, up to 3000 per female. This is an adaptation that permits survival in an aquatic habitat where predation is heavy. The mayfly sperm cells are somewhat atypical in that there is only one large acrosome, and the axoneme pattern is $9 + 9 + 0$, with no central pair of axonemes.

The double gonopores have been cited as evidence of the primitiveness of mayflies, because it can be assumed that ancestral arthropods bore their gonopores in pairs. But the double gonopores of mayflies are more properly interpreted as a reversed character state. Previous discussion showed that the ancestral insects probably had acquired the simple gonopores in both sexes, and that an ovipositor originated in insects before wings were acquired. The double gonopore of mayflies therefore is most probably the result of suppression of the median epidermal invaginations that resulted in the single gonopore of typical insects. The ovipositor must also have been lost in mayflies, because its origin must have been in the apterygote ancestor of all winged insects.

Mayflies are the only insects that have without question modified the coxal plates and styli of the ninth abdominal segment into male organs suited for clasping the female genital area during mating. Smith (1969) maintained with many previous authors that male insect clasping organs must all be homologized with ninth abdominal segment limb vestiges. This position is difficult to assume unequivocally. In the orthopteroid insects, the homologues of clasping organs of higher insects occur as simple or divided phallomeres developing from phallic lobes of the tenth abdominal segment *at the same time* that ninth segment vestiges of coxal plates and styli are apparent in the nymphs, particularly in grylloblattids, cockroaches, and mantids. The phallic rudiments of Caelifera (Orthoptera) even begin developing on the *embryonic* tenth abdominal segment, from embryonic limb buds.

Kristensen (1975) favored the idea that mayflies form a sister group to the Odonata and Neopterata because of some assumed primitive features. In the latter group, the basal wing tracheae arise from an alar arch, constructed by fusion of an anterior trunk and a posterior trunk in each winged segment, while in mayflies there is usually only the anterior trunk. Each leg is also supplied by two tracheal trunks which fuse, except in most mayflies. The presence of a complete alar arch and double tracheation of the legs of the mayfly genus *Epeorus* was considered a specialization. But because mayflies are really neotenous in many ways, the incomplete tracheal alar arch and lack of posterior leg tracheae most probably represent mayfly specializations. Kristensen also considered the absence of direct spiracular closing muscles (which are present in Thysanura, Odonata, and Neopterata), the lack of a nerve connection of the corpus alatum with the corpus cardiacum (also found in Thysanura), the presence of superlinguae and the presumed separation of Rs from R, represent primitive features not found in the Odonata and Neopterata. Since spiracular closing muscles do occur in the Thysanura (Maki, 1938), their absence in mayflies must be a specialization. The "superlinguae" also occur in Thysanura and Dermaptera, but these also are probably new and convergently acquired because no superlinguae occur in the entognathous orders. If Rs and R are occasionally separate in mayflies, this must be secondary since the normal branching of Rs from R occurs in *Triplosoba*, *Protereisma*, and *Lithoneura*, all ancient Paleozoic mayflies.

The primitive characters of mayflies are very few. They are still unable to fold their wings. The vein MA has persisted as a distinct wing vein. There is a regular alternation of convex and concave veins in the wings. The median caudal filament has remained, and the cerci are long and segmented. The tarsi are pentamerous. The lateral cervical sclerite is simple. When one considers the specializations, mayflies cannot any more serve as a model for the ancestral winged insect.

The practice of attempting to standardize the order names of insects with the Linnaean ending "-ptera" is deplorable, because it tends to produce nonsense names. For instance, the name Psocoptera Shipley, 1904, for Corrodentia Burmeister, 1839, suggests that the wings have a biting function. In the case of mayflies, Shipley chose to append "-ptera" to the stem of Ephemerida Leach, 1817. This makes the name Ephemeroptera to mean "wings of those which live for a day," and this is not as nonsensical as Psocoptera. But there is an older name that properly describes ephemerid wings: Plectoptera Packard, 1886, which refers to the plaited appearance, or the regular alternating creases of mayfly wings. The name Ephemerida is the oldest well-known name, and it is used here for that reason.

Subsection Neopterata

The ability to fold the wings seems to have arisen twice in the history of insects. Wing folding seems to have arisen in some of the Megasecoptera (Diaphanopterodea) but the line did not survive. The mechanism of wing folding in Megasecoptera is unknown, but the known basal wing sclerites appear to consist only of a fused mediocubital plate and a remnant of an anal plate.

All modern wing-folding insects share such a similar folding mechanism that there is little doubt that it came about as a monophyletic event independently of the Megasecoptera. A series of three axillary sclerites between the wing base and the notum occur as well developed structures. The anteriormost (first axillary, Snodgrass, 1935) is the anterior hinge plate of the wing base, articulating with the notum, second axillary, and vein Sc. The second axillary also bears the base of R, and by way of one or two median plates, the base of M, Cu, and 1A. It rests on the pleural wing process, acting as the wing pivot. The third axillary sclerite pivots anteriorly on the second axillary and posteriorly on the posterior notal process. It is associated with the bases of some of the anal veins. The stems of M, Cu, and the first anal (1A) articulate with sclerites generally known as median plates. The folding process involves a pleural muscle that pulls the outer arm of the third axillary upward and inward so as to crumple the posterior wing base, bringing the anal or jugal area under the wing and against the body as the wing swings backward. The details of folding differ in the various orders, but in all cases the revolution of the third axillary is involved in flexing the wing. Since the ancestral pteralia seem to have consisted of individual basal sclerites (Kukalova-Peck, 1974), the neopterous state appears to have resulted from the subcostal plate becoming the first axillary, the radial plate becoming the second axillary, the medial and cubital plates fusing into a medial plate, and the third representing the anal plate. The anteriormost anal vein became detached from the common anal base as an adaptation accommodating the flexion at the vannal fold (or plical furrow) between the first and second anal veins (Figure 69, B). The detached 1A vein is the one called empusal by Hamilton (1972a) and the postcubitus by Snodgrass (1935).

Several modifications occurred in the neopterous wings. Behind the anal field a sclerotized veinlike structure formed in the membrane, known as the jugal vein (Figure 69, B, j). The area bearing the jugal vein is separated from the anal area by a jugal fold. The jugal area is also known as the neala, or "new wing," because it is not found in the nonwing-folding insects. The fore wing folds so that the neala turns upside down against the wing proper at the jugal fold, thus differing from the hind wing in its method of flexing, particularly in the nonpupating insects. In the pupating insects (Endopterygotida) both wings fold at the jugal fold, possibly as a result of the narrowing of the anal area of the hind wing. In the nonpupating orders, often grouped as Exopterygota (a paraphyletic grouping), the hind wings fold along two main creases. The plical (or vannal) fold between 1A and 2A results in the area behind 1A (vannus) turning upside-down. The jugal fold between the last anal and the jugal vein results in the neala lying upward against the reversed vannal area in such wings.

Wing venation in the Neopterata (Figure 69, B), as exemplified by the

Comstock-Needham system, has changed from the hypothetical precursor pattern (Figure 69, A) in that vein MA has lost its identity. Vein M at its base is usually concave ($-$), and typically it is four-branched. The vein M often is neither concave nor convex. Some authors identify vein MA in those Neopterata whose media vein system has more than four branches (Carpenter, 1966), even though none of the branches of M is ever convex. Veins C ($+$), Sc ($-$), R ($+$), Rs ($-$), CuA ($+$), and CuP ($-$) are rarely confused, because these veins almost always are identifiable by their convex or concave positions. Vein MA has either been entirely suppressed, or it has joined the field of the radial sector and lost its convexity and its identity.

The pleural sclerite generally known as the trochantin probably originated as a new structure in the Neopterata. The trochantin typically is a slender straplike extension of the lower part of the episternum, articulating with the lower anterior border of the coxa. It has been assumed that the pterygote trochantin had its origin in the Thysanura (Matsuda, 1970). But the thysanuran trochantin is a specialization in that group, which permits the wide excursion of the horizontally movable coxa in oarlike fashion. Further, the pleural sclerites show no evidence of a trochantinal area (Figure 63) in the thorax of the well preserved Permian paleodictyopteran *Monsteropterum* (Kukalova-Peck, 1972), and the Odonata and Ephemerida seem never to have devloped a trochantin. The primitive neopteran trochantin is best exemplified in the Plecoptera, where it is an extension of the lower episternum. In the Orthopterodida the trochantin became separated from the episternum by a sulcus, or the separation became membranous and the trochantin is freely movable. The straplike trochantin was retained by most of the higher insects, but lost its identity in some orders, such as in Mallophaga, Anoplura, Coleoptera, Strepsiptera, Hymenoptera, Diptera, Siphonaptera, and Lepidoptera.

The median caudal filament was entirely suppressed in all the Neopterata. In addition, a new type of male genitalia evolved as the coxal plates and styli of the ninth abdominal segment merged with the ninth sternum and largely lost their identities. The only surviving neopteran insects that bear coxal plates and styli separated from the ninth abdominal sternum are the Grylloblattaria. The neopteran male external genitalia form through the growth of a pair of phallic lobes extending from the front border of the tenth abdominal sternum. As the phallic lobes grow, they are not penetrated by the mesodermal vasa deferentia, as in Ephemerida. Instead, an epidermal invagination forms between the phallomeres, becoming a median ejaculatory duct, which grows in to fuse with the ends of the vasa deferentia (Snodgrass, 1957). Since grylloblattids, cockroaches, and mantids still bear the ninth segment styli *and* at the same time the phallomeres form around the gonopore of the tenth segment, these phallomeres cannot have formed from modified "gonocoxae" and "gonostyli" of the ninth segment as is supposed by Smith (1969).

Infrasection Polyneopterata

The neopterous insects are represented by two sister groups, the Polyneopterata and the Phalloneopterata. Members of the first group share only a few advanced characters. The pleating of the anal field when the hind wings are folded consists of a convex fold at the vannal fold, at least one concave fold in the middle of the anal field and another convex fold at the jugal fold (Plecoptera), or the anal field is pleated in convex folds between adjacent anal veins. The members of the infrasection have not acquired the aedeagus or parameres such as in the Phalloneopterata. However, the ancestors of mayflies and Neopterata did acquire the ability of direct insemination in mating. The primitive neopterous method probably consisted of the eversion of the invaginated ejaculatory duct of the male, providing a simple everted penis. This method was not developed much further in the Polyneopterata.

The Plecoptera have remained primitive in that the median genital passage is largely unsclerotized. When it is everted during mating, the ends of the vasa deferentia, which join the genital passage separately, become the effective gonopores (Brinck, 1955), at the end of the everted ejaculatory duct. The phallomeres of the tenth abdominal segment in the Polyneopterata are best developed in the Orthopterodida, but even in this group, the phallomeres do not become functional male clasping organs.

The Polyneopterata have retained quite a number of ancestral insect features. The mouthparts consist of generalized biting organs. The maxillary lacinia is movably articulated with the maxilla. The palpi consist of three or more segments. Metamorphosis is gradual, and there is no pupa stage. The juvenile stages gradually acquire organs that continue their development through several nymphal stages. Thus the nymphs bear the functional beginnings of adult organs, such as cerci, ocelli, compound eyes, antennae, and developing wings for several instars before the final molt, and there is little or no hystolysis or loss of nymphal tissue. For instance, styli may form on the eighth and ninth abdominal segments of nymphal cockroaches, and these disappear in the adult female, or only the ninth segment styli remain in males of many orthropteroids. The abdominal nervous system is generally not concentrated forward and malpighian tubules are usually numerous. The ovipositor has been retained in modified form in the orthopteroid group, or entirely suppressed in various lines independently. The Polyneopterata mostly (except phasmids) bear typical insectan spermatozoa with two mitochondrial derivatives and the $9 + 9 + 2$ axoneme pattern.

Supercohort Plecopterida. This group forms a sister group to the Orthopterodida, and includes the orders Paraplecoptera, Plecoptera, and Embiidina. The derived features in common include the suppression of phallomeres and all abdominal styli, the development of accessory male clasping organs in the form of extensions of the tenth abdominal tergum and modified paraprocts, and the reduction of veins Rs and M to three or less neutral branches each. The Plecopterida have retained the primitive neopteran trochantin in that it is merely an extension of the epimeron, extending down to the lower front border of the coxa providing a secondary coxal articulation. The fore wings have not become leathery or sclerotized as in the Orthopterodida. The sister groups in the Plecopterida are the cohorts Perlodida and Embiodida.

The Perlodida are the only polyneopterous insects whose young are aquatic. Lateral tracheal gills in the form of filamentous tufts were acquired, variously positioned on the labium, neck, thorax, anterior abdominal segments, or near the anus. The adults may bear functional biting mouthparts, but they are more often reduced. The Perlodida, in contrast to the Embiodida, have retained the long, multisegmented primitive insectan cerci and lack the asymmetry of the male genitalia of the embiids. Two orders of perlodids are generally recognized. Paraplecoptera are extinct, with fossils found from the Carboniferous to the Triassic. They resemble primitive stoneflies, but true stoneflies are known from the Permian, so the Paraplecoptera appear to be a sister group to the Plecoptera. These insects reduced the tarsomeres to four segments, and the nymphs appear to have been aquatic, bearing numerous abdominal tracheal gills (Riek, 1970b) possibly derived from abdominal paranota. Vein M was reduced to three branches in many species. The ovipositor was well developed in other species.

The other order of this group is the Plecoptera. The derived characters which contrast with those of the Paraplecoptera are the lack of paranota on the prothorax, the suppression of the ovipositor, the loss of posterior lateral abdominal gills, the fusion of the testes into a single organ with retention of the two vasa deferentia, the reduction of branches in veins Rs and M, and the reduction of the tarsus to three tarsomeres. In contrast to the Embiodida, stoneflies bear a simple primitively undivided pair of lateral cervical sclerites.

The order Embiidina is the only one in the cohort Embiodida. The order name was changed to Embioptera by Shipley (1904) for the sake of uniform endings for the insect orders, and that name seems to have become popular. But the new name means "lively wings," an inappropriate name. The name Embiidina refers to the agility in running. The winged forms are poor fliers. The relationship of Embiidina to Plecoptera is suggested by the modification of the male tenth abdominal tergum and the paraprocts into copulatory claspers, the suppression of phallomeres, and the reduction in branching of veins Rs and M. Embiids otherwise are highly evolved in their own fashion, and apparently coexisted with the Permian Plecoptera. Silk glands occur on the basal fore tarsal segments, and the silk webbing is employed in building silken tunnels which they inhabit. The embiids are very agile runners, and are capable of backing up rapidly within their tunnels. The femora of the hind legs appear to be well adapted for rearward walking in that the flexor of the tibia is large and powerful.

The females are all wingless. The wings of males have a greatly reduced vannal field, and both wings are narrowed. The ocelli are lacking on the head and there is a gular bridge across the underside of the head. The lateral cervical sclerite on each side became fragmented into two pieces, convergently with the condition in the Orthopterodida and in beetles. The cerci are reduced to one or two segments, and in the male are asymmetrical and may be used as accessory clasping organs. The ovipositor has been suppressed in parallel with its suppression in recent Plecoptera.

Supercohort Orthopterodida. Hamilton (1971) erected the group name Pliconeoptera to replace the old name Orthopteroidea, because most of the winged members of this group fold the anal area into individual pleats between the anal veins. In general, the fore wings tend to be sclerotized, some to the extent that they resemble beetle elytra. In some extinct forms the fore wings appear not to be sclerotized (e.g., Caloneurodea). A distinct derived feature appears in the separation of at least the mesotrochantin from the epimeron by way of a membranous connection, so that the trochantin became secondarily movable. The ovipositor differs from the generalized type in that the prolonged ninth abdominal coxal plates, which serve as a sheath for first and second valvulae in the Phalloneopterata, have become sclerotized and serve in lieu of the reduced or suppressed second valvulae as penetrating organs for egg laying, unless the ovipositor has become nonfunctional and highly reduced. The lateral cervical sclerite consists of two pieces on each side of the neck in all the orthopteroid insects. Matsuda (1970) derived these sclerites from the presternum rather than from the preepisternum as in most other insects, because of their position relative to an apparently still existing prothoracic preepisternum. It may be that the posterior plate of the lateral cervical sclerites represents a fragment of the presternum that became inserted between the original anterior sclerite and the prothorax.

So many evolutionary diversions occurred among the Orthopterodida that it is difficult to characterize the group except to emphasize the ways they have remained primitive. The mouthparts are always of the generalized biting type. The cerci are generally short, but are always present, often as a single segment. In addition to cerci, adult males bear persistent abdominal styli on the ninth segment, except when they were eventually suppressed as in Dermaptera, Zoraptera, Acrididae, and Phasmida. None of the orthopteroid insects (except Zoraptera) have acquired a pterostigma. The group includes two sister groups, the Grylliformida and the Blattiformida.

Cohort Grylliformida.   This group includes those orthopteroid insects in which there is little in the form of advanced characters that apply to the whole group. In general, the cerci have become short and mostly of a single segment. The fore wings have acquired an expanded basal precostal area which in some forms is very prominent. The media vein and the cubitus have become associated, so that generally CuA and M arise from the same stem, with CuP arising separately. The sister groups comprise the Protorthopterida and the Orthopterida.

Subcohort Protorthopterida.   All the members of this group are extinct. Some species are among the oldest known winged insects, from the Carboniferous. A few forms survived into the Jurassic, but most became extinct in the Permian. Characterization of this group is difficult because only wings are known in numerous instances. The precostal field varies from almost nonexistent to prominent. The union of M with CuA may be as simple as through a strong oblique cross vein, or more like in the modern Orthoptera. The fore wings are weakly if at all sclerotized. The hind legs were elongated, and some species may have been saltatorial. Two lineages can be recognized. One is the catchall order Protorthoptera in which the large anal fan persisted, and the other line includes the Caloneurodea and Glosselytrodea, whose relationships are controversial. Tentatively these last two orders are regarded as orthopteroid, in which the anal field of the hind wing became reduced. Sharov's (1966b) claim that these are endopterygotes is tenuously based on obscure features of a few bodies.

Subcohort Orthopterida.   The advanced features of this group are several. In the fore wing, the precostal field is broadened and extended toward the wing tip. The precostal field usually does not lie flat, but is turned slightly downward so that there is a shallow fold along the C-Sc-R veins that lie very close together. The vannal fold of the fore wing is also a permanent fold, and the vannus lies right side up over the abdomen, with the remigium (the part of the wing ahead of the vannal fold) lying more or less vertically along the abdomen. In this group the fore wing is sclerotized, and is generally known as the tegmen. The cercus has become a single unsegmented organ, except in the family Tridactylidae, in which it is two-segmented.

Order Phasmida.   The walkingsticks are most obviously specialized in their elongate bodies and slender legs. The fore wings when present are much shorter than the hind wings and bear reduced venation and an almost unrecognizable remigium. The wings are entirely suppressed in many species. The head has become decidedly prognathous. Phasmids have become tree dwellers. Although some descend to the ground to lay their eggs, many merely drop them one at a time from their position in the tree. The ovipositor has become somewhat vestigial. The styli of the ninth abdominal segment have been suppressed. The phallic lobes of the male appear to have become suppressed, and copulation involves eversion of the expanded end of the ejaculatory duct. Phasmids differ from the other insects in general also in that their spermatozoa bear no mitochondrial derivatives, whereas most insect sperm tails bear two such derivatives.

Order Orthoptera. The order name is used here in its restricted sense, for the grasshoppers, katydids, and crickets. The advanced features in common include an expanded pronotum, and elongate hind legs fitted for leaping. The lateral edges of pronotum are turned downward, partly covering the propleura, and the hind edge extends partly over the mesonotum. The ovipositor typically consists of the first valvulae and the sclerotized third valvulae, which together provide an organ suited for digging into a substrate during oviposition. The second valvulae remain as vestiges. The male copulatory organ consists of the eversible terminus of the ejaculatory duct, which usually bears internal sclerotizations which are derived from the primary phallic lobes.

Cohort Blattiformida. This is the sister group to the Grylliformida. The pronotum typically is disclike, with varying degrees of lateral and fore-and-aft expansion. The postnota of the pterothorax are weakly developed or suppressed, and are not involved with the alinotum in arching upward at the wing base, to cause downward deflection of the wings as is the case in other opisthopterous insects. The wings articulate with the nota near the front edges, and the posterior notal processes are far forward of the small postnotum (if present). The extreme forward articulation of the wings is correlated with a strongly slanting pleural sulcus extending from the posteriorly displaced coxae up and anterior to the wing base. In this group there has been a change from the typical indirect flight muscles to a condition approaching the use of direct flight muscles (Figure 71, C). The dorsal longitudinal muscles have become small and weak, attached to highly reduced intersegmental thoracic phragmata. The downstroke of the wings is accomplished through the pull of strong basalar and subalar muscles originating on the coxae and lower pleuron. Their insertions on the basalare and subalare are lateral to the fulcrum, and the pull of these muscles is the chief action causing the downstroke. Elevation of the wing tip is accomplished by the pull of tergo-coxal and tergopleural muscles pulling down on the tergum near the wing base. The tergosternal muscles, which are so well developed as indirect wing muscles in the Ephemerida and most other Neopterata, are at best small weak muscles, and may be entirely suppressed.

The fore wing basally developed a costal expansion that consists of a widened area between the costa and the subcosta, imitating the precostal space of the Grylliformida. Only the Grylloblattaria have retained an ovipositor that is not reduced or retracted into a secondary genital chamber formed by the desclerotization and reversion of the ninth abdominal sternum. The two sister groups of blattiformids are the Dermapterida and the Dictyopterida.

Subcohort Dermapterida. The phylogenetic relationships of the order Dermaptera have been variously interpreted. They have been related to the Coleoptera (Matsuda, 1970), to the Grylloblattaria (Giles, 1963), to the Phasmida (Blacklith and Blacklith, 1968), and to the Blattarida (Blattopteroidea Hennig, 1969). The elytra of beetles and earwigs are so similar that even fossil forms have been confused, but these elytra cannot be other than convergent. Close relation to phasmids or grylloblattids was based on many primitive features in common. Hennig's conclusion took note of the fossil record, and more nearly approaches the viewpoint adopted in this book. In the Dermpaterida, the fore wing was shortened and became sclerotized, and ultimately became a veinless elytron in the Dermaptera. The main veins of the hind wing became crowded along the costal border, forming a narrow remigium. The vannus acquired additional anal veins, became greatly enlarged, and folded fanlike about a pivot point near the middle of the front border. The Permian representatives of the order Protelytroptera (Figure 73) suggest the changes that resulted in the evolution of the order Dermaptera. In the Protelytroptera, the fore wings of some species were flat and only slightly sclerotized, and retained obvious venation. In some species, the hind wing had not yet developed a fulcrum around which the anal fan could fold (Blattelytridae). The head was small, with large eyes, and the antennae were prominent, about half the length of the fore wings. The pronotum bore a small pronotal disc, much smaller than that of cockroaches. The legs bore pentamerous tarsi. The cerci were short but distinctly segmented. A short pointed ovipositor was borne by females (Carpenter and Kukalova, 1964). Since the Protelytroptera are known only from the Permian, and the fossil record of Dermaptera begins in the Jurassic (*Protodiplatys fortis* Martynov), there is no way to determine whether a species of Protelytroptera was ancestral to the Dermaptera, or the two orders had an unknown common ancestor. Besides *Protodiplatys*, which had pentamerous tarsi, segmented cerci, and typical dermapteran wings, a species in the Jurassic had transformed the cerci into forceps.

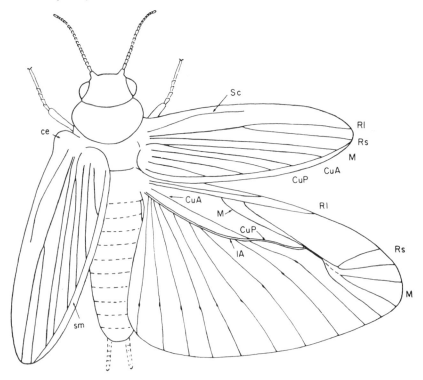

Figure 73 *Protelytron permianum* Tillyard (Protelytroptera), reconstruction. A, Anal vein; ce, costal expansion; CuA, anterior cubitus; CuP, posterior cubitus; M, media; R1, radius; Rs, radial sector; Sc, subcosta; sm, sutural margin of elytron. Reproduced from Carpenter and Kukalova, 1964. Used by permission of *Psyche: Journal of Entomology.*

Modern Dermaptera have highly reduced elytra devoid of veins and useless in flight, and which serve as wing covers for the double-folded hind wings. Some species are wingless. The thorax and flight musculature are remarkably like these of cockroaches. The dorsal longitudinal muscles are thin sheets, the phragmata are poorly developed, there is no postnotum, or the metapostnotum is small (Figure 74). The wings are attached on the front corners of the nota. The pleura are strongly slanted forward. The dorsoventral thoracic muscles are thin straps. But the tergocoxal and alary sclerite muscles are well developed and act as nearly direct wing muscles, as in cockroaches (Kleinow, 1966).

The venation of the hind wing consists of up to 10 radiating branches of 1A that converge on a fulcral pivot near the highly reduced remigium. The adult cerci have become robust unsegmented forceps, or are slender simple struc-

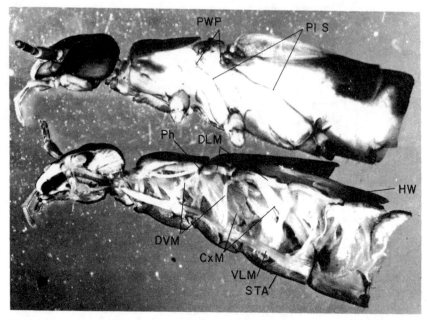

Figure 74   Head, thorax, and first abdominal segment of *Labidura riparia* (Dermaptera), ex-hibiting the blattiform type of thoracic musculature. CxM, coxal muscles; DVM, reduced dor-soventral muscles; HW, folded hind wing; Ph, reduced phragma; P1S, highly slanted pleural sulcus; PWP, pleural wing processes situated far forward on thoracic segments; STA, fused first abdominal and metathoracic sterna; VLM, ventral longitudinal muscle. Original.

tures in *Arixenia* and *Hemimerus*. The nymphal cerci are segmented in *Diplatys* and *Bormansia* (Giles, 1963). Matsuda (1976) maintained that ear-wig cerci are products of the periproct (twelfth "segment") because they lie behind the subanal lobes, and therefore are not homologous with eleventh segment cerci. But the embryonic eleventh and twelfth segments are not separate in most earwig embryos. It is likely that in the Dermaptera the cerci have shifted posteriorly, and are true cerci. Recent Dermaptera have reduced the tarsi to three tarsomeres.

The ovipositor was suppressed in almost all species. The genus *Echinosoma* bears a short ovipositor consisting of a pair each of dorsal and ventral valvulae protruding slightly from the tip of the abdomen. The middle valvulae are not apparent (Giles, 1963).

The male genitalia are most peculiar. Each nymphal paired phallomere splits into two distally, and remain fused basally. The adult phallic organ con-sists of a protrusible structure split into two main lobes. Each main lobe bears a small sclerotized pointed paramere, and a soft penis lobe or mesomere. The parameres, however, have no clasping function as in the Phalloneopterata. Clasping is performed by the cercal forceps. An epidermally derived duct invaginates into each penis lobe, and these join the simple ejaculatory duct, forming a forked passage into two functional gonopores. The two male gonopores of the Dermaptera cannot be considered as primitive because the vasa deferentia join the median ejaculatory duct in the typical insectan fashion. The ejaculatory duct appears to have split as a special derived character of the Dermaptera. In some species, the penis lobes are united, but the two secondary gonoducts open separately (Hemimerus), or one of the ducts is degenerate and there is a single gonopore (Forficula).

The trochantin of Dermaptera is clearly marked off by a suture from the episternum. This condition differs from the detached trochantin typical of all other Orthopterodida. It probably represents a secondary union of the trochanter with the pleuron. Finally, the first abdominal sternum has fused with the metasternum. This condition is signified in the ventral attachment of the muscles of the first abdominal segment on the posterior portion of the combined thoracic metasternum and first abdominal sternum (Figure 74).

Subcohort Dictyopterida.   The name Dictyopterida refers to the presence of numerous cross veins among the wing veins in the ground plan, persistent in cockroaches, mantids, and some termites. The cross veins generally form square or rectangular cells rather than polygonal cells as in the archedictyon of the Plagiopterata. The type of cross vein system in this group was called polyneurous (Hamilton, 1972c) because it is not a true archedictyon. The polyneurous condition was lost in the Zoraptera and most termites as the wing area was reduced.

A few derived features are additionally shared by members of the subcohort. The costosubcostal and costoradial cross veins became slanted, eventually to appear as slanting pectinate branches of Sc and R, and disappearing in the Zoraptera and most Isoptera. Veins Rs, M, and CuA in the fore wing had originally many branches, occupying the greater part of the remigium. The branches of CuA became directed posteriorly toward the rear border of the fore wing. Vein CuP characteristically is strongly concave and forms a strong arch at the base of the fore wing. The coxae have become conical, elongate, and backwardly directed.

Members of the group have retained a number of primitive ground plan features. The fore wings originally were weakly sclerotized. The cerci are still segmented, although short. The trochantin is a sclerite surrounded by membranous cuticle. The males bear styli on the ninth abdominal sternum. The phallomeres begin growth as a pair of lobes on the venter of the tenth abdominal segment, but never form an aedeagus and claspers. The phallomeres usually split each into two lobes of various sorts, but the functional penis is still nothing more than the everted median genital male passage.

Individual variations of the ground plan features are typical of individual orders included in this group. The two sister groups are the Protoblattarida and the Cursorida.

Infracohort Protoblattarida. This group is erected to include the extinct family Blattinopsidae and the order Grylloblattaria. The Blattinopsidae have variously been classified as Homoptera, Protoblattaoidea, or Protorthoptera (Kukalova, 1959) . The family is here regarded as the only representative of the order Protoblattaria (emended from Protoblattoidea Handlirsch, 1908). Most of the remains are of wings only. In these, the slanted c-sc and c-r crossveins are almost like the pectinate "branches" of Sc and R_1 of the primitive cockroaches, and vein CuP is a deeply concave, somewhat curved vein (Figure 75). The radial sector and CuA have many branches, as in the ground plan of the Cursorida. The group has a peculiar derived feature, in that there is a curving, cross fold extending from R, or Sc, to the posterior margin (Figure 75). *Blattinopsis martynovae* Kukalova, 1959, possessed an ovipositor very much like that of grylloblattids, but it was more blunt.

The order Grylloblattaria is considered to be a sister group to the order Protoblattaria chiefly because of the roughly similar ovipositors. The grylloblattids have been allied with the Grylliformida because of their prominent ovipositor, or with the Cursorida because of the segmented cerci. Both these characters are too primitive to be of phylogenetic significance. Giles (1963) thought the grylloblattids and Dermaptera were closely related, but their shared characters also are primitive rather than derived. Grylloblattids, even though they are wingless, possess a set of derived characters common to the cohort Blattiformida, which allies them to the group. These are the large conical hind coxae, set far back on the metathorax, the highly slanted pleural sulcus, the highly suppressed thoracic phragmata, the very thin weak tergosternal and dorsal longitudinal muscles, and the large coxal muscles (Figure 76). The asymmetrical male genitalia of grylloblattids may be synapomorphic with those of cockroaches and mantids, provided one may assume that the symmetry of male termites and zorapterans is a reversed state correlated with extreme reduction of the phallomeres.

Grylloblattids otherwise bear their own autapomorphies. These are the lack of wings, the small pronotal disc, the suppressed ocelli, and the development of eversible vesicles of the first abdominal sternum. The remainder of their features are mostly primitive. These are the persistent styli on the nymphal third valvulae, male styli on separate coxal plates on the ninth abdominal segment, the unreduced first abdominal sternum, the normal eighth abdominal

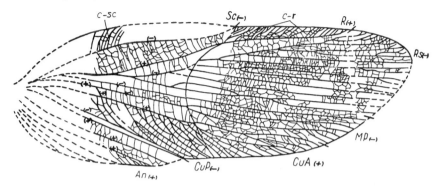

Figure 75 *Blattinopsis antoniana* Kukalova (Protoblattaria), Permian. Reconstruction of fore wing. An, anal veins; c-r, costoradial cross veins; c-sc, costosubcostal cross veins; CuA, anterior cubitus; CuP, posterior cubitus; MP, Posterior media; R, radius; Rs, radial sector. Reproduced from Kukalova, 1959, courtesy of the Czechoslovakian Academy of Sciences.

sternum of females, the large functional ovipositor, and the absence of an ootheca.

The male genitalia of grylloblattids are the strongest evidence against the notion that all male genitalia are homologous, as is maintained by Smith (1969). In these insects, the coxal plates of the ninth abdominal segment are prominent, separate structures that lie close together on the rear border of the ninth sternum. At the same time, phallic lobes (which in the Phalloneoopterata become claspers and aedeagus) grow from *behind* the ninth coxae. These lobes do not split as they do in cockroaches, mantids, and Phalloneoopterata, but become asymmetrical eversible structures (Walker, 1922). In all insects that have been studied, the phallic lobes grow from the venter behind any trace of sternal or coxal structures of the ninth abdominal sternum (Snodgrass, 1957; Matsuda, 1976). Thus the male claspers of the Ephemerida are clearly ninth coxal structures, and only their phallic lobes that become the penes are homologous with the phallic lobes of other insects.

There is no fossil record of grylloblattids. Their secretive habits and lack of wings would account for their absence as fossils. If they are a sister group of the Protoblattaria, their origin probably occurred during the Upper Carboniferous.

Infracohort Cursorida. This new group name is not equivalent to the order Cursoria Westwood, 1839, but includes the same orders as the superorder Blattopteroidea of Chopard (1949a). It is here considered as a sister group to the Protoblattarida, and includes the orders Zoraptera, Isoptera, Blattaria,

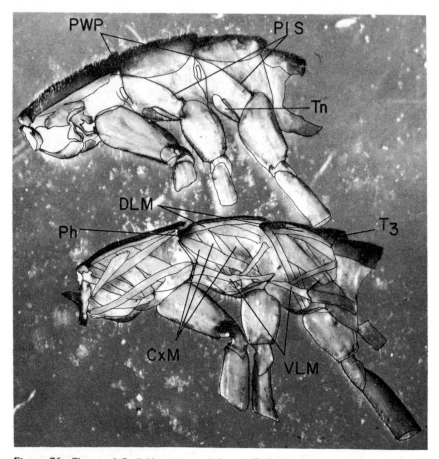

Figure 76 Thorax of *Grylloblatta campodeiformis* (Grylloblattaria), exhibiting the blattiform type of thoracic musculature. Retouched photograph. CxM, coxal muscles; DLM, reduced dorsal longitudinal muscles; Ph, reduced phragma; P1S, highly slanted pleural sulcus; PWP, pleurotergal terminus of pleural sulcus in extreme forward position; T_3, metanotum; Tn, trochantin; VLM, ventral longitudinal muscle. Original.

and Mantodea. The ground plan includes the conversion of c-sc and c-r cross veins of the fore wing into highly slanted veins, which cross veins appear to be forward pectinate branches of the subcosta and radius. The radial sector was reduced to just a few branches, leaving most of the remigium of the fore wing to contain the media and the anterior cubitus. The backwardly directed hind coxae hide a highly reduced first abdominal sternum. In this infracohort the sister groups are the Zorapterida and the Blattarida.

Subtercohort Zorapterida, Order Zoraptera. The Zoraptera have lost or modified most of their blattiformid characters, except for the wing motor system. Both wings are attached far forward on the terga, the alinota bear highly reduced postnota, the phragmata are small, the indirect flight muscles are reduced, and the coxae are conical. The Zoraptera have abundant autapomorphies, some in parallel with the Isoptera, some in parallel with the Acercarida, and others unique. Their convergent termite and psocid characters have led to opposing views about their relationships. The convergences with termites are roughly similar, but not sufficiently so to be homologous. Both bear vaguely similar moniliform antennae. Both cast off their wings after mating, but termites have a well marked basal wing fracture line beyond a sclerotized base, and zorapterans have no such line or sclerite. Both live in colonies, but zorapterans do not share each other's food as in the trophallactic termites, and zorapterans do not have a caste system. In both the cerci are short, but unsegmented in Zoraptera and multisegmented in primitive termites. Both groups have become neotenous in that some adult reproductives are wingless, but in termites the wingless neotenous state appears to result from trophallaxis with the queen. Trophallaxis and the cause of winglessness in zorapterans are unknown. Both groups have a tendency to live in excavated tunnels and cannot survive prolonged exposure to dry atmospheres. The two groups share some similarities in the small pleural areas with wide pleural membrane among the sclerites, but this condition is correlated with small size. If both groups fly employing mostly direct wing muscles, it is because of their blattiformid ancestry.

The characters acquired convergently with the Acercarida are the highly reduced wing venation, the two-segmented tarsi and the small number of malpighian tubules (six in Zoraptera, four or less in Acercarida). All but two abdominal ganglia have shifted to the metathorax in Zoraptera and at best only one abdominal ganglion remains in the abdomen in the Acercarida. None of the above characters shared by Zoraptera and Acercarida are trustworthy for phylogenetic purposes because all of them can be found in various lines, some distantly related. For instance, reduced tarsomery occurs in many modern insects whose ancient fossil relatives bore pentamerous tarsi (Dermaptera, Odonata, Plecoptera, Orthoptera, etc.). Reduced wing venation occurs in some ephemerids, cockroaches, most termites, some Diptera, Hymenoptera, and Lepidoptera. Malpighian tubules are reduced in number or even suppressed in the apterygote insects, ephemerids, termites, beetles, and the like. The abdominal ganglia have moved into the thorax in many unrelated orders, such as the higher Diptera, the Coleoptera, the Hemipterida, and so on. The Zoraptera have retained one outstanding cursorid feature, the backwardly directed conical hind coxae that hide a very small first abdominal sternum.

The zorapterid-derived characters that are fairly distinct from those of the sister group Blattarida are development of a pterostigma, highly reduced hind wings, lack of styli on the ninth male sternum, absence of anal veins in the wings, complete suppression of the ovipositor and male genitalia, and a very small body size of 3 mm or less. Although the apterous adults lack ocelli, the alates bear the usual three ocelli.

The mouthparts of Zoraptera are of the highly generalized biting type, with the lacinia firmly a part of the maxilla. This is quite different from the completely detached lacinia of all the Acercarida, in which in some forms the laciniae became the maxillary sucking stylets. As in other orthopteroid insects, the trochantin of the Zoraptera is a sclerite detached from the lower epimeron, and the lateral cervical sclerites consist of two pieces on each side. The blattiform specializations in direct flight musculature and in the neck and the thoracic structure clearly indicate that zorapterans are not transitional from orthopteroid to hemipteroid, but rather are a highly specialized group in the blattiform complex.

Subtercohort Blattarida.   This, the sister group to the Zoraptera, includes the termites, cockroaches, and mantids. The name Blattopteroidea (Hennig, 1969; Kristensen, 1975) was used for a superorder category in the same sense, but the "—oidea" ending is reserved for superfamilies. Blattarida is a new name for a new category, but includes the same insect groups. Chopard (1949a) used the name Blattopteroidea as a superorder for the group here named infrachohort Cursorida.

The derived features of the Blattarida are distinctive. The anterior tentorial arms evolved median processes on each side which grew together around the circumpharyngeal nerves, and the result is a seeming perforation in the corporotentorium, through which the nerves pass. The stomodaeum has a cone-shaped proventriculus in which there are strongly sclerotized teeth. The first abdominal sternum has become a tiny sclerite hidden by the hind coxae. There has been a return to the primitive position of the female gonopore on the rear border of the seventh abdominal sternum. The venter of the eighth abdominal sternal area formed a gutterlike passage from the gonopore in the Microcoryphia, and these ventral elements fused over the gutter to form the definitive female gonopore at the rear of the eighth abdominal segment in the Dicondylata, the passage having become a part of the median oviduct (vagina). In the Blattarida, the venters of abdominal segments 8 and 9 became desclerotized, and the vagina was lost in that the gutterlike invagination does not form. Thus the median gonopore has reverted to its former position on the hind border of sternum 7. With the loss of the secondary median genital passage, the definitive sternum 7 enlarged to form a large subgenital plate in females of this group, and a genital chamber resulted from the retraction of the ovipositor into the space above sternum 7. The radial sector branches from the radius much beyond midwing, and tends to be simple or sparsely branched.

The ground plan of the Blattarida includes some ancestral features that characterize the subcohort Dictyopterida and others that are older ancestral features. The retention of the styli on the ninth abdominal sternum of males, and ovipositor (reduced in Recent species), and segmented cerci are among the older primitive features that were retained, or modified in each line. The c-sc and c-r cross veins of the fore wing became highly slanted, so that the subcosta and radius veins appear to be highly branched (Figures 77, 78, 79). This condition was retained in some cockroaches (e.g., Blaberidae), a few mantids (e.g., *Choeradodis*) and was further modified in termites and some other cockroaches and mantids, so that in these the c-sc and c-r pectinate cross veins are suppressed. The two sister groups of the Blattarida are the superorders Isopterodea and Blattodea.

Superorder Isopterodea. The only order of this group is the Isoptera. Termites are regarded by Hennig (1969) and Kristensen (1975) as a sister group to the cockroaches, and the two together as a sister group to the mantids. The close relationship of cockroaches to termites is supposed to be indicated by the absence of a median ocellus, the course of the anal veins of the fore wings with respect to the cubital furrow and the shortened subcosta in both groups. The absence of a median ocellus is rather common among other insect groups, and must be suspect as a probable synapomorphy of termites and cockroaches. Grylloblattids, embiids, phthiriapterids (lice), beetles, stylopids, lepidopterans, and scattered families of neuropterids all lack median ocelli. Thus ocellar suppression appears to be a commonplace event occurring convergently in widely related groups. The ending of the first anal vein of the fore wing in the cubital furrow also occurs in primitive mantids (Figure 79). In termites, there are no obvious anal veins in the fore wings. The anal lobe of the fore wing of the *Mastotermes* is highly reduced and somewhat sclerotized without veins. The shortened subcosta then remains as the only possible synapomorphy of termites and cockroaches. This may also be a convergence, since cockroaches and mantids share some other derived characters not found in termites, and the subcosta extends beyond midwing in *Blaberus*.

Much has been made of the common occurrence of flagellate protozoa (*Oxymonas* and *Trichonympha*) in the hind gut of most termites and in the North American cockroach *Cryptocercus*. This cockroach is a wood-eater. The implication is that termites evolved from a wood-eating cockroach. Since only *Cryptocercus* among cockroaches bears the symbionts and is a wood-eater, and fossils of *Mastotermes* are known from North America, there is a

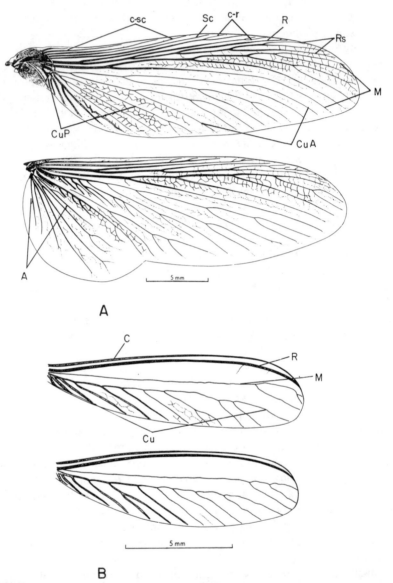

Figure 77 Wings of Isoptera. A, fore and hind wings of *Mastotermes darwiniensis*, a termite with primitive wing venation. Note the slanted costosubcostal (c-sc) and costoradial (c-r) cross veins, and the anal field of the hind wing. B, *Nasutitermes dixoni*, a termite without subcosta and anal veins. A, anal veins; C, costa; c-r, costoradial cross veins; c-sc, costosubcostal cross veins; CuA, anterior cubitus; CuP, posterior cubitus; M, media; R, radius; Rs, radial sector. From CSIRO: *The Insects of Australia*, by permission of *Melbourne University Press*.

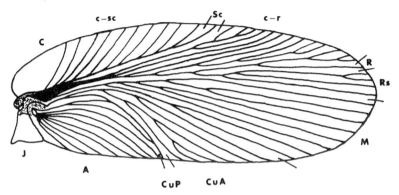

Figure 78   Fore wing of *Blaberus giganteus*, a cockroach with primitive wing venation. A, anal veins; C, costa; c-r, costoradial cross veins; c-sc, costosubcostal cross veins; CuA, anterior cubitus; CuP, posterior cubitus; J, jugum; M, media; R, radius; Rs, radial sector. Original.

strong possibility that the ancestor of *Cryptocercus* evolved the wood-eating habit in the presence of termites and secondarily acquired an adaptation to the symbionts from termites in the same habitat. The worldwide distribution of termites suggests that they originated in the Paleozoic before the breakup of Gondwanaland. The lack of known termite fossils before the Eocene is not evidence of their more recent origin. The lack of fossils may be because their specialized habitat has not permitted fossilization.

The production of a simple "ootheca" in *Mastotermes* has been described (Gay, 1970). To make such an ootheca homologous with that of cockroaches is forcing the issue. The cockroach-mantid ootheca results from the tanning of protein, and it contains organic calcium salts. The ootheca of *Mastotermes* is more comparable to that of the acridids, in that it consists of the dried secretions of the accessory glands, and merely forms a thin film over the eggs, which are laid up to 24 at a time in a double row in *Mastotermes*.

The following derived features of termites appear to be decisive for considering termites as a sister group to the cockroach-mantid group. The antennae are short and moniliform, and the median ocellus is suppressed. The jugum and anal veins of the fore wings are suppressed. The wings are broken off after mating at a special fracture line, with the wings sclerotized proximal to the fracture line. The subcosta of the fore wing extends only one-third the length of the wing, if it is at all present. The radial sector is simple. The costosubcostal cross veins and the costo-radial cross veins have shifted their origins basally in a few genera, and are extremely slanted, giving the appearance of a highly branched R, (Figure 77, A), or the subcosta and cross veins are all suppressed in other termites, leaving the simple radius as the first vein behind the

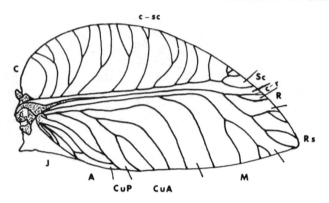

Figure 79   Fore wing of *Choeradodis* sp., a mantid with large fore wings. Lettering as in Figure 78. Original.

costa. The media (MP) may be highly branched (*Mastotermes, Zootermopsis*) to simple (*Kalotermes, Coptotermes,* etc.). The CuA vein is usually highly branched and posteriorly pectinate. The CuP vein is difficult to see in *Mastotermes*, and has lost its identity in the fore wings of other termites. Only *Mastotermes* retained the anal field and anal veins in the hind wings. A small ovipositor with three pairs of valvulae was retained by *Mastotermes*, but in other termites the ovipositor is highly reduced and vestigial. The male genitalia are highly reduced or suppressed. The malpighian tubules number from two to eight, and the first two abdominal ganglia have fused with the metathoracic ganglion. There are no dorsal longitudinal thoracic muscles. Termites have developed a highly organized social life, with the nymphs serving as workers, or with a special mostly sterile worker caste. The latter condition usually results from the suppression of adult sexual characters by secretions from the long-lived queen, which usually becomes highly distended with maturing eggs. Most termites harbor cellulose-digesting symbionts in the hind gut. The spermatozoa of some species of the Isoptera are atypical in being nonmotile.

The ground plan, exemplified by *Mastotermes*, includes the primitive features of a simple ovipositor, pentamerous tarsi, segmented cerci, and a pectinate R vein in the fore wing. Other termites bear four tarsomeres, vestigial ovipositors, and a simple R, and unsegmented cerci. There are no segmental arteries as in cockroaches and mantids. The pronotal disc is not enlarged. The males still bear styli on abdominal segment 9.

Superorder Blattodea (*Cockroaches and Mantids*).   The members of this group share the following derived features in the ground plan, with further modification of some of the ground plan features in each order. The head is highly mobile on a small neck. The antennae are long and filiform in contrast to the short moniliform antennae of termites. The pronotal disc is generally expanded laterally, but probably secondarily so. The prothoracic paranota were suppressed very early in the stem-line of the neopterous insects, as suggested by the Carboniferous and Permian mayflies. The fore wing has become slightly to well sclerotized, but retains a membranous jugum. Generally the first anal vein is reduced and does not reach the hind border of the wing, but usually ends against the concave CuP. Other anal veins also may terminate on the CuP without reaching the wing border. The male genitalia consist of asymmetric phallomeres of similar basic construction in both orders. Females produce an ootheca which sclerotizes around the eggs, and the ootheca contains organic calcium salt. The left collaterial gland produces the structural protein of the ootheca, and the right gland produces a glucosidase that releases the tanning substance (Neville, 1975). The abdomen and thorax bear lateral segmental arteries extending from the dorsal vessel. The malpighian tubules have become numerous, up to 100 in number. The orders Mantodea and Blattaria are the sister groups of the superorder.

Order Mantodea. Mantids bear large compounds eyes that give the head a triangular appearance. The mandibles are strongly toothed. The prothorax and procoxae are elongated, and the fore legs are strongly raptorial. These features are mantid adaptations to a carnivorous habit. Other derived features include a highly reduced ovipositor, the loss of segmental arteries in the thorax, the presence of calcium citrate in the ootheca. The ootheca is deposited, containing the eggs, while still soft. Hardening occurs after deposition.

Mantids have retained a few ancestral features. The pectinate Sc and R veins may bear numerous slanting cross veins to the costa (Choeradodidae, Figure 79), but in advanced forms the Sc and R veins extend near the tip of the fore wing without pectinate branches. The radial sector varies from fully four-branched to simple, arising from R distally. The disc of the pronotum does not cover the head as in cockroaches, but may bear thin to wide lateral expansions resembling paranota (*Choeradodis*, Deroplatyidae, and Orthoderidae). The head bears three ocelli.

Order Blattaria. In contrast to mantids, cockroaches bear the following derived features of their own. The body is depressed. The pronotal disc is expanded laterally and extends forward covering the hypognathous head. The median ocellus is suppressed (in parallel to the termites), and the lateral ocelli are poorly developed. The clypeofrontal sulcus has lost its identity, in parallel with the Machilidae. The thoracic terga and sterna are poorly sclerotized. The legs are adapted to rapid running, with strong internal muscles from the terga and pleura to the elongated conical coxae. The pleura have an extreme forward slant. The proventriculus has developed strongly sclerotized teeth. The ootheca is usually retained after formation in the genital chamber while it sclerotizes by a tanning process, and it bears calcium oxalate instead of calcium citrate as in mantids. The ootheca is usually deposited after sclerotization instead of before sclerotization as in mantids. The segmental abdominal arteries were suppressed in cockroaches.

The ancestral features of cockroaches include the highly pectinate Sc and R veins of the fore wing (Figure 78, *Blaberus*), but in most species the subcosta ends at midwing and bears no slanting pectinate cross veins from the costa. The nymphs of females bear styli on abdominal segment 8 and 9, on the rudimentary ovipositor. These styli are lost in adult females. The ground plan of the ancestral cockroach includes a well developed ovipositor, as is shown by the Permian *Kunguroblattina microdictya* (Figure 62) and a few others. Apparently the ovipositor was retained long after it became nonfunctional as an organ for digging into a substrate since fossil cockroach oothecae are known from older geological times. Modern cockroaches have reduced the ovipositor to small vestiges not visible externally.

INFRASECTION PHALLONEOPTERATA

The Phalloneopterata are here regarded as a sister group to the rest of the Neopterata (Polyneopterata). This group includes the Acercarida (which have been called Hemipteroidea or Paraneoptera) plus the endopterygote holometabolous insects. Hamilton (1971) named this group Planoneoptera but included the Plecoptera and Zoraptera. Henning (1969) called the group Eumetabola, but he also included the Zoraptera.

The outstanding derived character of the Phalloneopterata is a new type of male genitalia derived from phallic lobes of the venter of the tenth abdominal segment (Snodgrass, 1957). Smith (1969, 1970a) is one of the latest of many writers who considered all insect male genitalia to be derived from coxal plates, endites, and styli of the ninth abdominal segment. There is no question that the genital claspers of Ephemerida and the ninth coxal plates of Microcoryphia, Thysanura, and Grylloblattaria are homologous. Only in the Ephemerida, however, are these structures functional as the chief genital claspers. The phallic lobes of the tenth segment of Ephemerida are better candidates as the homologues of the genitalia of the Phalloneopterata. In the Polyneopterata, the pair of primary phallic lobes generally divide into median (mesomere) lobes and lateral (paramere) lobes during juvenile development.

But these tenth segment phallic lobes occur together with undoubted ninth coxal plates in grylloblattid males. In the other orthopterodid orders in which the phallic lobes divide, the ninth coxal plates fail to separate from the sternum, but in many males the remnants of the ninth styli are present. In these instances, the mesomeres fail to unite into an aedeagus, and semen is transferred through the everted median gonoduct.

In the Phalloneopterata, the rudiments of the future genitalia always begin their growth behind the ninth coxosternum, on the front border of the tenth coxosternum, just as the penis lobes begin their growth in ephemerids (Figure 65). The phallomeres are the source of both aedeagus and claspers in all members of the Phalloneopterata that have been studied (Snodgrass, 1957). Each phallomere splits into two. The median lobes, or mesomeres, fuse into a sclerotized organ whose lumen becomes continuous with the invaginated median ejaculatory duct, except in the Psocodida. Thus, the functional gonopore is a secondary one (phallotreme) connected to the primary gonopore by the lumen (endophallus) of the aedeagus resulting from the fusion of the mesomeres. In the Psocodida the mesomeres split off from the parameres, but do not fuse to form an aedeagus. Rather the penis results from extension of the ejaculatory duct (Matsuda, 1976). Both branches of the phallomeres serve as claspers in this group. The lateral lobes of other Phalloneopterata generally become the parameres, which act as genital claspers. The ground plan parameres are unsegmented, as in the Acercarida and Coleoptera, but become segmented into basimeres and telomeres in the Telomerida (the remaining endopterygotes). The assumption that basimeres are coxal plates (gonocoxae) and telomeres are gonostyli of the ninth segment ignores the ontogenetic origin of these structures from the tenth segment. Suppression of limb vestiges on the tenth abdominal segment appears to be an ancient insectan character, since there never are adult limb vestiges on that segment in all adult Euentomata previously discussed. Nevertheless, since all the tissues of any sexual organism are derived by mitosis from the zygote, the genetic code for "limb" must be present, but not "turned on" in all tissues, except in the thorax where the full array of the genetic code for limb is expressed in typical form for each species. It should not be surprising that partial expression of the limb genetic code could be achieved through the equivalent of a reverse mutation that would result in a superficial resemblance to a formerly well developed organ such as the coxal plates and styli of primitive insects. Such an example of the expression of limb genes is found in the mutant "labiopedia" of *Tribolium confusum* described by Daly and Sokoloff (1965). In this case the labial palps are replaced by fully formed legs similar to the prothoracic legs, in all the developmental instars and adults. The genetic basis for this phenomenon is a simple sex-linked recessive gene, apparently derepressing limb gene codes in the labial palpi. The segmented parameres of the higher insects may well be in part controlled by an ancient genetic code which was newly derepressed on the tenth abdominal segment as a result of mutations. These genitalia are not the result of continuous modification of former limbs, but seem to be new structures partly influenced by derepressed limb genes. Perhaps a good example of apparent continuous formation of primary phallic lobes from limb rudiments of the tenth abdominal segment occurs in the Orthoptera·(Matsuda, 1976). The primary phallic lobes appear very early on the tenth segment. This can be interpreted as an accelerated development resulting from the early expression of the phallic lobe genes, before the tenth embryonic limb buds are suppressed.

The Phalloneopterata initiate a new feature in metamorphosis, the delay of histogenesis of certain adult organs. Generally, there are no ocelli in the early stages. Ocelli develop during the preadult instar. When cerci are retained by the adult, as in Hymenoptera, Trichoptera, and Diptera, they form during the pupa stage. The external wing pads of the Acercarida generally are not progressively formed as in the polyneopterous nymphs, but first appear as small enlargements in a fairly late nymphal instar. Only in the last nymphal instar do the wing buds usually become prominent. The wing buds of the endopterygote insects begin their growth in late larvae as invaginated histoblasts, which develop into large pupal wings after eversion. The external genitalia make their appearance either during late nymphal instars or at the pupal stage.

The beginnings of a sternal furca first appear in the Phalloneopterata, in that the furcal arms tend to arise close together. The acercarids generally do not have the large median longitudinally invaginated cryptosternum found generally in the endopterygotes, but a median internal sternal ridge is formed in Corrodentia, Homoptera, Thysanoptera, and Heteroptera.

If the neala or jugal region of the hind wing is considered to be a ground plan character of the Neopterata, then this part has become highly reduced in the Phalloneopterata. It is represented only by a sclerotized jugal bar in the jugal area.

The primitive state of the neopteran trochantin was retained in the ground plan of the Phalloneopterata, except for some isolated major changes. Thus, the trochantin is mostly a narrow straplike extension from the lower epimeron, forming a secondary articulation with the coxae. The trochantin became secondarily detached and free at its upper end as in the Cicadidae (Homoptera), in parallel with the condition of the Orthopterodida, or lost its identity by fusing with the pleuron, as in most Coleoptera and Panorpida, in which the trochantin is barely recognizable at its articulation with the coxa.

While the ovipositor is reduced, modified or absent in various families of the Phalloneopterata, the ancestral ground plan included a typical ovipositor consisting of modified elements of the coxal plates of abdominal segments 8 and 9. The ovipositor is most nearly typical in the Thysanoptera, Homoptera and lower Hymenoptera, and only traces remain in beetles and most of the Neuropterida (Mickoleit, 1973).

The neck of the Phalloneopterata has retained the primitive state of having the lateral cervical sclerities consisting of single pieces. Exceptions are where the sclerites have become lost (some Anoplura, some beetles [Adephaga], Strepsiptera, and Hymenoptera), or have secondarily become double (Polyphaga, Coleoptera), or become three separate pieces (some Tipulidae, Diptera), or fused with a ventral sclerotization of the neck (Neuropterida).

SUPERCOHORT ACERCARIDA

This group name is used instead of Paraneoptera because the latter name is more recent, and it included the Zoraptera, which are here excluded. The name Acercarida refers to the total suppression of cerci in all instars. The ground plan includes other apomorphies restricted to the Acercarida. The lacinia is detached from the maxilla and is slender, often styletlike. The labial palpus has no more than two segments, and is often suppressed. The cylpeus is enlarged and serves as the origin for the large cibarial dilator muscles. The abdominal ganglia are shifted forward, with most or all united with the metathoracic ganglia. There are four or less malpighian tubules. The first abdominal sternum is highly reduced in all the Acercardia. The tarsi bear three or less tarsomeres, although the extinct psocidlike forms bore at least four-segmented tarsi. The aceracarid spermatozoa have evolved to the point of doubling the axoneme complex of the flagellum as a characteristic ground plan character. The Acercarida apparently have evolved along two lines represented by the sister groups Psocodida and Condylognathida.

Cohort Psocodida

In this group the lacinia has become a slender rod entirely detached from the maxilla except for the lacinial muscles. The lacinial function is obscure, and there is a tendency for its suppression in the Phthiriapterida. The lacinia is not involved in forming a sucking stylet as it is in the Condylognathida. The maxillary cardo is suppressed in the Psocodida, although the stipes may remain as part of the maxilla. The hypopharynx bears two modifications not known in other insects. On its upper part there is a sclerotized structure (oesophageal sclerite, sitophore, cibarial sclerite) into which a protruding part of the lower cibarium fits, in a "mortar and pestle" configuration. The sitophoral sclerite becomes reduced or lost in some of the parasitic lice. The other modification in common is seen as a pair of ovoidal sclerites on the lower border of the hypopharynx, consisting of a pair of oval sclerotized swellings, each with a sclerotized arm looping upward and joining the sitophoral sclerite. The function of these sclerites is unknown. These ovoidal sclerites also are lost in some of the parasitic lice.

Although the Psocodida bear phallic lobes that split into mesomeres and parameres, the mesomeres do not fuse to form an aedeagus (Matsuda, 1976). The mesomeres (internal parameres) together with the parameres act as claspers. The intromittent organ therefore is not a true aedeagus, but results from the evagination of the ejaculatory duct. In this sense, the Psocodida have suppressed the aedeagus and only appear to be primitive in lacking an aedeagus. The Psocodida consists of the sister groups Psocida and Phthiriapterida, commonly called psocids and parasitic lice.

Subcohort Psocida. This group consists of a single order, the Corrodentia. The commonly used name Psocoptera is unfortunate, because it is a nonsense name erected by Shipley (1904) to make all insect order names have a uniform ending. Psocoptera, translated "wings which gnaw," is highly inappropriate. The order Corrodentia originally included termites, embiids, and some Neuroptera. The order was restricted to the psocids by Comstock and Comstock (1895). The derived features of the Corrodentia not found in the Phthiriapterida are distinctive. The antennal flagellum is slender and setiform. The prothorax is reduced, and the phragmata between the thoracic segments are suppressed. The dorsal longitudinal mesothoracic muscles anteriorly are attached to the front of the highly arched mesoscutum. The mesopostnotum and metapostnotum serve as attachment points of dorsal longitudinal muscles (Matsuda, 1970). On the fore wing, the radial sector has two branches and the media has three branches. Veins M and Cu arise from a common stem, and M remains joined to CuA for some distance after the forking of CuP. Vein M often joins Rs for a short distance. There is only one anal vein. The hind wings are reduced and the venation is further simplified. The wings are coupled together in flight through a hook at the end of the forewing CuP.

The copulatory apparatus consists of two pairs of parameres and a penis apparently consisting of the eversible ejaculatory duct only. It appears that the primary phallic lobes divide, but the mesomeres do not fuse to form an aedeagus. The spermatozoa of the Corrodentia typically bear the $9 + 9 + 2$ (some with $9 + 9 + 1$) axoneme pattern, but the outer ring of nine axonemes consists of nine large fibrils. Each sperm bundle contains some spermatozoa bearing two flagella and others with a single flagellum each.

The primitive features include a retention of the maxillary and labial palpi, retention of the dorsal arms of the tentorium, a short but complete ovipositor, and the primitive simple lateral cervical sclerites. There are usually 10 pairs of spiracles, but in some the first abdominal spiracle is lacking. There usually are five or six nymphal instars.

Subcohort Phthiriapterida. The relationships of the parasitic lice have long been debated. The biting lice are generally assembled in the order Mallophaga, and the sucking lice in the order Anoplura. However, the two groups of lice represent a monophyletic assembly because they share derived features not found in the Psocida. These synopomorphies are as follows. The head is somewhat flattened and prognathous, placing the clypeus dorsal in part. The antennal flagellum is restricted to three or fewer segments, which are not as slender as in psocids. The ocelli are suppressed. The compound eyes are reduced to two or less ommatidia. The labial palpi are very small or suppressed, and the maxillae are reduced, either fused with the labium or not discernible. The lacinial stylets have become very small, and attached to them is a lacinial gland of unknown function. The stylets and gland are totally suppressed in some species. The supraesophageal ganglion has moved backward to lie over or behind the subesophageal ganglion. The abdominal and thoracic ganglia are united into a single thoracic mass. The dorsal tentorial arms are suppressed. The trochantin has been suppressed. There is no trace of wings. There is only one pair of thoracic spiracles, and usually only six abdominal spiracles are retained. Testicular follicles have been reduced to three or less per testis. The male genitalia are modified in that the split parameres arise from a common sclerotized basal plate. The ovipositor is vestigial. The juvenile states include only three instars. All are ectoparasitic on vertebrates. The eggs are glued to the hair or feathers of the host. The phthiriapteran sperm tail is a single piece, but with two sets of $9 + 9 + 2$ axial filaments, as in some psocids and heteropterans.

Although the biting lice are commonly placed in the order Mallophaga, it is clear the suborder Amblycera is a sister group to all the rest of the parasitic lice (Konigsmann, 1960; Clay, 1970), and that the suborder Ischnocera + Anoplura + *Hematomyzus elephantis* Piaget form the other sister group, which Konigsmann designated as Group A. In discussing these groups, the order Mallophaga is here restricted to include only the Amblycera. The order Anoplura in its original meaning included all the parasitic lice (Leach, 1815). As used here, the order Anoplura is nearly in its original sense, with the exclusion of the Amblycera, and corresponds to Group A of Konigsmann.

Order Mallophaga. With respect to the phthiriapterid ground plan, the Mallophaga, restricted to the Amblycera, retain mostly the ground plan features. The group is distinct from the Anoplura in only a few advanced features. The four- or five-segmented antennae bear somewhat swollen terminal and subterminal flagellomeres, and are ventrally located in a groove of the head. The first flagellomere (third segment) is pedunculate. Otherwise, the Mallophaga as here defined are still primitive phthiriapterids. There are two ommatidia in the reduced compound eyes, the maxillary palpi are four to five segmented, the labial palpi are still present, although highly reduced, the lacinial stylets are well sclerotized, the anterior tentorial arms are still fused with the posterior tentorium, there is no occipital apodeme or obturaculum, the feet are still two-clawed except for one family, the testes each bear three follicles, and the lateral cervical sclerites remained simple.

Order Anoplura. This group may be treated so as to include sister group taxa comprising the Ischnocera versus the Lipognatha, and the Lipognatha consists of the sister groups Rhynchophthirina and Siphunculata. The latter term is applied to the sucking lice only, as originally was done by Latreille (1825), and in the sense of many recent authors who use Siphunculata.

The order Anoplura as orginally used (Leach, 1815) included all the wingless parasitic lice. The name refers to the absence of tail bristles (cerci and terminal filament) such as are found in the Thysanura. Most entomologists employ Anoplura in a restricted sense, to include the sucking lice only. I propose that the name Anoplura be partially restored to near its original sense, to include the Ischnocera (usually classed as a suborder of Mallophaga), the Rhynchophthirina Ferris, 1931, with the single species *Hematomyzus elephantis* Piaget (elephant louse) and the sucking lice Siphunculata Latreille, 1925. The Siphunculata and Rhynchophthrina may be combined into the Lipognatha Börner, 1904. The name refers to the deficiency of ordinary insect mouthparts, in that the mandibles are reversed (elephant louse) or highly reduced and nonfunctional (Siphunculata), and the lacinial stylets are fully suppressed. The maxilla is so highly modified that it is barely recognizable.

Derived characters shared by the Anoplura as here defined were reviewed by Konigsmann (1960), who designated it as Group A of order Phthiriaptera. The anterior tentorial arms are reduced and do not reach the posterior tentorium (Ischnocera), or they are completely suppressed (Lipognatha). The lacinial stylets are highly desclerotized and still present in the Ischnocera, but they are totally suppressed in the Lipognatha. The mandibular articulations are rotated so that the axis of rotation of the mandibles is vertical. Both maxillary and labial palpi are suppressed. The last two flagellomeres bear peculiar saucer-shaped sensorial areas, each with a central raised area bearing radial ridges (Clay, 1970). The compound eyes are further reduced to only one facet, or, in the Siphunculata, there are no more eyes. The occiput of the head bears a prominent occipital apodeme which extents into the prothorax. In the back part of the head there is a mass of connective tissue that seems to act as a partial septum across the hemocoel, which is called obturaculum.

There has been a further reduction in the number of testicular follicles to two per testis, from the three-follicle state of the mallophagan testis.

While it seems clear the Anoplura as here defined are definitely a sister group to the Mallophaga as here restricted, the phylogenetic relationships within the Anoplura cannot be interpreted in the sense of Hennig to produce monophyletic groups. The Ischnocera do not appear to have acquired derived characters for their ground plan to appose them to the Lipognatha, although the Lipognatha share some synapomorphies, such as the absence of lateral cervical sclerites, the modified mandibles, suppressed laciniae, absence of anterior tentorial arms, and the well developed cibarial pump used for sucking blood. Thus, the Ischnocera appear to be a paraphyletic group in terms of the Lipognatha. However, the Lipognatha are more easily treated to include two sister groups. The elephant louse mandible appears to have rotated 180°, so that the teeth point outward, and biting in the ordinary sense is impossible. The pretarsal apodeme (tendon) extends through the tibia without the usual tibial muscles. Also, the usual tibiotarsal muscles appear to be lacking, and the tibia seems to be subsegmented. The posterior tentorial arms are very short, and do not meet to form a posterior tentorium. The Siphunculata, on the other hand, have evolved some new piercing stylets derived partly from the hypopharynx and partly from the labium, and a labial sac is formed which contains the stylets in repose. The mandibles of the Siphunculata are either mere vestiges or are gone. In addition, the tentorium is totally suppressed. The meso- and metathoracic terga of the Siphunculata have become very narrow, so that the pleura are extended dorsally, bringing the single thoracic spiracle to a dorsal position.

Cohort Condylognathida

This is the sister group to the Psocodida. The name refers to a special feature of the mouthparts. Both the mandibles and the maxillary laciniae have changed to form slender styletlike structures that are involved in feeding by making punctures into a substrate and sucking the fluids thus made available. The lacinial stylets consist of two pieces. The upper end forms a leverlike bar between the remnant of the stipes and the upper part of the stylet, so that the lacinial stylets may be protruded and retracted in the piercing act of feeding. Although the stylets are surrounded by a labral-labial sheathlike structure, the rostrum so formed is mostly labral in the Thysanoptera and labial in the Hemipterida. The maxillary galea is suppressed, and generally both the maxillae and labium are reduced or highly modified.

The fore wing of the Condylognathida bears a derived feature not found in other insects. The veins CuP and 1A (plical and empusal, Hamilton, 1972a) lie very close together or are fused, and the next two anal veins merge to form a "Y" vein, which then merges with CuP and 1A terminally. This arrangement forms a structure called clavus, somewhat set apart from the rest of the wing. The Condylognathida consists of the two sister groups Thysanopterida and Hemipterida.

Subcohort Thysanopterida. Only the order Thysanoptera is included in this group. The thrips suggest how the piercing-sucking mouthparts of the Hemipterida may have evolved, in the elongate mandible, the jointed maxillary stylet, and the rostrum enclosing the stylets. But the mouthparts of thrips have evolved past a stage that could have been ancestral to the Hemipterida. The stylets of thrips are enclosed by the labrum rather than the modified labium of hemipterids. The left mandible only remains, and appears incapable of protrusion and retraction, although it may be capable of making a puncture into which the lacinial stylets enter for sucking up liquids.

The thrips have characteristic wings in which the venation is highly reduced, the wings are very narrow and fringed with long setae. The clavus is highly reduced but still discernible only on the fore wing where it is generally called fore wing scale. Abdominal spiracles are reduced to one pair on the first segment and one on the eighth segment. The tarsi are two-segmented, and the claws are generally highly reduced. The abdominal ganglia are concentrated in the first segment.

The double sperm tail of thrips is in the form of two 9 + 2 complexes, or in a single disorganized double 18 + 4 complex, all in one flagellum, in contrast to the double tail typical of the ground plan of acercarids.

The life cycle of thrips is atypical in that the third instar is a nonfeeding stage (propupa). During this stage the coiled midintestine of the earlier nymphs is histolysed and reformed into a straight short tube, then is reconstructed into a short loop during the next nonfeeding stage (pupa), and finally returns to the coiled condition in the imago. The salivary glands degenerate during the propupal and pupal stages into long bodies that are restored in the adult. Fat body cells enlarge during the nymphal stages, and the stored material is used up during the quiescent stages in reconstruction of various organs. The nymphal position of the brain is in the thorax, and the brain shifts to its normal position in the head during the nonfeeding stages. Histolysis of nymphal head and thoracic muscles occurs during this propupal stage, and total regeneration of the head and thoracic musculature occurs in the pupal stage. Since the affinity of the thrips is with other acercarids, this metamorphic change apparently occurred as a convergence with the pupation of the endopterygote insects, and with the pupation of coccids and aleyrodids among the Homoptera.

The Thysanoptera bear only a few ancestral features. Ocelli were retained by thrips, and an ovipositor was an original character of the first thrips, although it was suppressed in one suborder (Tubulifera). The labial and maxillary palpi were retained, although shortened. The lateral cervical sclerites are still primitively simple.

Subcohort Hemipterida. This group is considered as a sister group to the Thysanoptera. The name is emended from the Linnaean name Hemiptera, is restricted to exclude the thrips, and the group is elevated in rank.

In this group the mandibles and laciniae are formed into slender interlock-ing stylets enclosed by the sheathlike labium. Feeding is performed by pierc-ing food sources and sucking the liquid food through a channel enclosed by the lacinial stylets. The Corixidae have modified this plan by shortening the gnathal cone, and appear to rasp their food from the surface of water plants, rather than to pierce and suck. In contrast to the thrips, the maxillary and labial palpi are suppressed. The musculature is better developed in the mesothorax than in the metathorax, resulting in the powering of flight mainly through the fore wings. The hind wings are generally coupled in flight to the fore wings by means of a gripping device near the apex of the clavus, which grasps the front border of the hind wing. The motion of the hind wings is largely passive following the beat of the fore wings, convergent with other mechanisms of wing coupling in the Corrodentia, Hymenoptera, and Trichopterida. The radial sector was reduced to two or fewer branches. The anterior veins of the fore wings have become somewhat consolidated, with Sc nearly obsolete or soon joining R. The veins CuP and 1A lie close together or have fused, forming the anterior boundary of the clavus, and veins 2A and 3A typically from a "Y" vein by fusion terminally. The abdominal ganglia have generally all moved forward to join the metathoracic ganglion in the thorax.

The phylogenetic relationships within the Hemipterida have been enigmatic. Hennig (1969) and Kristensen (1975) believed that there are three main sister groups, the Auchenorrhyncha, the Sternorrhyncha, and the Heteroptera + Coleorrhyncha line. In general, entomologists separate the Hemipterida into two groups, the Homoptera and the Heteroptera, regarded by some as orders, by others as suborders of the order Hemiptera. The fossil record is of little help because primitive homopterans and primitive heteropterans are known from the Permian. The controversial Peloridiidae, which some authors take to be primitive homopterans while others regard them as primitive heteropterans, do not appear in the fossil record until the Upper Triassic (China, 1962).

Although the Hemipterida can easily be defined by the synapomorphies listed above, the characters by which the Homoptera differ from the Heteroptera are essentially primitive: Homoptera lack a gula, have a well developed tentorium, typically fold their wings rooflike over the abdomen, bear a typical ovipositor, lack scent glands, have an exposed propleuron, and bear simple lateral cervical sclerites. Heteropteran synapomorphies include the following: presence of gula, absence of a posterior tentorium, wings fold flat and wings of both sides overlap, ovipositor reduced, scent glands present, pronotum covers propleura, the dorsum of the abdomen is flattened, and the lateral cervical sclerites have lost their identity.

It has been said (Hennig, 1969; Kristensen, 1975) that there are no derived features among the Homoptera that are not found also in the Heteroptera, thus suggesting that nonheteropterans do not form a monophyletic group. However, the presence of three tarsomeres in many of the Homoptera indicates that a reduction from five to three tarsomeres occurred in the hemipteran stem line, further reduced to two, one, or no tarsomeres in other Homoptera, and also in parallel in the Heteroptera to two tarsomeres. While this does not prove a homopteran synapomorphy, there are other features that suggest the unity of the Homoptera. The fore wing is larger than the hind wing in most Homoptera. The enlarged vannus in some Fulgoroidea is secondary. In addition, Hamilton (1972b) suggested that the vannal fold of the hind wing is displaced rearward, to occur among the anal veins in Homoptera, while it remains in the typical position just behind 1A (Hamilton's empusal) in the Heteroptera. I have examined a large number of homopterans, and can confirm that the fold is atypical in the Homoptera. In some, the fold crosses 2A (Hamilton's 1A) near the basal half, and in others, 2A appears to be ahead of the fold. This same type of fold occurs in Peloridium hammonorium according to China (1962). In the Heteroptera the fold is just behind 1A, as in other wing-folding insects.

The spermatozoa of the Acercarida have acquired a double set of axonemes, which at first were enclosed in a single sperm tail, and became separated into two flagella in various species. In the Homoptera, the biflagellar state has reverted mostly to a single flagellum bearing a single set of axonemes. In Ceresa (Membracidae) the flagellum is branched. In the Heteroptera the ground plan is represented by the two axoneme complexes in one flagellum, but some Heteroptera bear two flagella of unequal length on each sperm cell (Baccetti, 1979).

The so-called heteropteran type of wing coupling in Peloridium is not like the typical heteropteran coupling, in that the clasping device of many heteropterans is precisely similar to the condition in most Homoptera: A device consisting of two protruding bosses on the underside of the fore wing near the tip of the clavus grasps a raised ridge on the front upper surface of the hind wing during extension. This device on Belastomatidae (Heteroptera) and other heteropterans is remarkably similar to the structure of the Cicadidae and many other Homoptera. The peloridiid wing coupler then is a convergence on that of some heteropterans that bear bristles on the gripping device, and is not necessarily homologous.

The affinities of the Peloridiidae have been in dispute. Some relate this family to the heteropteran group because of the hemelytronlike fore wing folded flat, reduced antennal segments, fusion of CuP with the anal veins, flattened abdominal dorsum, and fusion of the anal cone (Kristensen, 1975). Others maintain that peloridiids are homopteran, because of the typical homopteran clavus, the cicadellid type of claws and unguitractor plate, the basally separate Sc, joining the unbranched radius, and the generally homopteran male and female genitalia (China, 1962). To this, one may add the homopteran speciliazation in which the hind wing folds behind the second anal vein. The so-called heteropteran characters are also found as autapomorphies among some Homoptera. For instance, the aphid genus *Monellia*, the Phylloxeridae and some cicadellids fold their wings flat and overlapping, and some fulgoroids (Tropiduchidae) bear a hemelytronlike fore wing. There can be no doubt that peloridiids are an early offshoot from the stem of the Hemipterida. There is generally the opinion among entomologists that the Homoptera have remained in many respects similar to the earliest hemipterids, and perhaps the Heteroptera orginated from a homopteran type before the acquisition by modern Homoptera of the peculiar wing folding of the hind wing behind the second anal vein. The Peloridiidae have a simple intestine forming a single loop, as in the Thysanoptera, and lack any kind of filter chamber. This could be correlated with their small size, as are the simple intestine and suppressed filter chamber of aphids whose small size has permitted the suppression of a complex intestine and filter chamber (Goodchild, 1966). The lack of a filter chamber in the peloridiids could also be a primitive condition, since the homopteran filter apparati appear to have evolved convergently in different lines of Homoptera: the fulgoroid type that consists of a coiled mesenteron ensheathed by a membrane; the cicadoid type involving the malpighian tubules joining the mesenteron; and the coccoid type with a helically twisted filter tube (mesenteron) embedded into a saclike thin walled rectum (Goodchild, 1966). It appears best to consider the peloridiids as a relict surviving offshoot of the early Homoptera. They have retained the typical posterior tentorium, but have their own autapomorphies: Paranota on the head and prothorax, development of ventral carinae forming a sheath around the proboscis (Coleorrhyncha refers to the ensheathed proboscis), specialization in becoming moss-feeders, small size, enlarged veins in the fore wing forming a pseudohemelytron, reduction in antennal segment number and number of tarsomeres.

In this work, I consider the Homoptera and Heteroptera as sister group orders, with the Peloridiidae as a sister group to all the other Homoptera. The derived features of the Homoptera include the following: the folding of the hind wing changed to occur behind or across 2A; the fore wing generally became larger and longer than the hind wing; tarsomeres were reduced to three in ground plan, and were further reduced in some lines; and biflagellar sperms reverted to a simple state. In contrast to the Heteroptera, the Homoptera are more primitive in that (1) the beak is attached to the rear part of the head, (2) there is a lack of gula, (3) the wings typically are folded roof-like over the abdomen, (4) there is a retention of a posterior tentorial bridge, (5) there is a well developed ovipositor, (6) they have not developed scent glands, (7) they retained the anterior gastric caeca (suppressed in Peloridiidae and some other Homoptera), (8) the pronotum does not descend downward to cover the propleuron, (9) the veins of the hind wing are branched (branching eliminated in Peloridiidae, coccids, some aphids), and (10) most have retained the simple lateral cervical sclerites.

The Heteroptera bear the following derived features that were not acquired by all the Homoptera: (1) a gula formed behind the labium, (2) the beak complex moved forward, (3) the wings fold flat over the abdomen in an overlapping fashion, (4) the posterior tentorial bridge is suppressed, (5) the ovipositor when well developed bears an enlarged gonangulum that is directly connected with the first valvulae, (6) all anterior gastric caeca are suppressed, with posterior caeca sometimes formed (Pentatomomorpha), (7) there are nymphal scent glands, (8) there is lateral and downward extension of the pronotum to cover at least a great part of the propleura, (9) there is lack of branching in the veins of the hind wing except CuA, (10) a hemelytron was acquired with a sclerotized basal portion and still membranous terminal part, (11) hind wings are wider than fore wings and bear a large vannal area, (12) there is total loss of lateral cervical sclerites.

The heteropteran sperm cell may be of two types. In some, the single flagellum bears two sets of 9 + 9 + 2 axonemes, as in the ground plan of Acercarida. Others bear two distinct flagella, one longer than the other. Biflagellar sperm are also found in some psocids, suggesting that the twin tails were acquired convergently.

PART 4
THE ENDOPTERYGOTIDA

This is the sister group to the supercohort Acercarida. It is typical of this group to delay greatly the development of imaginal organs, and to have an extended juvenile condition (larva) followed by an abrupt rapid development of the imaginal organs restricted to the last juvenile instar (pupa), accompanied by histolysis of larval muscles and the regeneration of imaginal muscles. The rapid histogenesis of imaginal organs from nonfunctional larval histoblasts includes these organs: compound eyes, ocelli, antennae, pentamerous adult tarsi, empodia, internal and external genitalia, wings, cerci, and so on. The wing histoblasts grow during the larval stages, first as slight thickenings of the epidermis, then they invaginate below the larval cuticle. These are everted and grow rapidly at pupation. The larvae never have ocelli or compound eyes that have a continuous development into the imago. Larval antennae have become small, with one or two segments on the flagellum. The larval tarsi are always simple.

Endopterygotes generally have lost the vannal fold of the hind wing and the anal area generally does not fold. Rather the hind wing folds along the jugal fold, as in the fore wings of all neopterous insects. The jugal area has been reduced and bears a simple jugal vein (Hamilton, 1971).

Sharov (1966a) argued that the endopterygotes (= Holometabola) evolved as an independent stock from an early flying insect that retained vestiges of abdominal limbs, these limbs becoming the larval prolegs. He thought erroneously that the first winged insects could fold their wings, so the ability to fold wings presented him with no difficulty. This concept was strongly based upon the mistaken notion that *Eopterum devonicum* was truly a Devonian winged insect.

The presence of abdominal prolegs on the larvae of some members of several of the endopterygote orders has led many to consider these abdominal structures as thoracic leg homologues, and therefore primitive. In view of the complications involving excessive convergence in this interpretation, it is best to consider abdominal larval limbs as secondarily derived. In any insect with thoracic legs, the gene code for limb must be present in all tissues, but is fully expressed only in the thorax. If a mutation occurs which causes derepression of limb code genes elsewhere, the embryonic abdominal limb buds can continue growth, but under the influence of new mutations which cause their growth into new structures, the prolegs. A good example of the derepression of the limb gene code is in the mutant "labiopedia" of *Tribolium confusum* (see p. 223) described by Daly and Sokoloff (1965). It is probable that the limblike abdominal appendages of some Gyrinidae and Dytiscidae (Coleoptera) on one hand, and abdominal limbs on the lower Hymenoptera, the Sialidae (Megaloptera), the Corydalidae (Megaloptera), the Mecoptera, the Trichoptera, and the Lepidoptera on the other hand, are in each case independently derived convergent structures resulting from early derepression of the abdominal limb buds. This interpretation requires convergent evolutionary events of a slightly different nature. In the case of the aquatic beetles, only a few genera bear lateral leglike structures possibly functioning as gills (*Coptotomus, Dineutus*) while other related genera and all other beetle larvae are without such structures. Hymenopteran larval abdominal legs develop continuously from abdominal limb buds, but under the influence of unique hymenopteran genes, since their structure is dissimilar to other insect abdominal prolegs. The same can be inferred about mecopteran abdominal prolegs, which are not fleshy, but which also develop continuously from derepressed limb buds of the abdomen. In the Trichoptera, only the terminal pygopods are developed, and there is evidence that these are highly modified larval cerci, changing to the adult cerci at pupation (Matsuda, 1976). In the Lepidoptera, the prolegs of Micropterygidae are convergently similar to those of Mecoptera, but have become fleshy and bear crochets in the higher Lepidoptera.

Since larval abdominal legs are known to develop continuously from embryonic abdominal limb buds where they occur in Megaloptera, Hymenoptera, Mecoptera, Trichoptera, and Lepidoptera, but not in Coleoptera (Matsuda, 1976), the derepression event may be a ground plan apomorphy for all the Endopterygotida except the Coleopterida. Dytiscid and gyrinid abdominal gills appear to be secondary outgrowths of the abdomen, restricted to a few unrelated genera. Assuming that the above is correct, then the subsequent second time suppression of abdominal limbs in the higher Hymenoptera, the Raphidiodea, the Neuroptera, the suppression of all but the pygopods in Trichoptera, partial loss of prolegs in some lepidopteran larvae (loopers) and the total suppression of all larval limbs in Siphonaptera and most Diptera represent at least six independent events.

This interpretation of the larval abdominal limbs permits a concept of homology between thoracic limbs and abdominal limbs, but in the sense that the larval abdominal limbs represent the expression of a previously repressed gene system in the abdomen. The phylogenetic system proposed in this book forces the conclusion that larval prolegs cannot have been directly converted from preexisting functional abdominal limbs of an ancestral preinsect. By the time insects acquired wings, adult vestiges of abdominal limbs were restricted to genitalia of the thysanurid type, and to cerci.

The endopterygotes have further extended the median invagination of the thoracic sternum, while the pleura have become extended ventrally fusing with the basisternum. The furcal arms tend to arise close together on each side of or on the internal median invaginated ridge. The coxae are thus very close together especially on the pterothorax.

In contrast to its sister group, the Acercarida, the endopterygote ground plan included some ancestral insect features. Cerci of the eleventh abdominal segment were retained in the adults, although in reduced form. These appear on the apparent tenth segment, but there has been fusion of the tenth and eleventh segments with the periproct, so that the cerci only appear to be on the wrong segment. In some cases the cerci are reduced to small setiferous bumps (fleas). The cerci are always simple. The ground plan also included pentamerous tarsi. These were secondarily reduced to fewer than five tarsomeres separately in various lines of Coleoptera and Strepsiptera. The ground plan mouthparts were of the usual biting type, subsequently modified or highly reduced in various orders of endopterygotes. All larvae except those of the Neuroptera bear biting mouthparts, except in some families such as the higher Diptera.

The sister groups of the Endopterygotida are the Coleopterida and the Telomerida. The Telomerida represent a new cohort including all the endopterygotes except the Coleoptera and Strepsiptera.

COHORT COLEOPTERIDA

This group differs from all the other Endopterygotida in the specialization of the metathorax only for flight, and in the partial desclerotization of the abdominal terga. The members of the group share derived features that appear to have arisen convergently in some members of other groups, such as the suppression of ocelli, the loss of an ovipositor, the simplification of the venation of the hind wings through loss of cross veins and reduction of branching of the main veins.

Members of the Coleopterida have not acquired the derived features that unite the other endopterygotes, such as the secondary separation of the parameres in the male into a basimere and telomere, and except for the Hymenoptera, the development of a meron in the coxa through a downward shift of the posterior basicoxal ridge. The spermatozoa are of the typical insectan type. The Coleopterida easily are separable into two sister groups, the orders Coleoptera and Strepsiptera.

Order Coleoptera

The beetles have evolved their own synapomorphies which have not appeared in the Strepsiptera. The sclerotized veinless elytra meet in a straight line medially, forming what is generally known as the suture. These sheathlike wings typically completely cover the hind wings in the folded position. The hind wings are remarkable in that they are usually folded both lengthwise and crosswise under the elytra, except when the folding has been modified, or the hind wings were shortened or lost. In the Myxophaga and the Archostemmata, the tips of the hind wings are rolled into a double spiral (Britton, 1970), perhaps best regarded as the primitive beginning of the double fold. The ocelli have become variously suppressed. Some dermestids bear only the median ocellus, other beetles lack the median ocellus but may have two lateral ocelli. The head of adult beetles is sclerotized ventrally (gula), seemingly as a result of the sclerotization and fusion of the ventral cervical sclerite with the extended lower borders of the postocciput, postgenae, and the basal part of the labium (Duporte, 1962). This is similar but not identical to the gula of raphidiids and megalopterans. In the latter two orders, the gula forms in the larvae without the incorporation of the ventral cervical sclerite, which remains as part of the neck containing single lateral cervical sclerites. The necks of beetles are further modified in addition to the incorporation of the ventral cervical sclerite into the gula. The lateral sclerites have either been lost (Adephaga) or have become fragmented into two sclerites on each side (Polyphaga).

The thorax of beetles consists of the immovably fused mesothorax and metathorax, more or less flexibly articulated with the prothorax. The male genitalia lie retracted into a sort .of genital capsule when not in use. In some beetles the simple nonarticulated parameres may be prominent, but in most the parameres are small, or are fused basally around the aedeagus to form a tube, and in some they are wholly suppressed. The primitive condition is seen in the Adephaga, in which the parameres are movable and articulated with the aedeagus. The first and second abdominal sterna are cryptic and membranous in beetles, and the first visible sternum is the actual third. The eighth and ninth abdominal segments generally are somewhat desclerotized and telescoped, lying hidden. The ovipositor has become reduced to a tiny pair of pointed first valvulae still remaining in the Carabidae, but lost in most beetles. The sheath (ninth coxal plates, labial lobes, third valvulae) remains at best as a pair of elongate ventral lobes with a stylus at the tip of each. These lobes are called ninth hemisternites by coleopterists, but they are more likely the remnants of the third valvulae (Mickoleit, 1973). Because of the similarity of these lobes with equivalent structures in Neuropterida, Mickoleit ventured to make the beetles a sister group to the Neuropterida. However, the Neuropterida are better regarded as a sister group to the Panorpida because of their shared synapomorphies. Furthermore, the vestigial ovipositor sheath does not protrude posteriorly with the upper surfaces of the valvulae fused, as is typical of the Neuropterida. The trochantin may persist as a thin detached or undetached strap, or may be merged with the pleuron. The caeca were reduced to a few (two to six). The pupae of Coleoptera are usually free in that they ordinarily do not form cocoons, and are never in a puparium.

Order Strepsiptera

The relationships of stylopids have been controversial. They are considered by some merely as a subtaxon of beetles, or as a separate order. As an order, they have been allied with beetles as a sister group, or allied with the mecopteroid groups, or with the Hymenoptera. Kinzelbach (1971) demonstrated that the Strepsiptera are probably best interpreted as the nearest relatives of the Coleoptera, based on extensive morphological data. Their own derived characters appear to have arisen subsequent to a split of the common ancestral stem of the beetles and the Strepsiptera, since the most primitive strepsipterans have features which could not have been derived from the beetles. Strepsipteran fore wings are highly reduced but are not sclerotized. The stubby fore wings are really membranous blood filled sacs, bearing vestiges of convex and concave veins. In some species (e.g.,· *Mengenilla kasabi* Kinz.) costa and subcosta are very short and close together and the radius is two-branched. Traces of the media remain in *Coriophaghus rieki* from Australia. The fore wings vibrate during flight, while those of beetles remain outstretched. The hind wings of males fold along the sides of the abdomen, but there is no sharp vannal fold, and only a small fragment of anal veins remains. The folding consists of a moderate amount of fluting in the field of M and Cu, with the wing folded lengthwise on itself. The veins of the hind wing consist of a fused C + Sc, a two-branched M *(Mengenilla)* generally reduced to one vein, a five-branched R *(Mengenilla)* variously reduced (two-branched in *Loania*), a two-branched CuA and a single CuP. A tiny trace of an anal vein persists in *Stylops*. The veins generally are not united basally, but the pupal wing veins show the primitive basal junction of the bases of Sc and R branches, and of M with CuA and CuP (Kinzelbach, 1971).

This group has been included with the beetles not only because of the presumed similarity of the wings, but because in many strepsipterans the tarsi bear four, three, or two tarsomeres. However, five tarsomeres are still found in the Mengenillidae, Mengeidae, and Corioxenidae, indicating that reduced tarsomery is a secondary feature within the Strepsiptera. The union of the first abdominal segment with the thorax has also been cited as a beetle or hymenopteran character. But in the free-living adult females of the Mengenillidae, the first abdominal segment is free from the thorax, and in the males, only the upper and lower parts of the first abdominal segment are overgrown by the metathoracic tergum and sternum, while laterally, the side of the first abdominal segment is exposed, bearing the first abdominal spiracle. Internally, the muscles of the first segment persist. In beetles the sterna of abdominal segments 1 and 2 lack internal musculature and are membranous in the adult, and the first visible abdominal sternum is that of abdominal segment 3. The Strepsiptera are said to possess a gula (Crowson, 1955), but the sclerotized undersurface of the head in male strepsipterans consists only of the postmentum of the labium, as shown by the musculature (Kinzelbach, 1971).

There are numerous derived features in the Strepsiptera which appear to have originated in their ancestral line after the beetles and the strepsipterans had separated from the coleopterid ancestral line. Those immediately following pertain to the adult male, the only winged state (Figure 80). The

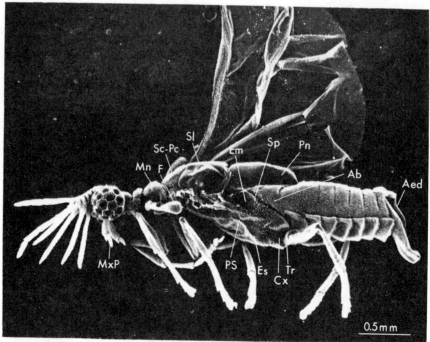

Figure 80 *Triozocera mexicana,* Strepsiptera, Mengenillidae. Scanning electron photograph. Left hind wing removed to show enlarged metathorax. Note the five-segmented tarsi. Ab, side of first abdominal segment; Aed, aedeagus; Cx, coxa; Em, epimeron; Es, episternum; F, fore wing; Mn, mesonotum; MxP, maxillary palpus; Pn, postnotum; PS, pleural sulcus; Sc-Pc, scutum plus prescutum; S1, scutellum; Sp, first abdominal spiracle; Tr, trochanter. Original.

antennal flagellum was reduced to five flagellomeres, the first one always flabellate. The second, third, and fourth flagellomeres may secondarily bear flabellae, making the antennae appear multibranched. The compound eyes bear few ommaditia (10 to 115) compared to the usual large eyes of other insects. The few ommatidia are separated from one another, and there are bristles in the spaces between facets. Ocelli are absent in the adult, although rudiments of median ocelli have been noted in the pupa. There are no lateral cervical sclerites. The postocciput and the tentorium do not develop in the adult head. The labium lacks palpi, and all that is left of the maxillae are the two-segmented palpi. The mandibles are reduced or undeveloped. The prothorax is a simple ring, freely articulated with the mesothorax. The mesothorax bears a pair of highly reduced club-shaped membranous wings that vibrate in flight. These seem to serve the same function as the halteres of flies, since strepsipterans whose fore wings were removed cannot fly (Kinzelbach, 1971). The venation of the fore wing is reduced to a fused C + Sc, a two-branched R, and a simple M. In some species there is further reduction of R and M. Both fore and middle legs lack a trochanter, and the coxae are elongated. The metathorax is greatly enlarged bearing the relatively large indirect flight muscles. The hind wings articulate with the notum only at the anterior notal process. There is no posterior notal process on the elongate postnotum. The subalare is usually very elongated, bearing strong muscles to the coxae. The hind coxae are firmly united with the metapleura in recent forms, but appear to be free in *Mengea tertiaria* (Menge) in Baltic Amber. The coxae are not as closely united in *Mengenilla*. The subalar-coxal muscles are mostly direct wing muscles, since the hind coxae do not move. The trochanter is the first free hind leg segment. The hind legs have moved far posteriorly, so that the pleuron is highly slanted forward and upward. The mesothoracic to metathoracic union is membranous, allowing some movement, quite different from the condition in beetles in which the pterothorax is a solid unit. The branches of the veins R, M, and Cu generally are detached from each other, and lie free in the membrane. Except for *Mengenilla* and *Stylops*, the anal veins are totally suppressed. Only a vestige of A remains in these. In the absence of a vannus, the hind wing folds in a longitudinally crumpled fashion into pleats among the M and Cu veins, and with a major fold so that the costal and cubital borders are down, the fold upward, while the folded wings lie vertically alongside the body. Beetle hind wings are folded horizontally over the dorsum of the abdomen. Only the sides of the first abdominal segment are visible, bearing the first abdominal spiracle. The tenth and eleventh segments remain unseparated from the periproct, forming a narrow anal tube. The male genitalia are unusual in that the ninth segment forms a subgenital plate with the simple aedeagus attached to the extended ventral part and folded upward in a groove of the subgenital plate. There is no trace of parameres. The development of the male genitalia has not been studied. The greatest number of abdominal spiracles is seven pairs on the first seven segments. The eighth pair is suppressed. In the more advanced species, only the first pair remain. The first abdominal spiracle is always larger than the others. The nervous system is highly concentrated, with the suboesophageal ganglion, thoracic ganglia, and one or more abdominal ganglia forming a ganglionic mass in the thorax. The nerve cords from the thorax to another ganglionic mass in the abdomen are generally fused into a single nerve. The posterior ganglionic mass contains abdominal ganglia that have not moved into the thorax. The malpighian tubes are highly reduced or absent. There is no free passage between the midintestine and the rectum. The testes are plump simple organs whose vasa deferentia are very short and thick, serving as seminal vesicles. The short ejaculatory duct is expanded as a sperm pump. There are no accessory glands.

In the most primitive families (Mengenillidae, Corioxenidae) the adult wingless females are highly neotenous and free-living, having a larval appearance. The thorax and head form a cephalothorax in the pupae. Neither sex feeds as an adult. The small mandibles are useful in escaping from the puparium. The female digestive tract is a simple flat epithelium and the midintestine ends blindly against the rectum or there is no rectum. Two or three bulbous short malpighian tubes remain at best. There are no ovaries as such. The eggs develop individually in the hemocoel of the female. In the Mengenillidae the median oviduct is blind internally and opens behind the venter of the seventh abdominal segment. The primary larvae escape by passing through the epithelium of the median oviduct (birth canal), which is lined with a thin cuticle bearing microtrichia. In other families the adult female remains in the puparium, exposing its front end through an intersegmental abdominal membrane of the host. The young escape through a series of dorsal invaginations similar to the median oviduct of Mengenillidae. These birth canals also have a thin cuticular lining bearing microtrichia. Fertilization of the female involves penetration of the aedeagus through the epidermis of the female.

The parasitic mode of existence of strepsipterans has allowed a high degree of neoteny to become established. The best developed male imaginal characters are those necessary for finding a mate: flabellate antennae, flight organs, and copulatory organs.

Pupation is in a puparium consisting of a late larval cuticle. When males emerge, they exuviate through pupal, prepupal, and larval cuticles.

There is nothing in the specializations of the Strepsiptera that must be considered as having an origin from any beetle ancestor. They resemble some parasitic beetles by virtue of convergently derived characters. The primitive insect characters of strepsipterans testify against their origin from the Coleoptera: absence of gula, jointed thoracic segments, membranous fore wings with vein vestiges, and a distinct first abdominal segment. Since they and the Coleoptera are the only endopterygote insects that use their hind wings for flight and have not developed the jointed parameres, the two together must be a sister group to all the other endopterygotes, and they bear a sister group relationship to each other. The other features which the Strepsiptera share with beetles are convergences with rather advanced species of parasitic beetles.

COHORT TELOMERIDA

This is the sister group of the Coleopterida, and includes all the remaining endopterygotes. The name refers to the jointed parameres of the male, which develop from lateral phallic lobes of the tenth abdominal segment just behind the coxosternum of the ninth segment. The basimere with its articulated telomere closely resembles the coxal plates and styli of the ninth abdominal segment in the Microcoryhia, Thysanura, Ephemerida, and Grylloblattaria. The reasons for considering the jointed parameres of the Telomerida as a derived feature were given earlier. Their resemblance to vestigial limb 9 of the lower insects appears to be the result of new mutations causing partial expression of limb gene codes that must be present in all the tissues of an insect. This derepression of limb gene codes also occurs elsewhere on the abdomen of the Telomerida, in which the embryonic limb buds of the pregenital segments and segment 11 fail to regress and become variously adapted abdominal prolegs, to be suppressed again convergently in the higher Hymenoptera, the Raphidiodea, Neuroptera, Diptera, and Siphonaptera, and partially suppressed in the Trichoptera and Lepidoptera. If derepression of embryonic limb buds is a ground plan character of the Telomerida, a secondary suppression of abdominal appendages must have occurred convergently in the Raphidoidea, Neuroptera, Diptera, and Siphonaptera. None of the Telomerida bear gastric caeca, in parallel with some other unrelated orders. The Telomerida consists of the two sister groups represented by Hymenopterida and the Meronida.

Subcohort Hymenopterida

The single order Hymenoptera makes up this group. The relationships of the Hymenoptera have long been debated. The members of the group are so highly specialized that it has been difficult to decipher their convergent features with other groups, and some of their autapomorphies are unique. The union of the first abdominal segment with the metathorax was cited as a character in common with Coleoptera and Strepsiptera. However, the first abdominal segment is distinct in the Symphyta, so fusion has secondarily occurred within the Hymenoptera. The production of silk by larval labial glands was cited as a feature in common with fleas, lepidopterans, and trichopterans. But these three orders all share derived features with other endopterygotes excluding the beetles, strepsipterans, and hymenopterans, such as the presence of a well developed meron and the lack of an ovipositor. The well developed ovipositor of Hymenoptera has been used to make them a sister group to all the other endopterygotes. But the ovipositor has been suppressed so many times convergently in nonpupating insects (Ephemerida, Odonata, Plecoptera, Blattiformia, Phthiriapterida, aphids, etc.) that one must suspect that the loss of an ovipositor in the pupating insects has also occurred several times.

The derived features peculiar to the Hymenoptera in contrast to their sister group the Meronida are as follows. The fore wings are the principal organs of flight. The hind wings are reduced and the musculature of the metathorax is weak. The hind wings derive most of their function in flight by virtue of their being coupled to the fore wings through a series of hooks (hamuli) grasping the hind border of the fore wings. The wing venation is reduced and highly modified, in that many cross veins mimic the main veins, obscuring the primitive venation. Male haploidy occurs in all the Hymenoptera. Females have haplo-diploid parthenogenesis. Fertilized eggs hatch into females, while unfertilized eggs become males. The lateral cervical sclerites have been lost in contrast to their presence in the Meronida. The femur of the more primitive Hymenoptera basally bears a distinct subsegment known as the trochantellus, mimicking a second trochanter. The higher Hymenoptera have mostly lost the trochantellus. The male genitalia are modified, in that the parameres and aedeagus arise from a solid phallobase, and the parameres split so that a median pair of accessory claspers (volsellae) develop in addition to the two-jointed parameres. There is a suggestion that the accessory glands of all female Hymenoptera produce venom (Hennig, 1969), but its significance for the phytophagous species is unclear.

In some features, the Hymenoptera have kept a few ancestral features. The ovipositor is fairly generalized in the primitive Hymenoptera. The third valvulae (coxite 9 plus stylus in some) normally ensheath the piercing first and second valvulae, which slide on each other on a tongue-and-groove device. There is no meron on the thoracic coxae. The subalar-coxal muscles insert on the upper rear edge of the middle and hind coxae, above the basicoxal sulcus. Abdominal prolegs occur on most of the phytophagous Symphyta, but are suppressed in the higher Hymenoptera. The fused tenth and eleventh abdominal segments bear generally a pair of simple cerci (socii). The spermatozoa are typical of insects in general. There is usually a moderate number of malpighian tubules, which are fewer in the small parasitic species.

Subcohort Meronida

The presence of a coxal meron distinguishes this group. Normally the tergocoxal remotor and the subalar-coxal muscles insert on a narrow internal flange at the rear upper rim of the coxa. This inner ridge externally forms the basicoxal sulcus. In the Meronida, the flange above the basicoxal sulcus has become enlarged while the sulcus has moved distally behind the coxal articulation, so that the sulcus appears to divide the coxa into an anterior eucoxa and the posterior meron. The meron serves as a wide surface for the insertion of the large subalar-coxal and tergocoxal remotor muscles. A superficially similar condition occurs in the orthopteroid insects, in which a sulcus divides the coxa into an anterior and a posterior piece. But the basicoxal flange is still at the upper border of the coxae in these, and the coxal muscles insert on the upper edge of the coxa. A close approximation of a meron occurs in the Strepsiptera. The hind coxae have become merged with the pleuron, and a large subalar-coxal muscle inserts broadly on an apparently expanded coxal meron, even showing a sort of coxo-meral sulcus. This must be secondary within the Strepsiptera, because the extinct Baltic Amber genus *Mengea* bears elongated free hind coxae with no trace of a meron (Kinzelbach, 1971). In the Meronida, the thoracic sterna between the legs are deeply invaginated, elevating the sternal apodemes on the median vertical part. Thus, the furcal pits lie close together and may not be clearly evident. The legs of the Meronida thus lie very close together with hardly any space between them. Although the same is the case in the higher Hymenoptera, the primitive state is seen in the lower forms in which the furcal pits lie separated on the mesosternum (Matsuda, 1970). Therefore, the close apposition of the legs of the Hymenoptera is autapomorphic in the group. The highly reduced, almost nonfunctional, ovipositor is another derived feature of the group. The second valvulae are entirely suppressed, and the first valvulae are very small, if at all present (Megaloptera, *Sialis*). The third valvulae when present are largely membranous, with a pair of slender lateral sclerites and a short stylus remaining. In addition, the malpighian tubules are reduced to fewer than 10 in the Meronida.

The ancestral ground plan of the Meronida included functional hind wings as important as the fore wings in flight, still retained in the Neuropterida, Trichopterida, and the Mecoptera. In all, the lateral cervical sclerites are primitively simple and not suppressed as in the Hymenoptera, although they may fuse with the ventral cervical sclerite (Neuropterida) or become divided (tipulids). The Meronida fall into two fairly distinct sister groups, the Neuropterida and the Panorpida.

Infracohort Neuropterida. The members of this group have retained a somewhat highly modified ovipositor in the ground plan, but it is variously reduced in the different groups. The third valvulae have become descleroLtized and fused dorsally, but bear internal muscles (Mickoleit, 1973). A narrow lateral sclerite occurs on each side of this structure, and these sclerites appear to be vestiges of the third valvulae, since they articulate with the ninth abdominal tergum. Usually a pair of styli remain on the posterior tip of the organ. The second valvulae appear to be totally suppressed. Only vestiges of the first valvulae remain in some species of *Sialis*, and these do not serve as a piercing organ. The costosubcostal area tends to be somewhat expanded, bearing many cross veins. The lateral cervical sclerites of the Neuropterida have become fused with the ventral cervical sclerites, forming a sclerotized structure reaching from one occipital condyle to the other around the ventral portion of the neck, except in the Neuroptera and Sialidae in which the ventral median part is secondarily desclerotized. The postlabium is sclerotized and united with the ventrally extended postgenae, but the gula so formed does not incorporate the ventral cervical sclerite as in beetles. The adult and larval maxillae usually bear reduced laciniae and galeae, except in the order Neuroptera in which the larval maxilla is highly modified. The labium generally lacks terminal muscles in the prelabium. There is a median dorsal diverticulum on the stomodaeum of the members of the Neuropterida.

The Neuropterida appear to be a good monophyletic group, but the relationships of the three generally recognized orders, the Raphidiodea, Megaloptera, and Neuroptera, have not been easy to decipher. The latter two orders are clearly monophyletic groups because each has derived features exclusive of the other. The Raphidiodea, however, have been variously treated. Achtelig and Kristensen (1973) have reexamined the various suggestions of raphidian relationships, and could not decide whether to consider them as a sister group of either the Megaloptera or the Neuroptera, or whether the Raphidiodea belong in the Megaloptera. It seems best to consider the Raphidiodea as a sister group to the Megaloptera + Neuroptera in two subtercohorts, the Raphidida and the Sialida.

Subtercohort Raphidida. This group contains the single order Raphidiododea. For the sake of uniformity of order names, the group has been called Raphidioptera. This is another nonsense name, because the stem is Greek for "needle," referring to the long ovipositor. Externally the elongate ovipositor of raphidians resembles that of the Dilaridae (Neuroptera), but in the Raphidiodea the ventral median part of the eighth abdominal sternum is prolonged into a pointed structure that extends into the space formed by the dorsally fused elongated third valvulae. This structure is provided with two grooves that receive two ventrally directed carinae on the median sides of the third valvulae, imitating the tongue-and-groove union of the first and second valvulae of normal ovipositors, but forming a closed passage for the eggs between the prolonged eighth sternum and the third valvulae. Such a structure must represent a secondary specialization of the Raphidiodea, because the second valvulae are totally suppressed in all the Neuropterida (Mickoleit, 1973). The dilarid ovipositor is typical of the Neuropterida in lacking any trace of the first valvulae, and in the highly reduced styli on the third valvulae. Styli are well developed on the raphidian ovipositor.

In addition to the peculiar long ovipositor, raphidians departed from the stem line of Neuropterida by developing a long head, neck, and prothorax quite unlike the elongated prothorax of mantispids (Neuroptera). The postocciput is extended behind the head, imitating a neck, but the postoccipital sulcus which is clear in the larva is lost in the adult head. The head is ventrally sclerotized behind the labium, forming a gula in both larvae and adults. The posterior tentorial pits of the adult are quite close together, in contrast to their position far apart in the Sialida. The dorsally membranous neck is nearly as long as the prothorax, but is covered by forward extension of the pronotum. The pronotum is also extended laterally and downward, hiding the pro-pleuron. Ventrally the neck bears an elongate sclerite which Matsuda (1970) called the presternum. This plate joins the greatly elongated narrow lateral cervical sclerites, forming a unit ahead of the prothoracic basisternum, and differs from the condition in Sialida in its great length. The raphidian pro-coxae are at the extreme rear of the prothorax. The tarsi differ from those of Sialida in that the raphidian tarsi bear a bilobed third segment followed by a very short fourth tarsomere. The fifth tarsomere bears a pair of strong claws but not an empodium.

Otherwise, the Raphidiodea appear to have remained primitive in many features. The terrestrial larvae are not much different from the adults, as is the case in the Sialida. The adults retained the three ocelli on the head. The wing venation is primitive in that the subcosta ends in the front border of the wing closer to midwing than to the tip, while in the Sialida Sc joins R or ends free near the tip of the wing. If a pterostigma appeared in the ground plan of the Neuropterida, it has remained pigmented in raphidians while it lost the pigmentation (but not the blood sinus) in the Sialida (Kristensen, 1975). The maxillae of both larvae and adults are not modified as in the Sialida, retaining a somewhat normal short stipes. In males, the genital claspers are rep-resented by the larger basimere bearing a simple pointed telomere.

If the raphidians are truly an early offshoot of the neuropterid stem, as in-dicated by their primitive characters, then they seem to have evolved some structures in parallel with the Megaloptera on one hand, and with the Neuroptera on the other. Both Megaloplera and Raphidiodea bear a gula, telotrophic ovarioles, and have similar synchronous antennal cleaning behavior (Kristensen, 1975). Both Raphidiodea and Neuroptera have malpighian tubules joining the rectum, show some terminal bifurcation of the radius and media, and have precocious disjunction of the sex chromosomes during cell division. None of these characters is unique to the groups in-volved, and so the possibility of convergence is there.

Subtercohort Sialida. As a sister group to the Raphidida, this group bears some distinctive advanced features. The subcosta ends in the radius instead of in the front border of the wing. The subcosta appears to have secondarily lost its union with R_1 in some families. In the Coniopterygidae, Sc ends at the wing tip, but there is a great suppression of branching and of cross veins in this family, correlated with the small size. In some other families Sc may continue as a highly branched vein to the wing tip. At any rate, the course of Sc is different in this group than in the Raphidida. In the latter, Sc is simple and ends just past midwing against C. In the Sialida, Sc goes past the third quarter of the wings before either ending in R_1 or continuing to the tip of the wing as a highly branched vein. A pigmented pterostigma similar to that of raphidians seems to have been lost in the Sialida, but the stigmatal blood space persists in the absence of pigment (Kristensen, 1975). Stigmalike pigmented spots appear to have secondarily been acquired by some of the Myrmeleontidae (Neuroptera) since these spots are in the terminal branches of the long Sc. The Sialida bear another derived feature not found in raphidians. The stipes of the maxilla is slender and elongated in the adults. In raphidians the stipes is short in all stages. Judging from the condition in the Megaloptera and most of the Neuroptera, the third valvulae were highly reduced in the ground plan of the Sialida, and the styli were reduced to small structures or eventually lost. Lengthened third valvulae approaching the condition in raphidians were redeveloped by the Dilaridae, but in these the styli are gone. The Berothidae and Mantispidae bear prominent third valvulae, but not as prominent as in raphidians.

The Sialida probably bore ground plan ancestral characters that were variously changed. For instance, all of of them except the family Sialidae bear unadorned cylindrical tarsomeres. The fourth tarsomere has become short and bilobed in Sialidae. In contrast to the Raphidida, the arolium below the tarsal claws was not lost. It remains as a thin-walled bladderlike structure, or it may be sclerotized, bearing bristles. The suppression of the first valvulae of the ovipositor may have occurred three times in the Neuropterida. There is no trace of the first valvulae in raphidians. Some species of Sialidae bear nonfunctional, tiny, paired, pointed vestiges of the first valvulae, so there must have been at least a second suppression in the Megaloptera. If one considers the Neuroptera (Plannipennia) a sister group of the Megaloptera, the absence of the first valvulae in the former group would appear to be a third-time suppression of the first valvulae.

Other ancestral ground plan features of the Sialida are absence of prothoracic paranota, lateral cervical sclerites fused with ventral cervical sclerite (as in raphidians), short neck, and probably terrestrial larvae. The Sialida are easily separable into two sister groups, the order Megaloptera and the order Neuroptera.

Order Megaloptera. The two families Corydalidae and Sialidae comprise this order. This group differs from the Neuroptera in that all the members are aquatic as larvae. In becoming aquatic, the derepressed embryonic limb buds become vaguely leglike structures into which presumably respiratory tracheae penetrate. The larval cerci in the Corydalidae persist, but became modified into a pair of leglike structures bearing terminal hooks. These pygopods are equivalent to similar cercal legs in the Mecoptera, Lepidoptera, and Trichoptera (Matsuda, 1976). There is a larval gula similar to that of the Raphidiodea, but the gula persists in the adult only in the Corydalidae. The larval maxillary stipes is elongated, similar to the condition of the adults. The wings have lost the pigment in the pterostigma, although a stigmatal blood sinus is said to remain. If the terminal bifurcation of the main veins is a ground plan ancestral neuropterid character, then the Megaloptera have eliminated this character. Terminal venational bifurcations occur in raphidians and in the Neuroptera. Within the Megaloptera, each of the two families has unique derived features. The Corydalidae bear an enlarged jugal area in the hind wing, with a prominent jugal bar. The hind wing folds among the posterior anal veins rather than at the typical vannal fold of the nonpupating insects, and different from the jugal fold typical of the endopterygotes and of the Sialidae. In the Sialidae, the adult gula has secondarily become desclerotized, along with the desclerotization of the midventral area of the fused lateral cervical sclerites and the ventral cervical sclerite. The prothoracic sternum is also highly desclerotized in Sialidae, leaving a small median area bearing the furca. The lateroventral cervical plate is a solid continuous structure in the Corydalidae, and the prothoracic sternum is fairly large. While sialids have lateral abdominal appendages, they have lost the pygopods which remain in the Corydalidae. The fourth tarsomere of sialids is very short, strongly bilobed, and followed by an elongated fifth tarsomere. In Corydalidae the tarsomeres are all cylindrical. In contrast to the Corydalidae, the Sialidae have no ocelli. The male claspers of the Corydalidae are segmented, bearing the telomere, but the claspers have become secondarily simple and small in the Sialidae.

Order Neuroptera. This order as erected by Linnaeus included the Odonata, Ephemerida, Mecoptera, Trichoptera, Raphidiodea, and Megaloptera. With the removal of these orders from the original Neuroptera, the remaining families form a well defined taxon, which Latreille (1817) named Plannipennes, later changed to Plannipennia. Since it is desirable to preserve Linnaean names, and because of priority, the name Neuroptera in its restricted sense is retained for this group.

The derived features of this order are such that the group is clearly monophyletic. The larval mouthparts have become peculiarly adapted for a piercing and sucking type of feeding as predators. The long, generally sickle-shaped larval mandibles are grooved below. This groove is closed by the close apposition of a similarly elongated flat lacinial stylet. In some families there are two grooves, one through which a poison is injected into the prey, and another which serves as a sucking tube for the liquified insect prey after it has been impaled on the pointed mandibles. The cardo and stipes are very short in the larvae. Although the larvae are typically terrestrial, the larval metathorax is without spiracles and the larval prolegs are suppressed. The atypical family Sisyridae, the spongillaflies, have secondarily become aquatic as larvae, and have reacquired tracheated proleglike structures on the abdomen. In the Neuroptera the larval midintestine is closed, and the proctodaeal malpighian tubules become silk-secreting glands. The silk issues from the anus in forming a silken cocoon for the pupa. Suppression of the ocelli of the adult has occurred within the order, since three ocelli are found on the Osmylidae. The adult head has secondarily desclerotized the neuropterid gula, since the larvae generally bear a gula. The median sclerotization of the ventrolateral cervical plate has been lost in parallel with a similar condition in the megalopteran Sialidae. The lateral cervical sclerites so formed are generally free in the neck membrane, so that the head does not articulate directly with the cervical sclerites. The ovipositor of Neuroptera consists of only the remnants of the third valvulae, mostly without styli. This structure is secondarily elongated in the Dilaridae. The wings of the Neuroptera are primitive endopterygote wings in that the jugum is not enlarged as in the Megaloptera. But the venation has become specialized, in that Rs has become typically many-branched, with the branching pectinate along the stem of Rs. The bifurcation of the tips of the veins has become very elaborate in some species, while the venation was secondarily simplified in the Coniopterygidae. The male genitalia of the Neuroptera are specialized in that the telomeres are suppressed in the parallel with the condition in the Sialidae.

Infracohort Panorpida.   This is the sister group to the Neuropterida. It has long been known as the Panorpoidea, or the Panorpoid Complex. There are few obvious derived features to define this group. One such is the further invagination of the thoracic sterna, so that there is little left externally of the mesosternum and metasternum. The furcal arms arise near the upper part of the ridge that represents the invaginated sternum. Veins 1A and Cu of hind wing join together then separate near bases. In all the Panorpida the first abdominal sternum is highly reduced or membranous medially in the adults. All members of the group have entirely or almost entirely suppressed the usual ovipositor, and a subsitute ovipositor has evolved consisting of the modified terminal abdominal segments which usually telescope within each other.

Some members in each of the orders here included pupate in a silken cocoon secreted by larval labial (salivary) glands. It appears that silk production by the salivary glands is a ground plan feature of the Telomerida, secondarily lost in many Diptera, such as in the mosquitoes and the cyclorrhaphous flies. The Panorpida have reduced the malpighian tubules to six or fewer, while in the Neuropterida the ground plan number is eight malpighian tubules.

The other derived features common to the group are not easily seen because to observe them requires some dissection or other special preparation. The spermatozoa differ from most insect sperm in that there is only one mitochondrial derivative. Some of the higher Diptera, however, bear the two unequal mitochondrial derivatives in the sperm tail. Since the single condition is found in all the rest of the group, including some primitive Diptera, the double condition in the higher Diptera may be secondary. According to a summary by Kristensen (1975), in the Panorpida the first axillary sclerite bears a muscle originating on the pleuron; the usual tergo-coxal remotor of the pterothoracic coxae is suppressed, leaving only the subalar-coxal remotor; and stipito-lacinial and stipito-galeal muscles are lacking in the larvae. In addition,. the larvae of the Panorpida all bear a divided stipes, and the clypeus is fused to the frons, eliminating the larval frontoclypeal sulcus.

The Panorpida are represented by two sister groups, Mecopterida and Trichopterida. The Mecopterida includes the orders Mecoptera, Siphonaptera, and Diptera. The Trichopterida includes the orders Trichoptera and Lepidoptera.

Subtercohort Mecopterida. Hennig (1969) used the name Antliophora (pump-bearers) for the group consisting of the Diptera and the Mecoptera. He excluded the Siphonaptera from the group because he thought the fleas did not bear a sperm pump. Kristensen (1975) indicated that the group should include the fleas, because these insects also bear a sperm pump at the base of the aedeagus, modified from the ejaculatory duct. The details of the structure of the sperm pump differ in all three orders, but all may have been derived from a simple ground plan. The name Mecopterida is preferred because of its simplicity, and because it is formed from the stem name of one of the included orders. No morphological studies have explored the possibility of homology in the sperm pumps of the three orders. These actually may not be strictly homologous. A sperm pump consisting of a muscular development of the short ejaculatory duct occurs in the Strepsiptera (Kinzelbach, 1971). Hennig (1969) stated that only in these three is the semen transferred as a fluid in mating, and there is no spermatophore. But spermatophores are common in the lower Diptera (Leppla et al., 1975). The absence of spermatophores in the higher Diptera must be secondary, so the absence of a spermatophore in Mecoptera and Siphonaptera does not necessarily relate them to each other. In fact, the spermatozoa are transferred in a mass wrapped around the tip of the penis lobe in fleas, analogous to a fork full of spaghetti (Rothschild, 1965), and therefore are not transferred in a fluid. The unity of the group is borne out by other apparently shared derived features. The malpighian tubules were reduced to four, in contrast to the presence of six malpighian tubules in the Trichopterida. In the Mecopterida the larval labium lacks the lateral labial retractor muscles (Hinton, 1958). The larval cardo tends to fuse with the basistipes. The cranial promotor of the cardo is suppressed. The larval postmentum is desclerotized. The larval labial palps are reduced to two or fewer segments, and they are not musculated. Other usual insect muscles that are lacking are the hypopharyngeal retractor and the ventral muscle of the salivarium. The larvae undergo four or fewer larval instars, while the Trichopterida typically live through six instars. Larval mecopterids are more primitive than larval trichopterids in that in the former, silk is produced only at pupation, forming a cocoon, while in the latter, the early instar larvae can produce silk before pupation.

In all three orders of Mecopterida, some loss of the outer doublets of axonemes of the sperm tail has occured, reducing the axoneme pattern to only one ring of nine doublets. Apparently all the fleas and the Mecoptera have returned to the primitive arthropod 9 + 2 pattern (Baccetti, 1970). In the Diptera, *Plecia nearctica* (Bibionidae) bears the 9 + 0 pattern (Trimble and Thompson, 1974). The central doublet is lost. If the 9 + 2 pattern was a ground plan feature of the Mecopterida, then the higher Diptera regained the outer ring of doublets in the sperm tail, because such patterns as 9 + 9 + 0, 9 + 9 + 1, 9 + 9 + 2, and 9 + 9 + 3 occur in various Diptera (Baccetti, 1979). The sperm tails of Mecoptera and Siphonaptera are peculiar among insects in that the axonemes spiral around the single mitochondrial derivative. In the Mecoptera, the spiral is extended also around the nucleus (Baccetti, 1970). This coiling feature tends to relate the fleas to the scorpionflies, but it is not unique among arthropods. The 9 + 2 axoneme bundle coiled around single mitochondrial derivative is also found in a pseudoscorpion (Arachnida) and in Mystacocarida, Cirripedia, and Branchiuria (Crustacea) according to Baccetti (1979). That this could be a convergence between fleas and scorpionflies is indicated by other features shared by fleas and Diptera, but not found in the Mecoptera, discussed below.

The relationships of fleas to either the Mecoptera or the Diptera is still controversial, but the Mecoptera and the Diptera appear to be clearly separate monophyletic groups. The discussion that follows is in favor of a sister group relationship between the Mecoptera, and the Diptera + Siphonaptera. The Mecoptera have had an apparently much longer evolutionary history separate from the other two orders. The two sister groups are here designated as the Mecopterodea and the Haustellodea.

Superorder Mecopterodea. The only order in this group is the Mecoptera, or scorpionflies. The derived features that separate the Mecoptera from the Haustellodea are as follows. (1) The head is elongated, bearing a long labium, and the mandibles are apical. (2) The labrum, clypeus, and frons of the adult form a single piece not well marked off by sulci. (3) The tentorial mandibular adductor is suppppressed. (4) There are no hypopharyngeal muscles. The hypopharynx lies on the extended labium and is not separate as in the Haustellodea. (5) The adult antennae are long and filamentous, usually as long as the wings (secondarily shortened in the Bittacidae). (6) The larvae have developed lateral eyes similar in structure to compound eyes, but these are histolyzed at pupation and adult compound eyes develop from separate histoblasts. (7) The larval tarsus bears no extensor muscle. (8) The pretarsus is without the pretarsal flexor. These last two characters may have been in the mecopterid ground plan, but flea and fly larvae are apodous. (9) Although there are 10 pairs of larval spiracles, the metathoracic spiracle is not functional. (10) The ninth abdominal segment of males forms a sclerotized ring without demarcation into separate tergum and sternum. (11) The spermathecal opening is separate, from the gonopore. The ovipositor of scorpionflies is highly reduced. The eighth sternum extends in the female as a subgenital plate, bearing two lobes which may represent vestigial first valvulae. The highly reduced third valvulae are behind the membranous ninth segment, with the gonopore in front of this structure (called medigynium), and the opening of the spermatheca is near the distal end.

It has been pointed out that the larval Mecoptera bear reduced thoracic legs, with the trochanter, femur, and tibia remaining unseparated (Hinton, 1958). However, the larva of Nannochorista was figured with all the podomeres clearly visible (Riek, 1970c). Therefore, the lack of separation of podomeres in larval thoracic legs probably occurred within the Mecoptera, and convergently with thoracic leg reduction in other endopterygote larvae such as in Siphonaptera, Diptera, some Lepidoptera, and some beetles. The full leg segmentation of larval Nannochorista suggests that the podomeres may be musculated, but this has not been investigated.

In terms of the sister group Haustellodea, the scorpionflies have remained primitive in some features. The wings are typically fully developed (except in Boreidae, Apterobittacus, and Apteropanorpa). Both wings are essentially similar in size and function. The pupa escapes from the pupal cocoon employing its pupal mandibles (decticous, exarate). There are still traces of an ovipositor. The larvae bear abdominal limbs functional in locomotion. The last pair of abdominal limbs (pygopods, hooklike in the Nannochoristidae) become the adult cerci.

Superorder Haustellodea. The name Haustellata was used by Clairville (1798) for insects with an obvious sucking beak. In addition to the Siphonaptera and Diptera, he included the Lepidoptera and Hemipterida. The name is here emended and restricted to include only the Siphonaptera and Diptera.

The derived features restricted to the Haustellodea in contrast to the

Mecopterodea are as follows. (1) The ground plan adult piercing-sucking mouthparts consist at least of a pair of maxillary lacinial stylets, possibly a pair of mandibular stylets, an elongate epipharynx (labrum), a pointed hypopharynx enclosing the exit of the salivary duct, and dorsally involved with the formation of a cibarial sucking pump, plus an elongate labium basally enclosing the stylets. The fleas and flies each have further modified this ground plan. (2) The pupae escape from the pupal cocoon through the use of cephalic cocoon cutters in those forms that spin a cocoon (adecticous pupae), or otherwise *not* with the aid of pupal mandibles. The inflatable ptilinium of the cyclorrhaphous Diptera is a speciality of the higher forms. (3) Hind wings are reduced. The fleas bear no trace of hind wings at any stage. The hind wings of Diptera are reduced to halteres, and aid in flight by serving as gyroscopic sense organs. (4) The metanotum is much smaller than the mesonotum. (5) The adult antennae are shortened, and generally are never as long as the wings, in contrast to the long antennae of scorpionflies. (6) Thoracic legs are suppressed in the larvae. (7) Larval eyes are suppressed.

In addition to the seven haustellate synapomorphies listed, there is another feature of flies and fleas worth noting. Sharif (1935) described some peculiar mesothoracic outgrowths on the pupae of ceratophylloid fleas (absent in pulicoid fleas) that he thought were wing buds. Rothschild (1975) denied that these could be wing buds because they grew out from a position on the upper pleuron below the future pleural arch. This does not necessarily rule out the possibility that these are vestigial wing buds.

The first sign of the nymphal wing buds in some Odonata appears low down on the side of the nymphal thorax, and with subsequent molts the wing buds shift upward, eventually to be in the proper place lateral to the small tergum, but above the pleuron (Matsuda, 1970). Since fleas are totally wingless as adults, the position of incipient wings is not important to the adult, and the pupal position of mesothoracic wing buds in fleas is probably merely a siphonapteran peculiarity of development. At any rate, the emphasis on the mesothoracic wings in Diptera matches the presence of wing buds only on the mesothorax of fleas. This emphasizes the good possibility that the haustellate stem line had already initiated the suppression of hind wings before the fleas and the Diptera separated.

The ground plan of the Haustellodea apparently included some primitive features that were modified in the higher forms. Holopneustic larval spiracles (10 functional pairs) are typical of flea larvae, and bibionid larvae are holopneustic in late larval stages. Other Diptera have variously become either apneustic, metapneustic, propneustic, or amphipneustic as larvae. In contrast, larvae of Mecoptera are peripneustic in that the metathoracic spiracle is nonfunctional. Larval proleglike structures occur on the abdomens of dipterous blepharocerids and some other unrelated genera. These may be secondary acquisitions in these forms, since abdominal prolegs were probably suppressed in the ground plan of the Haustellodea (Hinton, 1955). But the terminal leglike structures on some Diptera are apparently the precursors of adult cerci (Matsuda, 1976).

Order Siphonaptera. The features of fleas that seem to relate them to the Mecoptera rather than to the Diptera can be interpreted either as generally primitive insect characters or as convergencies. Rothschild (1975) enumerated several such characters shared by fleas and scorpionflies: (1) The spirally coiled axonemones of their spermatozoa are not unique. This condition also occurs in pseudoscorpions and many crustaceans, and so could be convergent in fleas and scorpionflies also. (2) The axoneme pattern in which one of the outer nine rings of doublets is missing also occurs in some Diptera. *Plecia nearctica*, a bibionid fly, bears the 9 + 0 axoneme pattern (Trimble and Thompson, 1974). The absence of one outer ring of axoneme doublets may be a ground plan character of the Mecopterida, and restoration of the outer ring of doublets in the higher Diptera represents another instance of reverse mutation. Other Diptera have various axoneme patterns, particularly in the central axonemes. (3) The thick glycogen coat also is found in the sperm of Lepidoptera, and the outer nine secondary fibers rich in glycogen occur commonly in insects (Baccetti, 1970). (4) The presence of resilin in the pleural arch of fleas does not relate them to the Mecoptera, because resilin is a common cuticular substance in the pterygotes. (5) Multiple sex chromosomes are also found in the Diptera as well as in some fleas and scorpionflies. (6) Acanthae, very fine cuticular bristles, occur in the proventriculus of both fleas and Mecoptera. These are very thin structures each secreted by one cell, which form a velvety lining in the proventriculus of Mecoptera, and which are not so numerous and are backwardly directed in the fleas. These acanthae are elsewhere known among insects only in the beetle genus *Rhynchophorus* (Curculionidae), (Hepburn, 1969; Richards and Richards, 1969). I have seen acanthae in the proventriculus of the following insects: *Labidura ripara* and *Euborellia annulipes* (Dermaptera); *Agulla* sp., larva (Raphidiodea); an unidentified mantispid (Neuroptera); *Corydalus cornutus*, larva and adult (Megaloptera); *Calosoma alternans* (Coleoptera, Carabidae); *Odontotaenus disjunctus* (Coleoptera, Passalidae); *Dibolocelus ovata* (Coleoptera, Hydrophilidae); *Cybister fimbriolatus* (Coleoptera, Dytiscidae); *Sitophilus zeamais, Sphenophorus pertinax, S. cariosa, Scyphophorus yuccae, Rhodobaenus tredecimpunctatus, Rhynchophorus* sp., and *R. cruentatus* (Coleoptera, Curculionidae). Since they are unknown in the Diptera, their unusual presence in both Siphonaptera and Mecoptera strengthened ideas of close relationship between fleas and Mecoptera (Rothschild, 1975; Kristensen, 1975).

Since acanthae occur in other insects, such acanthae must be a convergence with those of fleas and Mecoptera. This leaves open the question of convergence in the latter two orders, since fleas are here regarded as sister order to the Diptera. Possibly acanthae belong in the ground plan of the Mecopterida. The various species of Mecoptera feed on a variety of food, such as plant material, dead or injured organisms, or as active predators. *Panorpa* spp. are frequently observed feeding on entrails of field dressed game animals, such as deer and squirrels. I have watched a species of *Panorpa* bite into a live pupa of laboratory reared *Trichoplusia ni*, insert its probiscis, and eventually suck out the entire contents in about one day. The entire head would enter the pupa in the last stages of feeding. The scorpionflies would regurgitate a brownish fluid on the surface of a bit of dog food pellet, and suck up the material after a while. Hepburn (1969) reported the same behavior, suggesting that digestion begins extraorally, with the liquid food containing only tiny particles. The acanthae seem to serve as a strainer when regurgitating digestive fluid. It is possible that the hausteliate line leading to Siphonaptera and Diptera may also have had similar food habits before becoming adult fluid feeders. The Siphonaptera retained the proventricular acanthae, even though they are not efficient as filters to hold back red blood cells. In the Diptera, the absence of acanthae may have resulted from evolutionary suppression, since their loss would not be a disadvantage in fluid feeding. (7) The sexual dimorphism in the abdominal nervous system of males also occurs in some Lepidoptera (Kristensen, 1975), but not in *Boreus*, a mecopteran. This also could be a convergence. (8) The presence of a prothoracic transverse interfurcal muscle in fleas and some scorpionflies is not unique. This condition also occurs in the Mallophaga and the Heteroptera (Matsuda, 1970), and tranverse interfurcal muscles are found also in the pterothorax of Blattaria, Phasmida, Ephemerida, and Hymenoptera. Such muscles are common in insects whose spinasternum is suppressed, and probably are furco-spinasternal muscles that have become secondarily transverse (Matsuda, 1970). Their presence is not necessarily indicative of close relationship because of the good probability of convergent evolution.

The derived features of the Siphonaptera are quite diagnostic. The wings are totally suppressed in the adult. The basalar-coxal and subalar-coxal muscles of the hind leg have become important leg muscles, and do not have a double function (direct flight muscles) as in the winged insects. The meron of the coxae has nearly lost its identity. The only trace of a meron is an internal vertical ridge in the middle and hind coxae. Adult fleas are ectoparasitic blood sucking insects, primarily of mammals. Their presence on a few bird hosts and on a snake seems to be secondary. On the head, there are only vestiges of compound eyes. The antennae of the male serve as accessory clasping organs during mating. The antennae are usually very short and lie in a groove of the head. However, antennae of the nematoceran type were present on the head of a cretaceous flea (Riek, 1970d). This flea also lacked the typical ctenidia or combs found on the head and thorax of most modern fleas. The body is well adapted for moving among the hair or feathers of their hosts. The body is laterally compressed, and there are usually backward-pointing ctenida on the head and thorax. The head is also compressed in modern fleas, but in the cretaceous flea reported by Riek (1970d), the head was of normal shape, suggesting that the flattening of the head occurred after the fleas separated from the haustellate stem line. Riek's cretaceous flea bore rather long legs compared to modern fleas, and the hind legs did not appear to have the enlarged femora so well adapted for leaping in modern fleas. In fleas the second through seventh abdominal terga overlap the sclerotized sterna, in contrast to the Diptera, whose abdominal sterna generally join the terga by way of arthrodial membrane. Although the metanotum is shorter than the mesonotum, the metaepimeron is expanded behind, covering the sides of the first abdominal segment. Large muscles adapted for leaping originate on this enlarged structure.

In contrast to the Diptera, whose larvae are either aquatic or associated with damp environments, the larvae of fleas live in dry situations without need for moisture. The first two larval instars are peripneustic, but become holopneustic in the third instar (Hinton, 1958). The holopneustic larval tracheal system bears the usual spiracular closing apparatus.

Order Diptera   Unlike fleas, the Diptera typically are flying insects, but use only the fore wings as organs of flight. The hind wings are reduced to small club-shaped halteres, whose vibrations during flight serve as gyroscopic sense organs. The mesothorax is highly developed in the Diptera, consistent with the restriction of flight function to the fore wings. The phragma between mesothorax and metathroax is greatly developed, serving as attachment for the powerful indirect flight muscles. The metathorax is highly reduced. Dipteran mouthparts are typically piercing and sucking. Unlike fleas, the mandibles of flies are elongate, styletlike structures, and the hypopharynx typically is elongate and styliform. The epipharynx (labrum), mandibles, laciniae, and hypopharynx form a bundle of six stylets, usually ensheathed by the elongated prelabium. The food canal is bordered anteriorly and laterally by the epipharynx and posteriorly by the crossed mandibular stylets. the salivary canal penetrates the elongated hypopharynx. In contrast, the fleas lack mandibular stylets, and the short hypopharynx conveys saliva to a channel in each lacinial stylet. The labial palps of adult flies are modified into the labella at the end of the labium. The labella may serve as a sponging device in some Diptera. Within the Diptera, the higher forms (Cyclorrhapha) have lost the piercing stylets, and are restricted to sponging up liquid food. However, some muscoid flies, such as *Stomoxys* and *Glossina*, have reacquired a piercing and sucking ability through the stiffening and elongation of the labium, which has become a piercing organ.

Larval Diptera generally inhabit aquatic or semiaquatic environments, or live in moist media. The tracheal system is primitively holopneustic (Bibionidae), but secondarily within the Diptera there has been a reduction in functional larval spiracles, and none bear any sort of closing apparatus.

The males of Nematocera bear the primitive jointed parameres, but in the Cyclorrhapha the basimeres have fused (Griffiths, 1972). The basimeres are fused together ventrally and extended upward to meet on the dorsum, obliterating the tergum of the ninth segment. This structure, formerly called epiandrum, was renamed periandrum. The telomeres remain as the only functional claspers in the Cyclorrhapha. In many Diptera, the male genitalia have rotated during the pupal stage so that the morphological venter is dorsal. In others this inversion is further rotated so that a complete 360° circumversion occurs at pupation. Since the inversion is not present in various unrelated species, the rotation of the genitalia probably occurred convergently many times in the Diptera (Griffiths, 1972).

Subtercohort Trichopterida. This, the sister group to the Mecopterida, includes the orders Trichoptera and Lepidoptera. Hennig (1969) called the group Amphiesmenoptera (dressed-up wings). The group name Trichopterida is preferred for the same reason as the name Mecopterida. The cohort suffix -ida appended to Henning's name would form an eight-syllable awkward word: Amphiesmenopterida. Either name refers to one of the derived features of the group. The wings are abundantly clothed with setae, both on the main veins and on the membrane between, often forming distinctive pigmented patterns. The fore wing venation is modified, so that the first anal vein is joined by the other anal veins, forming a double-Y vein. The ground plan method of wing coupling seems to be through the extension of the jugal area of the fore wing into a jugum, which fits under the front border of the hind wing. This method of wing coupling occurs in some Trichoptera and the lower Lepidoptera. Other coupling devices employing setae of various sorts have evolved independently in the Trichoptera and the Lepidoptera. In the two groups, a dense area of fine setae occurs on the metanotum near the subalare. These setae are hooked and are modified as a wing holding device for the hind wings at rest in the Lepidoptera (Kuijten, 1974). The bristles hook into an area of aculeate microtrichia on the undersurface of the base of lepidopteran fore wings. On the prothorax of both orders there is a pair of setigerous protuberances, which are enlarged in the higher Lepidoptera into large lobes called patagia. In the thorax, the trichopterid furca is large, with anterior lateral arms that fuse broadly with the epimeron. On the fifth abdominal sternum, a pair of ventral glands is often present in members of both orders, probably representing a ground plan feature. A pair of sternal apodemes often extend inward on the eighth and ninth abdominal segments. These apodemes were lost in some families of each order (Ross, 1967). The larval midintestine often contains goblet cells, suggesting that these were an early acquistion in the trichopterid ground plan. The Trichopterida are the only insects in which there is female heterogamy. Normally, the XY sex chromosome pattern is in the male, rather than the female. Oogenesis is achiasmatic, the reverse of the usual condition. The chromosome number is elevated in the Trichopterida (n = 30 or 31), in contrast to the smaller number of chromosomes in the Mecopterida. The spermatozoa of trichopterids are peculiar in lacking an acrosome, and in having the outer ring of axoneme doublets rather stout and containing abundant glycogen. The lower cervical muscles come together and insert together in the middle of the corporotentorium.

The ability to spin by larval labial glands has been extended in the Trichopterida to early larval instars. Silk webbing or cement is spun by trichopteran larvae (with exceptions), and many lepidopteran larvae spin silk in early instars. If this is a ground plan character of the Trichopterida, this ability was secondarily lost independently in both orders.

The Trichopterida have some primitive features compared to the Mecopterida. The larvae generally have six instars instead of four. There are six malpighian tubules instead of only four in Mecopterida. The spermatoza bear the usual 9 + 9 + 2 axoneme pattern. It is usual to consider the Trichoptera to be the closest relatives of the Lepidoptera, and these bear a sistergroup relation to each other.

Order Trichoptera. The Trichoptera have evolved into secondarily aquatic forms as larvae, so the larvae bear many derived features associated with the aquatic habitat. These are lack of open larval spiracles, the development of the anal prolegs, epidermal respiration through various epidermal evaginations containing air-filled trachae, and the use of silk in various ways (protective cases, nets for capturing prey). The larval antennae are much like papillae. The adult has very poorly developed mandibulate mouthparts. The adult mandibles are nonfunctional. The labium is shortened, and the hypopharynx has joined the labium forming a protrusible sucking tube. The galea is reduced and the lacinia is suppressed. Compared to the trichopterid ground plan, the adults are not otherwise highly evolved. They have a fairly generalized wing venation, but have developed the original jugate wing coupling in various ways in different families, such as the loss of the jugum with a simple overlap, or with specialized bristles that hook the wings together. The male genitalia may be quite complicated, but the usual telomeres and basimeres have been retained. The abdomen bears a pair of simple cerci on the fused tenth and eleventh abdominal segments. According to Matusda (1976) the cerci are transformed larval pygopods. Exit from the cocoon is made by the use of pupal mandibles operated by adult mandibular muscles. Three ocelli are present pn the head except in some advanced species.

Order Lepidoptera. The primitive moths of the family Micropterygidae appear to have digressed least from the trichopterid ground plan, so that they have been variously regarded as more closely related to the Trichoptera, or as a separate order, the Zeugloptera. Kristensen (1971, 1975) has shown that they are truly Lepidoptera. The setae on the wings have become transformed into overlapping scales. The metathorax is usually smaller than the mesothorax. The median ocelli have been suppressed, leaving only the lateral ocelli on the adult head. The genital claspers of the male have lost the telomeres, so that the parameres are simple when present. The females lack cerci on the abdomen. The mesothoracic prescutoscutal sulcus is suppressed. The tenth abdominal tergum tends to be bilobed. The apical labial palpal segment bears sensory setae in an invagination. Except for butterflies and a few moth families, a dense patch of hooked aculeae (spinelike setae) forms a "holding area" laterally on the metascutum near the subalare, which intermingle with a similar patch of ac uleae on the underside of the fore wing base. When the wings are folded, these setae apparently couple the hind wing to the thorax (Kuijten, 1974). Kristensen (1971, 1975) listed other lepidopteran derived features of internal anatomy of the adults. These include the following: the presence of a medial posterior process on the corporotentorium; the lack of a tentorial adductor of the mandible; the presence of a separate slender cranio-stipital muscle; the lack of antagonistic muscles in the maxillary palp; the lack of a dorsal muscle of the salivarium; exit of labial nerve close to the frontal ganglion; connectives between abdominal ganglia form a solid cord rather than the usual paired nerve cord; and the recurrent nerve is enclosed by the dorsal aorta.

Lepidopterous larvae generally bear four to seven pairs of abdominal prolegs plus a pair of pygopods. Judging from the condition in the primitive Micropterygidae, the ground plan state of the prolegs was in the form of a protuberance with an apical spine, similar to the prolegs of Mecoptera. These legs became more fleshy and musculated, and developed crochets in the higher Lepidoptera. The larvae are peripneustic in that the metathoracic spiracle is nonfunctional in the higher Lepidoptera, but the primitive Micropterygidae have functional metathoracic larval spiracles, suggesting that the peripneustic condition is convergent with larval Mecoptera.

Not all lepidopterous larvae can spin silk in stages earlier than the last larval instar. Some spin silk trails, others form larval cases, and others produce a web around the leaves on which the colony feeds in various trees. This early spinning ability may have been secondarily lost within the Lepidoptera, perhaps several times, since larval spinning does not appear to be very common except at pupation.

CHAPTER 8

Phylogenetic Classification of Insects

This chapter is a summary of the phylogenetic table presented at the beginning of Chapter 7, with a condensed argumentation from the longer discussion of that chapter and Chapter 6. The derived features (synapomorphies) that characterize the ground plan of each taxon are those that are presumed to have been acquired by the common ancestor of each group. The synapomorphies of each group become the ancestral characters (symplesiomorphies) of the descendant sister groups. It should be noted that the synapomorphies of each taxon are usually further modified in descendant taxa, or may become further modified in different members of each taxon (autapomorphy). The synapomorphies of each group are ground plan characters that became established in each line subsequent to its split from its superordinate sister group, and before the line split further into subordinate sister groups. The plesiomorphies of each group are these characters that had presumably not yet changed to a further derived state shortly after the split, but that may have changed in descendant forms.

The presentation is based on the phylogenetic system of Hennig (1966), in which each taxon is presumed to consist of a monophyletic group, identified by unique shared derived features in each group. The ranks are not determined, as Hennig would do, by the presumed geological age. The categories are ranked in the sequence presumed to have occurred in the evolution of the groups using the highest available superordinal ranks necessary. This method produces some monotypic superordinal taxa. Monotypic taxa are frowned on by purists, but it is felt that they are useful in the argumentative method of Hennig.

Phylogenetic classifications for each order have not been attempted, except for a new treatment of the order Anoplura. The tabular presentation assumes a dichotomous split in each stem line, and corresponds with the cladogram of Figure 53. The descending ranks of categories here employed are class, subclass, infraclass, subterclass, supersection, section, subsection, infrasection, supercohort, cohort, subcohort, infracohort, subtercohort, superorder, and order.

In most instances, the oldest available name is used. Well-known synonyms are indicated. Each group has been given a standard ending as explained in Chapter 7. The geologic distribution of each order is shown, but it is not intended to represent the geological age. It is very likely that the actual age of each order is greater than its oldest fossil form.

Class Insecta Linnaeus, 1758, s.s., = (Hexapoda Latreille, 1825)

Synapomorphies. (Contrasted with Myriapoda). Head of four metameres plus acron; locomotor tagma (thorax) consisting of the fifth, sixth, and seventh metameres and their locomotor appendages; reduction of trochanters from two to one; thoracic pleuron formed by association of pleural anapleurite plus coxal catapleurite; maxilla evolved from appendage of third head metamere, consisting of cardo and stipes (two coxomeres) and palpus (telopodite); maxillary coxal endites transformed into galea and lacinia; maxillary palps evolved from maxillary telopodites; labium evolved by fusion of the two appendages of the fourth metamere, consisting of the postlabium (postmentum, labial cardo) and the prelabium (prementum, labial stipes) and palps (telopodites); labial endites became glossae and paraglossae; reduction of first through ninth abdominal limbs to coxal base plus a two-segmented telopodite (stylus); suppression of tenth abdominal limbs; telopodites (styli) of eleventh abdominal segment converted to segmented cerci; ecdysis involving a dorsal longitudinal split of cuticle; loss of digestive function of midintestinal caeca; spermatozoa with an extra outer ring of axoneme doublet fibers, forming a $9 + 9 + 2$ axoneme pattern in the flagellum; two mitochondrial derivatives extending into the sperm flagellum.

Plesiomorphies. Maximum number of somites 18, plus acron and periproct; acondylar coxa-body joint; intersegmental tendon system primitive; simple rolling mandible with a single articular condyle; compound eyes; three ocelli on head, two lateral, one median; musculated antennomeres; at least 11 pairs of postcephalic spiracles (three thoracic, eight abdominal); retention of intersegmental groove in head (postoccipital sulcus); presence of tergal paranota on thorax and abdomen; unsegmented tarsus; prominent simple pretarsus; no closing apparatus on spiracles; eversible coxal vesicles on first through ninth abdominal segments; anamorphosis; no extraembryonic amnion; meroblastic yolky eggs; postimaginal molting.

1. Subclass Entognathata (Hennig, 1953); name emended from Entognatha.

Synapomorphies. Overgrowth of facial integument over mandibles and maxillae, fusing with edges of labium, resulting in entognathy; posterior prolongation of posterior tentorial arms, supporting the maxillae; protrusible mandibles; reduced or lost compound eyes and ocelli; highly reduced malpighian tubules, gastric caeca, paranotal lobes, and palpi; suppression of postoccipital sulcus.

Plesiomorphies. Individually musculated antennomeres; simple tarsus; obvious pretarsus; absence of amnion and sinking into yolk in embryos; absence of external genitalia; rolling mandible with single articulation; eversible coxal vesicles; transverse tendon system; molting continued in adults; ecdysis of head without median split of cuticle; indirect transfer of spermatophore (no internal copulation); absence of cervical sclerites; abdominal telopodites in form of styli; simple pleuron with anapleurite and catapleurite; three-lobed hypopharynx; cerci; spermatozoa with two mithochondrial derivatives and 9 + 9 + 2 axonemes in flagellum.

A. Infraclass Diplurata (Hennig, 1953); (= Diplura Hennig, 1953).

Order Diplura Börner, 1904; (= Endotropha Grassi, 1889; = Entognatha Imms, 1925). No fossil records. Campodeids, japygids.

Synapomorphies. Union of terminal abdominal segments into definitive tenth segment; two pretarsal claws; biting toothed mandibles; abdominal coxal plates fused with sterna; seven pairs of abdominal spiracles (sometimes no abdominal spiracles); two to four pairs of thoracic spiracles; gonopore of both sexes on abdominal segment 9; oral folds not meeting or close together on underside of head, separated by labium; eyes suppressed; anterior tentorium suppressed; 9 + 2 axonemes in sperm tail of Japygidae.

B. Infraclass Ellipurata (Börner, 1910); emended from Ellipura.

Synapomorphies. Cerci and abdominal spiracles suppressed; antennae reduced or suppressed; postmentum very narrow, oral folds nearly meeting behind prementum; small size, 8 mm or less; gonopore terminal; spermatozoa coiled, immotile.

(A). Subterclass Oligoentomata Berlese, 1909.

Order Collembola Lubbock, 1873. Devonian to Recent. Springtails.

Synapomorphies. Collophore from eversible sacs of first abdominal segment; compound eyes reduced; ocelli vestigial; maxillary palpus one-segmented, lacking muscles; unsclerotized sternum and propleuron; one pair prothoracic spiracles, or no tracheae; neoteny; failure of incorporation of catapleurite (upper part of coxa) into pleuron; extensive tendinous struts anchoring cuticle against blood pressure; tibia and tarsus unseparated; pretarsus with one accessory claw; neotenous abdomen of five somites; third abdominal limbs develop into retinaculum; leaping organ developed from abdominal fourth limbs; blood pressure employed in extending leaping organ for jumping; secondary total cleavage of eggs.

(B). Subterclass Myrientomata Berlese, 1909.

Order Protura Silvestri, 1907. No fossil record.

Synapomorphies. Antennae vestigal (pseudoculi); mandibles solid stylets, protrusible but nonrolling; laciniae styletlike; suctorial feeding; tentorium fused medially; hypopharynx suppressed; eyes suppressed; catapleurite movable against anapleurite; coxa articulated with sternum; spiracles reduced to two pairs on thorax, or none; abdominal styli suppressed except on first abdominal segment; coxal vesicles lost; gonopores double in male, on a protrusible phallic complex; malpighian tubes, caeca, and intestinal muscles suppressed; digestion intracellular; spermatozoa with axonemes secondarily 9 + 0, 12 + 0, or 14 + 0, disc-shaped.

II. Subclass Ectognathata (Hennig, 1953); emended from Ectognatha.

Synapomorphies. Head with clypeus and postocciput; anterior tentorial pits near clypeofrontal sulcus; postoccipital sulcus with posterior tentorial pits; posterior tentorial arms fused together across back of head; prothoracic spiracles suppressed; ovipositor acquired; female gonopore simple on front of eighth abdominal segment (secondarily on rear of eighth or ninth); male gonopore simple on tenth abdominal segment; tergum of eleventh abdominal segment (epiproct) prolonged into a segmented median caudal filament; periproct a distinct structure only in the embryo; cerci with subsegments; median dorsal ecdysial line extended into the head.

A. Infraclass *Archentomata*. New infraclass.

Order Monura Sharov, 1957. Upper Carboniferous, Lower Permian (extinct).

Synapomorphies. Tergum of tenth abdominal segment larger than other terga; caudal filament as long as body; cerci suppressed (?); prothoracic tergum reduced.

Plesiomorphies. Subocular traces of separate terga of mandibular, maxillary, and labial segments (?); compound eyes well developed; trunk segments all similar; thorax distinct only in bearing fully developed legs; maxillary palpus leglike, five-segmented; catapleurite still part of upper coxa, not incorporated into pleuron; tarsus a simple segment; pretarsus distinct, elongated, without claws; abdominal styli two-segmented; ovipositor simple, consisting of enlarged coxal plates on venter of eighth and ninth abdominal segments; paranota with lateral margins covering side of body.

B. Infraclass Euentomata Berlese, 1909.

Synapomorphies. Distinct clypeus, separate from frons by clypeofrontal sulcus; anterior tentorial pits in clypeofrontal sulcus or nearby; dorsal remotor muscle of mandible enlarged; ventral mandibular muscle of mandible enlarged; ventral mandibular muscles partly shifted to originate on anterior tentorium; thorax a distinct tagma; catapleurites implanted into pleuron; pleural apodeme formed in catapleurite; tarsi subsegmented; pretarsus reduced; claws articulating on last tarsomere (not part of pretarsus as in Entognathata); two thoracic and eight abdominal pairs of spiracles; abdominal styli reduced to unsegmented pegs, suppressed on first and tenth abdominal somites; cerci became many-segmented; male gonopore simple, on tenth abdominal sternum, opening through fused phallomeres; female gonopore simple through union of two gonoducts with invaginated eigth abdominal sternum; ovipositor acquired with conversion of eighth and ninth abdominal sterna.

(A). Subterclass Archeognathata (Börner, 1904); (= Monocondylia Hennig, 1953); emended from Archeognatha.

Order Microcoryphia Verhoeff, 1904; (= Machiloidea, Handlirsch, 1903). Triassic to Recent. Bristeltails.

Synapomorphies. Head reduced in size, retracted partly under pronotum; antennae and compound eyes moved close together near midline; maxillary palpus enlarged, subsegmented, appearing eight-segmented; mandible with molar area and slender picklike terminal lobe; leaping developed by (a) short hops using the thoracic legs in phase (as oars) and (b) high leaps aided by strong abdominal push on substrate; styluslike pegs on coxae of second and third legs on some species; abdominal muscles in twisted groups, providing for high leaps by slapping abdomen to ground; anapleurites membranous, on inner faces of paranota.

Plesiomorphies. Ocelli retained; hypopharynx three-lobed; labium typical;

mandible vertical, monocondylic, rolling; intersegmental tendons extensive in head, thorax, and abdomen; anterior tentorial arms separate from each other and from posterior arms; three tarsomeres; femur-tibia joint barely dicondylic; abdominal styli and coxal sacs present; abdominal coxae separate from sterna; genitalia simple; primary mesodermal female gonopore behind seventh abdominal sternum; vagina represented by groove on eighth sternum; molting continued in adult; amnion not yet acquired; lateral cervical sclerites not yet acquired.

(B). Subterclass Dicondylata (Hennig, 1969); emended from Dicondylia.

Synapomorphies. Mandibles with secondary antero-ventral condyle, with axis tilted forward toward horizontal, typically biting (variously modified); dorsal transverse tendons suppressed; ventral mandibular muscles attached to tentorium or suppressed; anterior tentorial arms fused to each other; hypopharynx consolidated into a single lobe; maxillary palpi reduced; femur-tibia joints with two condyles; tarsi pentamerous; origin of use of tip of tibia for weight bearing, tarsi held flat on ground; pleuron variously modified; ovipositor with gonangulum; tracheal system complex with intersegmental and transverse tracheal trunks.

A. Supersection Apterata (Linnaeus, 1758), s.s.; (= Zygentoma Börner, 1904); emended from Aptera.

Order Thysanura Latreille, 1796, s.s.; (= Ectotropha Börner, 1904, s.s.; = Lepismoidea Handlirsch, 1903). Oligocene Baltic Amber to Recent. Silverfish, firebats.

Synapomorphies. Reduction of compound eyes; coxae greatly enlarged, with shift of coxa-body axis toward vertical, and with dorsal concavity receiving femur when in extreme flexion; body somewhat depressed; pleuron with anapleurite, catapleurite and an additional sclerite detached from the catapleurite, the pseudotrochantin; spermatozoa unite in pairs.

Plesiomorphies. Ocelli, pentamerous tarsi, eight pairs abdominal styli, abdominal coxal sacs, and abdominal coxae not totally fused with sterna are primitive features retained by the Lepidothricidae. The apterous state is primitive. Atelurids and nicoletiids still bear a few coxal sacs. The abdominal styli became restricted to three pairs in most and the abdominal pregenital coxae are united with the sterna, and are not individually discernible. Tendinous endosternum was retained ventrally in variously modified ways. The segmented ovipositor valves became shortened and more robust in the Lepidothricidae. The paired phallic lobes joined to form a simple hollow penis on the tenth abdominal segment of the male. There is yet no internal copulation, and sperm transfer is indirect. Molting is continued past sexual adulthood. The lateral cervical sclerites not yet acquired.

B. Supersection Pterygota Brauer, 1885.

Synapomorphies. Wings acquired by modification of paranota; anterior tentorial arms united with posterior arms; pleuron formed by union of pleurites into epimeron and episternum, with elongate pleural sulcus; thoracic and abdominal tendons suppressed, replaced by sternal and pleural apodemes and dorsal intersegmental thoracic phragmata; all abdominal eversible sacs suppressed; epicuticle waterproof; pregenital styli suppressed; blades of ovipositor stiffened, improving penetrating ability of ovipositor; molting of imago suppressed; sinking of embryo into yolk further developed so that embryonic blastoderm encloses embryo completely, forming an amnion and chorion. (?) Establishment of a lateral cervical sclerite on each side of neck.

(A). Section Plagiopterata (Lemche, 1940);(= Orthomyaria Schwanwitsch, 1956); emended from Plagioptera.

Synapomorphies. Prolonged median caudal filament reduced or suppressed; wing beat performed mostly by tergosternal muscles (upstroke) and lateral dorsoventral muscles (downstroke); wing veins tend to curve posteriorly at tips; posterior edges of abdominal and thoracic terga with hollow filamentous outgrowths.

Plesimorphies. Nymphal wing pads extended laterally (modified in Odonata); adult wings incapable of folding over abdomen (modified in some Megasecoptera); wings with well developed archedictyon (reduced in some); wing venation usually full including both anterior and posterior branches of media; legs all alike; fore and hind wings similar; spermatozoa (judging from Odonata) typically insectan with two mitochondrial derivatives and 9 + 9 + 2 axonemes in flagellum.

1. Subsection *Paleodictyopterata.* New subsection.

Synapomorphies. Mouthparts modified into a sucking proboscis; well developed prothoracic and abdominal paranota extended laterally; median caudal filament suppressed.

Plesiomorphies. Cerci long, multiarticulate; both MA and MP wing veins in wings, both primitively branched; tarsi pentamerous; tergal filaments present.

i. Order Paleodictyoptera Goldenberg, 1854, including Protohemiptera Handlirsch, 1908. Carboniferous to Permian (extinct).

Synapomorphies. Prothoracic paranota very well developed, with venation similar to wings.

Plesiomorphies. Abdominal paranota; archedictyon. Typical genera: *Stenodictya, Eugereon.*

ii. Order Megasecoptera Brongniart, 1893; including Protohymenoptera, Archodonata, Diaphanopterodea. Carboniferous to Permian (extinct).

Synapomorphies. Paranota reduced or lost; tergal filaments well developed; branching of wing veins reduced; MA simple; wings basally narrowed; archedictyon reduced or suppressed. Ability to fold wings acquired by Diaphanopterodea (autapomorphy convergent with Neopterata).

2. Subsection Odonatopterata (Martynov, 1938); emended from Odonatoptera.

Synapomorphies. Prothoracic paranota weak; abdominal paranota suppressed; cerci short; paraprocts prominent at least in juveniles; compound eyes large, close together dorsally; MA vein unbranched, basally associated with vein R; MP vein suppressed; CuA vein reduced; archedictyon reduced to numerous cross veins; tergal filaments not elaborate, or reduced; basal wing sclerites fused into a large radio-anal plate.

i. Order Meganisoptera Martynov, 1932; (= Protodonata Brongniart, 1893, in part). Upper Carboniferous to Lower Triassic (extinct).

Synapomorphies. Tarsomeres reduced to four or less; prothoracic paranota reduced; body size very large; vein CuA very short, soon joining CuP; large precosta in wing base; median caudal filament reduced to a two-segmented epiproct. Includes *Meganeura monyi* Brongniart, the largest known insect, wingspread 70 cm.

Plesiomorphies. Wings without pterostigma, arculus, or triangle; base of MA not fused to R; thoracic terga wide as in most insects; abdomen not slender; tergal filaments of abdomen retained, forming a pilose border *(Meganeurula);* cerci segmented.

ii. Order Odonata Fabricius, 1792. Lower Permian to Recent. Dragonflies, damselflies.

Synapomorphies. Antennal flagellum reduced to a tiny bristle; larval labium modified into a prehensile organ; nymphs aquatic; nymphal respiration through posterior tracheal gills; nymphal wing pads reversed, elevated over abdomen with dorsal surfaces against abdomen and costal borders medial; lateral cervical sclerite fragmented into three pieces; adult prothorax very small, without paranota; tergal filaments reduced to small spines in rows on posterior borders of abdominal terga; mesonotum and metanotum reduced; pteropleura strongly tilted backward, with mesepisterna extended ahead of mesonotum, meeting in midline between prothorax and fore wings; vein Sc ending abruptly proximad of middle of wing at a nodus; pterostigma acquired; vein CuA imperceptible; base of MA joined to R, leaving R to form an arculus with cross vein m-cu; strong intercalary vein in Rs field between R3 and R4; trochanter immovably subsegmented; tarsi reduced to three tarsomeres; walking locomotion abandoned in adult, legs prehensile or perching; dorsal longitudinal thoracic muscles reduced or absent; flight through direct wing muscles; strong tergosternals elevate wings, leg and alary muscles depress wings; cerci and epiproct unsegmented; epiproct and cerci of zygopteran larvae modified as gills; cerci used as claspers in males, grasping neck of females; abdomen narrow, slender, with terga nearly meeting across very narrow sterna; male accessory copulatory organs on abdominal segment 2; male gonopore shifted to abdominal sternum 9, covered by highly reduced ninth coxal plates; phallic lobes eliminated; ovipositor when, retained somewhat reduced, but bearing styli on ninth coxal plates.

(B). Section Opisthopterata (Lemche, 1940); (= Chiastomyaria Schwanwitsch, 1956); includes Ephemerida; emended from Opisthoptera.

Synapomorphies. Nymphal wing pads positioned over abdomen with costal border lateral, as in folded position of adult wings; wing beat downstroke largely effected through powerful dorsal longitudinal muscles arching the pteroterga, lifting the wing base (basic pattern modified in Blattiformida); dorsal longitudinal muscles attached to large internal intersegmental apodemes (phragmata); upstroke effected by powerful tergosternal and tergopleural muscles lowering the terga and wing bases; internal copulation with direct transfer sperm from male gonopore to female gonopore.

1. Subsection Ephemerata (Handlirsch, 1908); emended from Ephemeroidea.

Synapomorphies. Enlarged styli of abdominal segment 9 used in clasping of female during mating; larvae aquatic; ovipositor lost; rich intercalary venation in wing; spermatozoa with only one mitochondrial derivative and with a 9 + 9 + 0 axoneme pattern.

Plesiomorphies. Both MA and MP veins retained; neala not developed; unable to fold wings at rest over abdomen as in neopterous forms, but wings held over body with upper surfaces appressed; long segmented cerci and median caudal filament; trochantin not developed; abdominal styli 9 prominent in male.

i. Order Protephemerida Handlirsch, 1908. Carboniferous (extinct). *Triplosoba pulchella* Brongniart.

Not sufficiently known. Features mostly primitive: fore and hind wings similar in size and venation; MA not associated with R; MA simple.

ii. Order Ephemerida Leach, 1817; (= Plectoptera Packard, 1886; = Ephemeroptera Shipley, 1904); includes Permoplectoptera Tillyard, 1932. Upper Carboniferous Syntonopterodea Laurentiaux, 1953 to Recent. Mayflies.

Synapomorphies. Fore wings larger than hind wings in recent spp.; MA branched and associated with Rs vein; wings of last nymphal instar (sub-imago) well developed and functional in flight; antennal flagellum reduced (11- segmented in Permoplectoptera); genitalia secondarily reverted to "double gonopore" state by suppression of invaginated median sternal gonoduct; ovipositor suppressed; employment by male of abdominal coxal styli 9 as genital claspers; adult mouthparts suppressed; aquatic larvae with paired lateral abdominal gills; anterior tentorial pit not in clypeofrontal sulcus as in other pterygotes; males with elongated clasping fore legs; imaginal life very short, gonads mature during nymphal stage.

Plesiomorphies. The only surviving pterygote order with a long segmented median caudal filament. Persistent MA vein; fluted wings; inability to fold wings; cerci long and segmented; pentamerous tarsi; simple lateral cervical sclerites.

2. Subsection Neopterata (Martynov, 1925); emended from Neoptera.

Synapomorphies. Wings at rest folded over abdomen with costal border lateral, through action of muscle of third axillary sclerite; fore wings fold along a basal-jugal fold; hind wings fold along a vannal fold (modified in En-dopterygotida); jugum (neala) developed, or secondarily reduced; MA vein suppressed, MP retained; first anal vein (= empusal) basally detached from other anal veins; phallic lobes of male form phallomeres on tenth abdominal segment; median caudal filament suppressed (convergent with Entognathata and Plagiopterata); development of trochantinal arc from episternum, forming a secondary ventral coxal articulation. Ancestral pentamerous tarsi convergently modified to fewer tarsomeres in various orders.

(1). Infrasection Polyneopterata (Martynov, 1925); emended from Polyneoptera.

Synapomorphies. Hind wings typically larger than fore wings because of enlarged anal area, folding along vannal fold and jugal fold; male intromittent organ by eversion of ductus ejaculatorius; ovipositor either suppressed, or third valvulae become part of penetrating system.

Plesiomorphies. Mouthparts typically biting; lacina articulated with maxilla; labial palpus with three or more segments; metamorphosis generally gradual, without pupa stage, in that the alatoid juveniles bear adultoid cerci, ocelli, compound eyes, antennae, wing histoblasts, for several instars before the final molt, organs that continue development without histolysis (contrast with Phalloneopterata); nervous system generally not concentrated forward; malpighian tubules usually numerous; ovipositor originally well developed, but may be reduced; juvenile male phallomeres never developed into claspers (parameres) and aedeagus (mesomeres) (contrast with Phalloneopterata); spermatozoa typical of insects, with exceptions in Phasmida and Isoptera.

1. Supercohort *Plecopterida* (= Panplecoptera Crampton, 1917, excluding Dermaptera). New supercohort.

Synapomorphies. Phallomeres highly reduced or absent; male intromittent organ from everted median gonoduct; accessory male clasping organs from various sources; M and Rs reduced to two or three neutral branches; abdominal styli suppressed.

Plesiomorphies. Trochantins still attached to epimera without sulcus between (contrast with detached trochantins of Orthopterodida);fore wings not sclerotized as in Orthoptera; spermatozoa with two mitochondrial derivatives and 9 + 9 + 2 axonemes; lateral cervical sclerites primitively simple.

(1). Cohort Perlodida (Handlirsch, 1908); emended from Perloidea.

Synapomorphies. Juveniles aquatic with lateral tracheal gills in tufts; vannus of hind wing very well developed; veins Rs and M neutral (not plus or minus).

Plesiomorphies. Long multiarticulate cerci; male genitalia symmetrical (may include variously developed paraprocts and epiproct).

i. Order Paraplecoptera Martynov, 1938; (= Protoperlaria Tillyard, 1926); includes Miomoptera Martynov, 1927. Carboniferous to Triassic (extinct).

Synapomorphies. Tarsomeres reduced to four; nymphal tracheal gills numerous (? = paranota of abdomen?).

Plesiomorphies. Paranota on prothorax; wings similar to those of Plecoptera; ovipositor well developed in female. *Lemmatophora typa* Sellards, *Protembia permiana* Tillyard.

ii. Order Plecoptera Burmeister, 1839. Permian to Recent. Stoneflies.

Synapomorphies. No paranota (? except *Peltoperla*, extinct); ovipositor suppressed; tarsomeres reduced to three; adult mouthparts highly reduced; nymphal gills suppressed posteriorly; testes united; ejaculatory duct expanded and eversible, functioning as a penis.

(2) Cohort Embiodida (Handlirsch, 1908); emended from Embioidea.

Order Embiidina Enderlein, 1903; (= Embioptera Shipley, 1904). Oligocene Baltic Amber to Recent (from Permian, *Sheima sojanesis* Martynova?). Embiids.

Synapomorphies. Females wingless; neala, vannus, and venation reduced; cerci short, one- or two-segmented; silk gland in fore tarsi; male genitalia mostly asymmetrical (some spp. not); gular bridge; ocelli suppressed; left cercus aids in copulation; lateral cervical sclerites of two pieces.

2. Supercohort Orthopterodida (Handlirsch, 1908); (= Panorthoptera Crampton, 1917; = Pliconeoptera Hamilton, 1971); emended from Orthopteroidea.

Synapomorphies. Trochantin separated from episternum by a sulcus or a membrane; fore wings somewhat sclerotized; phallomeres may split each into two lobes, but median lobes (mesomeres) do not fuse to form sclerotized aedeagus (some fusion in Dermaptera), nor do lateral lobes (parameres) become functional claspers; in addition to vannal and jugal folds, vannus of hind wings forms pleats when folded; addition of lateral fragments of prothoracic presternum to lateral cervical sclerites, forming two cervical sclerites on each side of neck.

Plesiomorphies. Persistent abdominal styli on ninth abdominal segment of males (on eighth and ninth in some female nymphal cockroaches, suppressed altogether in Dermaptera, Zoraptera, acridids, and phasmids); spermatozoa typical with two mitochondrial derivatives and 9 + 9 + 2 axonemes.

(1). Cohort *Grylliformida.* New cohort.

Synapomorphies. Ovipositor modified, third valvulae (= sheath of others) functional in oviposition, second valvulae reduced; cerci relatively short with few segments; prominent precostal field.

Plesiomorphies. In contrast to Blattiformida, male genitalia symmetrical; wing motion through usual indirect musculature.

a. Subcohort *Protorthopterida.* New subcohort.

i. Order Protorthoptera Handlirsch, 1908. Carboniferous to Permian (extinct).

An assembly of various forms, some cursorial, some seemingly saltatorial; bearing prothoracic paranota; exserted orthopteroid ovipositor; short cerci (segmented, or unsegmented and cheliform in *Chelopterum perigrinum* Carpenter); four or five tarsomeres; fore wings weakly if at all sclerotized; hind wing with obvious neala and large vannus.

ii. Orders Glosselytrodea Martynov, 1938, Permian to Lower Jurassic, and Caloneurodea Martynov, 1938, Upper Carboniferous to Permian. Several families of uncertain affinities, provisionally regarded as highly evolved protorthopteroid insects. Imperfectly known.

b. Subcohort *Orthopterida*. New subcohort.

Synapomorphies. Fore wing somewhat sclerotized, not generally used in flying; cerci simple or two-segmented; lack of well developed phallomeres.
Plesiomorphies. Large pleated vannus, in contrast to some Protorthopterida..

i. Order Phasmida Leach, 1815; (= Cheleutoptera Crampton, 1915). Lower Triassic to Recent. Walkingsticks.

Synapomorphies. Ovipositor weak; eggs laid free, individually; thorax and legs elongate. Some forms with reacquired abdominal paranotumlike structures and flattened legs; fore wings shortened, nonfunctional; styli suppressed in males; spermatozoa without mitochondrial derivatives.

ii. Order Orthoptera Oliver, 1789; s.s.; (= Saltatoria Latrielle, 1817). Upper Carboniferous (Oedischiidae) to Recent. Grasshoppers, katydids, crickets.

Synapomorphies. Paranota of pronotum expanded laterally and turned downward to cover propleuron, and pronotum extended posteriorly over mesonotum; hind legs adapted to leaping (leaping ability suppressed in mole crickets); ovipositor well developed, inserting eggs into substrate (piercing ability of some tettignoiids lost, eggs laid on surfaces).

(2). Cohort Blattiformida (Handlirsch, 1908); emended from Blattaeformia.

Synapomorphies. Pronotum disclike; postnotum reduced; loss of action of flexible meso- and metanota in indirect wing motion; loss of action of tergosternal thoracic muscles for indirect wing elevation and dorsal longitudinal muscles for indirect wing depression; reduction of thoracic phragmata; forward shift of wing fulcrum to fore edge of thoracic nota; extreme forward slant of pleura; wing elevation through depression of wing base by tergopleural and tergocoxal muscles; wing depression through pull of basalar-coxal and subalar-coxal muscles lateral to wing fulcrum.

a. Subcohort Dermapterida (Martynov, 1925); emended from Dermapteroidea.

Synapomorphies. Fore wings reduced and sclerotized; hind wing with main veins reduced and crowded near costal border; anal fan of hind wing enlarged, and folded double in repose; styli suppressed in males.

i. Order Protelytroptera Tillyard, 1931; (= Prototcoleoptera Tillyard, 1924). Permian (extinct).

Synapomorphies. Not clearly different from Dermaptera, essentially suggesting early stages of dermapteran evolution. Fore wing sclerotized, with evident veins; hind wing with very large anal field; cerci short.

Plesiomorphies. Tarsi pentamerous; cerci segmented; ovipositor present.

ii. Order Dermaptera Leach, 1815. Jurassic to Recent. Earwigs.

Synapomorphies. Cerci large, unsegmented, forcepslike (Miocene and younger forms); *Protodiplatys fortis* Martynov, Jurassic, with segmented cerci, pentamerous tarsi; fore wings when present without venation, short, not useful in flight; flight muscles as in other Blattiformida; ovipositor suppressed or vestigal; first abdominal sternum united with methathoracic sternum; trochantin partially reunited with epimeron leaving a partial suture at union; median mesomeres of males partly or completely fused, traversed by forked extension of ejaculatory duct resulting in secondarily double functional gonopores; parameres weak, nonclasping; Miocene and younger forms with three tarsomeres.

b. Subcohort Dictyopterida (Leach, 1815); expanded and emended from Dictuoptera.

Synapomorphies. Fore wing with CuP simple and forming a strong arch; polyneurous dictyon; veins R, M, and CuA of fore wing originally highly branched, forming greater part of remigium; CuA with numerous posteriorly directed branches to hind edge of fore wing; radio-costal and subcosto-costal cross veins appearing as slanting pectinate branches of Sc and R; coxae conical, elongate and backwardly directed.

(a). Infracohort *Protoblattarida.* New infracohort.

Synapomorphies. Known mostly from remains of fore wings only, with features not much different from those of the subcohort; radial sector of fore wing with many branches.

Plesiomorphies. Head of nymph (*Protoblatiniella* Meunier) not covered by pronotum; nymph with prominent ovipositor; fore wings not sclerotized; cerci segmented.

i. Order Protoblattaria (Handlirsch, 1908), s.s; restricted to Blattinopsidae; emended from Protoblattoidea. Upper Carboniferous to Permian (extinct).

Synapomorphies. Fore wing with a "tranverse line" across main veins.

ii. Order Grylloblattaria (Brues and Melander, 1932); emended from Grylloblattodea. Recent. No fossil record.

Synapomorphies. Wings completely suppressed; pronotal disc reduced; ocelli suppressed; eversible vesicle on first abdominal sternum; asymmetrical male genitalia.

Plesiomorphies. Ovipositor prominent, bearing styli on third valvulae of nymphs; abdominal sternum 8 not prolonged nor forming a subgenital plate; eggs not laid in oothecae; ninth abdominal coxal plates prominent in males, separate from sternum 9, and bearing styli; first abdominal sternum not reduced; malpighian tubules not numerous.

(b). Infracohort *Cursorida*. New infracohort.

Synapomorphies. Veins Sc and R_1 with numerous anteriorly directed pectinate branches to front edge of fore wing; Rs forwardly pectinate or reduced to few branches; remigium of fore wing consisting of M and CuA (ground plan of fore wing modified in Zoraptera, Isoptera, and Mantodea); backwardly directed conical hind coxae hiding the very small first abdominal sternum.

a. Subtercohort *Zorapterida*. New subtercohort.

Order Zoraptera Silvestri, 1913. No fossil record.

Synapomorphies. Small body size (up to 3 mm); simplification of wing venation; loss of anal field; small hind wings; antennae short, moniliform; cerci simple, short; ovipositor suppressed; subgenital plate from abdominal sternum eight; loss of styli in males; phallic lobes highly reduced; neoteny (some wingless, poorly sclerotized adults with no ocelli); wings broken off after mating; malpighian tubules reduced to six; first through sixth abdominal ganglia fused with metathoracic ganglia; semicolonial in rotten wood; loss of ability to withstand dry atmosphere; tarsi two-segmented.

Plesiomorphies. Three ocelli; blattiformid wing motor mechanism with highly reduced thoracic dorsal longitudinal and tergosternal muscles, use of coxal and tergopleural muscles for "direct" wing motion; wing fulcrum on fore edge of tergum; undeveloped postnotum; orthopteroid mouthparts; detached trochantin; lack of phallomeric aedeagus.

b. Subtercohort *Blattarida*. New subtercohort.

Synapomorphies. Union of anterior tentorial arms ahead of circumesophageal nerves, forming a foramen; reduction of abdominal sternum 1 to tiny sclerite, hidden by hind coxae; abdominal sternum 7 enlarged to form subgenital plate; female gonopore secondarily between seventh and eighth sterna through desclerotization of abdominal sternum 8 and 9 and formation of genital chamber; conical proventriculus with a ring of sclerotized teeth.

Plesiomorphies. Retention of styli on abdominal 9 of males; ovipositor in ground plan, later reduced in each line; pectinate Sc, R, and CuA; reduced radial sector in fore wing; segmented cerci.

(a). Superorder *Isopterodea*. New superorder.

Order Isoptera Brulle, 1832. Eocene to Recent Termites.

Synapomorphies. Antennae short, moniliform; median ocellus suppressed; anal field of fore wing reduced (*Mastotermes*) or suppressed; jugum of fore wing suppressed; autotomy of wings at special fracture line, with basal sclerotization; subcosta suppressed in wings (retained with three "branches" in *Mastotermes*); fore wing with no anal veins, and mostly with a simple R vein; ovipositor highly reduced (*Mastotermes*) or suppressed; male genitalia highly reduced; two to eight malpighian tubules; first two abdominal ganglia fused with metathoracic ganglia; thoracic dorsal longitudinal muscles suppressed; development of social life, caste system, trophallaxis (resulting in neoteny); specialized wood eaters; with symbiotic intestinal organisms for wood digestion; long-lived physogastric queens.

Plesiomorphies. Pentamerous tarsi (*Mastotermes*), but reduced to four tarsomeres in others; pectinations of R vein fore wing (*Mastotermes*) but reduced to simple R in most other termites; anal field present in *Mastotermes* hind wing, but no anal veins in other termites; pronotal disc not enlarged; male styli; cerci multisegmented to simple; no segmental arteries; spermatozoa typical, but some species have nonmotile sperm.

(b). Superorder Blattodea Brunner, 1882; (= Cursoria Westwood, 1839; = Oothecaria Verhoeff, 1903; = Blattiformia Werner, 1906).

Synapomorphies. Highly mobile head on small neck; pronotal disc expanded laterally; M of fore wing with a few branches; fore wing sclerotized; fore wing with large membranous jugum; vein 1A of fore wing ending in groove of

CuP, and other anal veins often ending against CuP; male genitalia asymmetric, with well developed phallomeres; eggs laid in exposed proteinaceous ootheca sclerotized after deposition of eggs into it; ootheca containing organic calcium salt; malpighian tubules numerous; segmental arteries in abdomen and thorax; antennae long, filiform; flight power weak.

i. Order Mantodea Burmeister, 1838. Triassic to Recent. Mantids.

Synapomorphies. Prothorax and procoxae elongated; fore legs raptorial; carnivorous insects; compound eyes enlarged; ovipositor reduced; ootheca hardened after depositon; calcium citrate in ootheca; mandibles strongly toothed; Rs of fore wing reduced; branching from R distally; segmental arteries lost in thorax; lateral cervical sclerites further fragmented to four pieces each.

Plesiomorphies. Pectinate Sc and R retained in some (Choeradodidae), but mostly the Sc extends to tip of wing without branches, and R lacks pectinations in most forms; anal veins mostly terminate in posterior border of fore wing, but in some the first A terminates in CuP before reaching border; pronotal disc never covering head as in cockroaches, but pronotum with weak to prominent lateral expansions (*Choeradodis*, Deroplatyidae and Orthoderidae); head with three ocelli; segmental arteries retained in abdomen.

ii. Order Blattaria Burmeister, 1829. Upper Carboniferous to Recent. Cockroaches.

Synapomorphies. Body depressed; pronotum large with laterally expanded sides, and protruding forward over the hypognathous head; thoracic sterna and terga poorly sclerotized; legs adapted to rapid running; median ocellus suppressed; lateral ocelli poorly developed; suppression of clypeofrontal sulcus; ootheca sclerotized before deposition while in genital chamber; calcium oxalate in ootheca; development of proventricular "gizzard"; extreme forward slant of pleura; segmental arteries of abdomen suppressed.

Plesiomorphies. Highly pectinate Sc and R in fore wing (Blaberidae), but Sc reduced to single vein ending before midwing in most species; female nymphs bearing styli on eighth and ninth abdominal segments, on developing ovipositor; ovipositor well developed in ground plan (Permian *Kunguroblattina microdictya* and others), but reduced to vestiges in modern forms.

(2). Infrasection *Phalloneopterata.* (= Planoneoptera Hamilton, 1971, *s.s.*, excluding Plecoptera and Zoraptera). New infrasection.

Synapomorphies. Complete suppression of coxal plates and styli of ninth male abdominal segment; new male genitalia evolved from phallomeres of tenth abdominal segment: paired phallomeres divide into mesomeres and parameres, mesomeres fuse to form a sclerotized functional aedeagus, parameres assume clasping function (or secondarily suppressed); histogenesis of imaginal structures variously delayed into last preimaginal instar, such as ocelli, wing buds, abdominal cerci, rudiments of external genitalia; thoracic sternal apodemes moved closer together; jugum (neala) generally reduced.

Plesiomorphies. Trochantin not usually detached or separated basally from epimeron, usually elongate if present (in contrast to detached condition of trochantin in Orthopterodida); ovipositor with piercing blades consisting of the first and second valvulae; third valvulae employed as sheath for ovipositor blades; spermatozoa originally typical, variously modified (Acercarida, Mecopterida); lateral cervical sclerites originally simple, retained as such or modified.

1. **Supercohort Acercarida** (Börner, 1904); (= Hemipteroidea Handlirsch, 1908; = Paraneoptera Martynov, 1925, excluding Zoraptera); emended from Acercaria.

Synapomorphies. All stages lack abdominal cerci; lacinia elongated and detached from the maxilla, with slender chisel-shaped styliform or toothed tip; labial palpus reduced to two or less segments or suppressed; clypeus enlarged housing large cibrarial muscles; abdominal sternum 1 reduced; nervous system strongly concentrated in thorax; not more than four malpighian tubules; not more than three tarsomeres (four in extinct psocids): generally anteromotoral in modern forms; spermatozoa basically biflagellate or with doubled axial filament complexes in one flagellum.

Plesiomorphies. Trochantin generally retained, narrow and not basally separate from epimeron, and articulating terminally with coxa (as in Plecopterida); anal fan not pleated when wing is folded.

(1). **Cohort Psocodida** (Weber, 1933); (= Psocopteroidea Jeannel, 1945); emended from Psocoidea.

Synapomorphies. Lacina detached from maxilla, serving as a "pick," and protrusible and retractible; ovipositor reduced; esophageal (cibarial) sclerite; ovoid sclerites on hypopharynx; no cardo; aedegus not formed by fusion of mesomeres; parameres and mesomeres both clasping.

Plesiomorphies. Maxilla generally retaining galea, mandibles biting, dicondylic, or suppressed.

a. Subcohort *Psocida.* New subcohort.

Order Corrodentia Burmeister, 1839, *s.s.;* (= Psocoptera Shipley, 1904). Lower Permian to Recent. Bark lice, book lice, psocids.

Synapomorphies. Antennal flagellum slender, setiform; Media of fore wing three-branched; Rs forked; wings coupled at rest by blunt projection of stigma of fore wing; Cu and MP united at base; metathorax small; prophragma and mesophragma suppressed or reduced; dorsal longitudinal muscles insert on strongly arched mesonotum; spermatozoa biflagellate and uniflagellate.

Plesiomorphies. Ocelli retained; maxillary palpi retained; labial palpus one or two-segmented; (Permian Delopteridae with short cerci. Considered by some to be related to Embiidina); ovipositor short; lateral cervical sclerites simple.

b. Subcohort Phthriapterida (Weber, 1939); emended from Phthiriaptera. Parasitic lice.

Synapomorphies. Head flattened dorsoventrally and prognathous; antennae with three or fewer flagellomeres; ocelli suppressed; compound eyes reduced to two or fewer ommatidia; labial palpi reduced to one segment or suppressed; maxillae highly reduced; lacinial stylets small, internal, connected to lacinial gland; supraesophageal ganglion (brain) moved backwards; dorsal tentorial arms eliminated; trochantin absent; wings suppressed; one pair thoracic and usually six pairs abdominal spiracles; ovipositor eliminated; testicular follicles reduced to three or fewer per testis; vertebrate ectoparasites with eggs laid singly, glued to hair or feathers of host; life cycle reduced to three nymphal stages; spermatozoa with two axial filament complexes in one flagellum.

i. Order Mallophaga Nitzsch, 1818, s.s.; restricted to Amblycera Kellog, 1896. Unknown as fossils. Biting lice, in part.

Synapomorphies. Antennae four or five segmented, with the last two flagellomeres somewhat swollen; first flagellomere (third segment) pedunculate; antennae hidden in a groove under the head; mesonotum generally smaller than metanotum; head prognathous but mandibles articulated about a horizontal axis.

Plesiomorphies. Compound eyes with two ommatidia; maxillary palpus four-segmented; three testicular follices per testis; pretarsi generally two-clawed; lacinial stylet sclerotized; no obturaculum or occipital apodeme as in Anoplura; lateral cervical sclerites simple.

ii. Order Anoplura Leach, 1815, s.l.; including Isochnocera, Rhynchophthirina, and Siphunculata;(= "Group A," Konigsman, 1960). Oldest fossil an ischnoceran in Pleistocene ice on a frozen ground squirrel.

Synapomorphies. Antennae with saucerlike sensoria on last flagellomeres; maxillary palpi suppressed; only one ommatidium on eye (or eye suppressed); lacinial stylets desclerotized or absent; occiput of head with occipital apodeme protruding into thorax; connective tissue septum (obturaculum) nearly occluding the occipital foramen; mandible (if not suppressed) rotated to operate about a vertical axis; anterior tentorial arms small, not reaching posterior tentorium, or absent; testicular follicles reduced to two per testis.

Plesiomorphies.   Antennae exposed; flagellomeres filiform; lateral cervical sclerites simple.

Suborder Ischnocera Kellogg, 1896. Biting lice, in part.

Synapomorphies.   Mandibular axis rotated to the vertical; laciniae weak and desclerotized; anterior tentorial arms vestigal.

Plesiomorphies.   Essentially those of the ground plan for Phthiriapterida.

Suborder Lipognatha Börner, 1904.

Synapomorphies.   Suppression of anterior tentorial arms; suppression of lacinia; mandibles modified or suppressed; development of large cibarial pump plus pharyngeal pump; lateral cervical sclerites suppressed; pretarsal claw single; blood sucking ectoparasites of mammals.

Infraorder Rhynchophthirina Ferris, 1931. Includes one species. *Haematomyzus elephantis* Piaget, the elephant louse.

Synapomorphies.   Prolongation of head into a snoutlike structure, with mandibles terminal and mandibular teeth turned outwards; mandibles lying outside the "mouth"; pretarsal apodeme extending into femur without tibial muscle bundle; suppressed tibiotarsal muscles; posterior tentorial arms very short, not meeting.

Infraorder Siphunculata Latreille, 1825; (= Anoplura Auct. s.s.). Sucking lice.

Synapomorphies.   New piercing stylets developed from hypopharynx and labium; labial sac acquired, enclosing stylets in repose; mandibles vestigal or suppressed; mesothoracic and metathoracic terga reduced, narrow longitudinally, with pleura extending dorsally; thoracic spiracle on dorsal part of pleuron; tibia modified for grasping hair of host.

(2).   Cohort Condylognathida (Börner, 1904); emended from Condylognatha.

Synapomorphies.   Lacinia slender, pointed, styletlike, articulated basally with a barlike structure to the remnant of the maxilla; maxillary galea suppressed; mandibles slender, styletlike; stylets ensheathed in protruding labral-labial complex (rostrum); union of CuP and 1A to form upper boundary of "clavus" in fore wing.

a. Subcohort Thysanopterida (Weber, 1933); emended from Thysanopteroidea.

Order Thysanoptera Haliday, 1836. Permian to Recent. Thrips.

Synapomorphies. Right mandible suppressed; beak (labrum-labium) a broad cone; mandible and lacinal stylets ensheathed by labrum; preimaginal instar an inactive, nonfeeding "pupa" with histolysis and reconstruction of intestine; wings straplike with edge fringe of long setae; clavus vestigal; abdominal spiracles reduced to two pairs (segs. 1 and 8); two-segmented tarsus; claws suppressed; no caeca; abdominal ganglia concentrated in abdominal segment 1; spermatozoa with two 9 + 2 complexes, or 18 + 4 disorganized double complex.

Plesiomorphies. Labial and maxillary palpi present; ocelli present.

b. Subcohort Hemipterida (Linnaeus, 1758) s.s.; emended from Hemiptera, excluding Thysanoptera.

Synapomorphies. Mandibles and laciniae slender stylets, enclosed in elongated three to four segmented sheathlike labium; piercing-sucking mouthparts; labial and maxillary palpi suppressed; mesothorax more powerful in flight than metathorax; wings generally coupled together in flight (anteromotorial); radial sector simple or two-branched; abdominal ganglia concentrated in thorax.

i. Order Homoptera Leach, 1815. Upper Carboniferous to Recent. Cicadas, plant hoppers, aphids, scales, whiteflies, and so on.

Synapomorphies. Fore wing generally larger than hind wing; (vannus secondarily enlarged in Fulgoridae); two or three tarsomeres; folding of hind wing across or behind 2A; spermatozoan filament complex reduced to a single 9 + 9 + 2 complex, some with a branched flagellum.

Plesiomorphies. (In contrast to Heteroptera). Beak attached to posterior part of head; no gula between labium and neck; wings folded rooflike over abdomen (except some aphids, cicadellids, and peloridiids with wings flat and often overlapping over abdomen); posterior tentorial bridge retained; ovipositor common, but sometimes suppressed (e.g. aphids); no scent glands; some with one caecum, others no caeca; pronotum not covering propleuron.

ii. Order Heteroptera Latreille, 1810. Permian (*Paraknightia magnifica* Evans) to Recent. Bugs.

Synapomorphies. Beak attached forward on head, with distinct gular area sclerotized between labium and neck; lateral cervical sclerites suppressed; wings folded flat and overlapping over abdomen; fore wings typically sclerotized basally (corium), membranous at tip, smaller than hind wings; posterior tentorial bridge suppressed; pronotum enlarged, the sides crowding out the propleura; third valvulae of ovipositor reduced, or ovipositor suppressed; no caeca in midintestine except near the posterior midintestine in some; the double flagella of the spermatozoa of unequal length, sometimes only one flagellum with two axoneme complexes.

2. Supercohort Endopterygotida (Sharp, 1898); (= Oligoneoptera Martynov, 1925; = Holometabola Auct.); emended from Endopterygota.

Synapomorphies. Pupating insects with extreme late rapid histogenesis of histoblasts becoming imaginal organs during the pupal stadium, such as, larval invaginated wing histoblasts everted at pupation, development of compound eyes, ocelli, antennae, many muscles, pentamerous adult tarsi, genitalia, empodia, and so on; histolysis of larval organs, such as "prolegs," muscles, some mouthparts, alimentary canal, and the like; many larval organs different from similar imaginal organs, such as larval eyes generally not compound and not continuing into pupal stage, tarsus and pretarsus simple; cerci absent on larvae or larval cerci modifid as anal legs; larvae generally vermiform; anal field of hind wing highly reduced, jugal vein simple.

Plesiomorphies. Larval and adult cerci usually present, but reduced or modified; tarsi pentamerous; ground plan mouthparts biting; ocelli in ground plan.

(1). Cohort Coleopterida (Handlirsch, 1908); emended from Coleopteroidea.

Synapomorphies. Fore wings not useful as flight organs; flight through motion of hind wings; metathorax enlarged serving the hind wings; hind wings with simplified venation; venation of fore wings highly reduced; dorsum of abdomen less sclerotized than venter; ovipositor suppressed; ocelli reduced to two or fewer.

Plesiomorphies. Parameres (clapers of male) simple, not two-segmented as in Telomerida; spermatozoa typical, with two mitochondrial derivatives and a 9 + 9 + 2 axoneme pattern; regression of embryonic limb buds of abdomen before hatching (except cerci).

i. Order Coleoptera Linnaeus, 1758, s.s.; excludes Dermaptera, Blattaria, and Orthoptera. Permian to Recent. Beetles.

Synapomorphies. Elytra (fore wings) sclerotized, with longitudinal ridges and pits, venation unrecognizable; elytra folded covering hind wings, with anal borders contiguous; elytra held motionless and outstretched in flight; hind wings folded rolled or double folded above abdomen under elytra; posterior tentorial pits elongated ventrally and forward, enclosing a gula posterior to labium; gula formed from ventral cervical sclerite, postgena and base of labium; lateral cervical sclerites either lost (Adephaga) or fragmented into two pieces (Polyphaga); male genitalia retracted and hidden when not in use; abdominal segments 8 and 9 telescoped; first and second abdominal sterna membranous, not visible externally; gastric caeca reduced to six or less; one, two, or no ocelli; trochantin merged with pleuron; pleuron heavily sclerotized; mesothorax immovably fused with metathorax; trace of nonfunctional ovipositor (third valvulae) remaining.

Plesiomorphies. Tarsi pentamerous, but reduced variously; mouthparts typically biting; pupa free.

ii. Order Strepsiptera Kirby, 1813. Oligocene Baltic Amber to Recent. Stylopids, twisted-winged parasites.

Synapomorphies. Endoparasitic in insects; males only winged; fore wings reduced to membranous sacs serving as balancers in flight; veins of fore wing highly reduced but persistent; hind wings without vannus or anal veins, or vannus reduced to alulae; branches of veins independent, venation radial; hind wings folded longitudinally, vertical and lateral to abdomen; antennae seven-segmented with a flabellum on the third (first flagellomere), flabella often also on the fourth, fifth, and sixth; compound eyes of males berrylike, with bristles between the facets; free-living females with reduced compound eyes; vestiges of ocelli only in male pupae; lateral cervical sclerites lost; postocciput suppressed; tentorium vestigal; labium without palpi; maxillae reduced to two-segmented palpi; adult mandibles used only for ecdysis; prothorax of males reduced to a small ring; fore and middle legs of males without trochanters; all legs without trochanters in females and larvae; hind coxae nonmovably joined to metapleuron in males; middle and hind legs of males far apart; metathorax greatly elongated with pleural sulcus almost horizontal; hind wing without postnotal articulation; abdominal segment 1 exposed laterally bearing large spiracles and overgrown dorsally and ventrally by metathorax; abdominal segment 8 without spiracles in both sexes; male genitalia consisting of an exposed aedeagus without parameres; ovipositor completely suppressed; nervous system condensed into three masses, the brain, the thoracic mass including the subesophageal and some abdominal ganglia, and an abdominal mass; malpighian tubules highly reduced; midintestine not opening into hind intestine; male genital tract shortened, plump, without accessory glands; ejaculatory duct formed into a sperm pump; ovaries lost, the eggs developing in the female hemocoel; viviparous females; gonopore of female opens into a blind invagination (reduced median oviduct of primitive species) at rear border of abdominal segment 7, modified into a birth canal (suppressed in higher forms, which acquired dorsal segmental birth canals); extreme neoteny in females; partial neoteony in males; pupation in a puparium; alimentary canal not used in feeding; copulation through female epidermis.

Plesiomorphies. (Contrasted with Coleoptera). Freely articulated thoracic segments; pentamerous tarsi (variously reduced in higher forms); male genitalia exposed; absence of gula; fore and hind wings beat in unison; presence of tegula; primary larvae with 11-segmented abdomen bearing cerci; larval compound eyes.

2. Cohort *Telomerida.* New Cohort.

Synapomorphies. Male genitalia with parameres of tenth abdominal segment jointed (basimere plus telomere secondarily simplified in some families of various orders); derepression of abdominal limb buds, variously secondarily suppressed; Gastric caeca suppressed; silk from labial glands.

Plesiomorphies. Pentamerous tarsi; well developed ovipositor; simple lateral cervical sclerites.

a. Subcohort Hymenopterida (Handlirsch, 1908); emended from Hymenopteroidea.

Order Hymenoptera Linnaeus, 1758. Triassic to Recent. Sawflies. wasps, bees, ants.

Synapomorphies. Fore wings are the principal flight organs; small hind wings coupled to fore wings by hamuli; wing venation modified, with enlarged cross veins obscuring the primitive venation; metathorax reduced; lateral cervical sclerites suppressed; base of femur with a trochantellus, lost in higher forms; parameres and aedeagus arise from a solid phallobase; parameres with median accessory claspers (volsellae); males haploid; haplo-diploid parthenogenesis; female accessory glands producing venom.

Plesiomorphies. Ovipositor generalized, but modified in some (e.g., sting); absence of meron; abdominal prolegs (lost in higher forms); cerci in adult; mouthparts mandibulate; spermatozoa typically insectan; malpighian tubules generally not highly reduced.

b. Subcohort *Meronida*. (= Petanoptera Brauer, 1885, s.s.). New subcohort.

Synapomorphies. Origin of coxal meron (not same as orthopteroid "meron"); thoracic sterna deeply invaginated, furcal pits adjacent to each other; ovipositor highly reduced or suppressed, with third valvulae vestiges at best; malpighian tubules less than 10.

Plesiomorphies. Hind wings not subordinate to fore wings in flight (except in Haustellodea); lateral cervical sclerites simple; male claspers with telomeres, secondarily simplified in some.

(a). Infracohort Neuropterida (Handlirsch, 1908); emended from Neuropteroidea.

Synapomorphies. Ovipositor with third valvulae fused dorsally, desclerotized, bearing internal muscles; second valvulae suppressed; first valvulae reduced and incapable of piercing; costal space expanded, with cross veins; lateral cervical sclerites fused ventrally with ventral cervical sclerite; gula formed from postmentum and postocciput, free of ventral cervical sclerites; maxillae with reduced galea and lacinia (modified in Neuroptera); labium without terminal prelabial muscles; stomodaeum with median dorsal diverticulum; wing vein tips with bifurcations.

Plesiomorphies. Basisternum not reduced; larval maxillary and labial muscles not lost; first abdominal sternum normal; spermatozoa normal; ocelli normally present.

a. Subtercohort Raphidida (Handlirsch, 1908); emended from Raphidioidea.

Order Raphidiodea Burmeister, 1839. Lower Permian to Recent. Snakeflies.

Synapomorphies. Ovipositor (third valvulae) highly elongated, with secondary tongue and groove union with prolonged eighth abdominal sternum; first valvulae suppressed; head elongated posteriorly; posterior tentorial pits close together; neck long, membranous dorsally, covered by anterior extension of pronotum; pronotum extended downward hiding propleuron; neck bearing elongate ventral cervical sclerite joined to long lateral cervical sclerites; third tarsomere bilobed; empodium lost; larval abdominal prolegs suppressed.

Plesiomorphies. Terrestrial larvae little different from adults; three ocelli; subcosta ending in costa; pterostigma pigmented; larval and adult maxillae not elongate as in Sialida; genital claspers with simple telomere; prominent styli on third valvulae.

b. Subtercohort Sialida (Handlirsch, 1908); emended from Sialoidea.

Synapomorphies. Subcosta joining the radius terminally near tip; pterostigma with blood sinus but not pigmented; adult stipes elongated; styli of third valvulae small.

Plesiomorphies. All tarsomeres cylindrical, except fourth tarsomere in Sialidae bilobed; empodium retained; short first valvulae in Sialidae, lost in Corydalidae and Neuroptera; no thoracic paranota; larval prolegs (but lost in most Neuroptera); neck short; tentorial pits not close together.

i. Order Megaloptera Latreille, 1802, *s.s.*, excluding Raphidiodea. Permian to Recent. Dobsonflies, alderflies.

Synapomorphies. Larvae aquatic; abdominal prolegs respiratory; larval stipes elongated as in adult; tips of wing veins without bifurcations.

Plesiomorphies. Larvae with biting mouthparts; alimentary canal open; three ocelli; radial sector not excessively branched; pupae not in silk cocoons, males with telomeres (secondarily suppressed in Sialidae); abdominal prolegs; neuropterid gula (secondarily lost in Sialidae).

ii. Order Neuroptera Linnaeus, 1758, *s.s.*; (= Plannipennes Latreille, 1817). Lower Permian to Recent. Excluding Odonata, Ephemerida, Mecoptera, Trichoptera, Raphidiodea, and Megaloptera. Spongillaflies, antlions, aphidlions, lacewings, and so on.

Synapomorphies. Larvae with grooved elongate mandibles covered by elongate maxillary lacinal lobe, forming piercing-sucking tubes; larval metathorax without spiracles; larval prolegs suppressed (reacquired in aquatic Sisyridae); larval midintestine closed; malpighian tubules secrete silk at pupation; gula secondarily desclerotized; median cervical sclerotization lost; cervical sclerites free of head; Radial sector with multiple pectinate branches; telomeres suppressed in male claspers; ovipositor (third valvulae) with highly reduced styli.

Plesiomorphies. Ocelli in ground plan, but suppressed in most; terminal bifurcation of wing veins (secondarily reduced in some, very extensive in others); tarsomeres cylindrical, without lateral lobes.

(b). Infracohort Panorpida Handlirsch, 1908; (= Mecopteroidea Hennig, 1969, plus Siphonaptera); emended from Panorpoidea.

Synapomorphies. Ovipositor suppressed; functional ovipositor (oviscapt) derived from modification of terminal abdominal segments; extreme invagination of thoracic sterna, with pits of sternal apodemes (furca) hidden; 1A of hind wing curved forward temporarily fusing with Cu; first axillary sclerites with pleural muscle; mesothoracic spiracle shifted to prothorax; tergocoxal remotor of pterothoracic coxae suppressed; larval stipito-lacinial and stipito-galeal muscles absent; larval stipes divided; larval clypeofrontal sulcus obliterated; spermatozoa with only one mitochondrial derivative; pupation in silken cocoons secreted by labial glands (secondarily lost in some Diptera); six malpighian tubules. (Neuropterida, eight).

Plesiomorphies. Lateral cervical sclerites simple; parameres with basimere and telomere (changed in Diptera and Lepidoptera); tarsi pentamerous; larval mouthparts biting; cerci retained (except in female Lepidoptera).

a. Subtercohort *Mecopterida.*(= Siphonaptera plus Antliophora Hennig, 1969). New subtercohort.

Synapomorphies. Sperm pump at base of aedeagus; loss of outer ring of axonemes in sperm tail (reacquired in higher Diptera?); absence of lateral larval labral retractor muscles; larval cardo fused to basistipes; suppression of cranial promotor of cardo; desclerotized labial postmentum; larval labial palp reduced to two segments and without muscles; larval hypopharynx without retractor muscle; larval salivarium without ventral muscle; four larval instars; four malpighian tubules; acanthae acquired in proventriculus.

Plesiomorphies. (In contrast to Trichopterida). Male normal, heterogametic; wings not densely setose; acrosome present in sperm cells.

(*a*). Superorder *Mecopterodea.* New superorder.

Order Mecoptera Packard, 1886; includes Protomecoptera Tillyard, 1917. Lower Permian to Recent. Scorpionflies.

Synapomorphies. Head elongate, beaklike, with reduced terminal mouthparts; labrum, clypeus, and frons a single sclerite, sulci obscure; no tentorial mandibular adductor; no hypopharyngeal muscles; hypopharynx fused with adult labium; larval eyes similar to compound eyes, but histolyzed and replaced at pupation with adult eyes; no larval tarsal extensor or pretarsal muscle; larval metathoracic spiracle nonfunctional; male abdominal segment 9 a ring without demarcation between tergum and sternum; ovipositor highly reduced, with nonfunctional vestiges only remaining; spermathecal opening separate from gonopore; 9 + 2 axial filaments in sperm tail coiled around single mitochondrial derivative.

Plesiomorphies. (In contrast to Haustellodea). Wings fully developed, with little difference between fore and hind wings and between mesothorax and metathorax; wings with costal cross veins; pupae decticous and exarate; traces of ovipositor; larval abdominal limbs; adult cerci fairly prominent; adult mouthparts biting; proventricular acanthae.

(b). Superorder Haustellodea (Clairville, 1798), s.s.; emended from Haustellata, excluding Lepidoptera and Hemipterida.

Synapomorphies. Adult mouthparts piercing and sucking with styletlike epipharynx and laciniae, pointed free hypopharynx containing salivary duct, the stylets ensheathed by elongate prelabium; pupae adecticous, primitively exiting from cocoon through use of cephalic cocoon cutters; reduced hind wings; small metanotum; shortened adult antennae; larval thoracic legs suppressed; larval eyes suppressed; no trace of ovipositor.

Plesiomorphies. Holopneustic biting larvae (becoming apneustic, metapneustic, propneustic or amphipneustic in various Diptera).

i. Order Siphonaptera Latreille, 1825; (= Suctoria DeGeer, 1778). Lower Cretaceous to Recent. Fleas.

Synapomorphies. Adult wings suppressed; meron suppressed; adult compound eyes vestigal; male antennae serve as accessory clasping organs; adult mandibles suppressed; salivary channels in lacinial stylets; body laterally compressed; abdominal terga overlap the sclerotized sterna; metepimeron enlarged accommodating leaping muscles; ectoparasites of mammals and birds; larval abdominal prolegs suppressed; spiral axonemes.

Plesiomorphies. Larvae terrestrial, with biting mouthparts; larvae holopneustic, with spiracular closing apparatus; ground plan antennae similar to nematocerous Diptera, but later shortened further; retention of proventricular acanthae.

ii. Order Diptera Linnaeus, 1758. Permian to Recent. Flies, gnats, mosquitoes, midges, and so on.

Synapomorphies. Hind wings reduced to halteres; metathorax small, fore wings only used in flight; mesothorax enlarged, with large mesopostphragma; adult mandibles elongate, styletlike; hypopharynx styletlike, bearing salivary duct; labial palps modified into a labellum; larvae aquatic or semiaquatic; larval spiracles without closing apparatus; suppression of proventricular acanthae.

Plesiomorphies. Ground plan larval mouthparts biting, progressively suppressed in higher forms; ground plan larval spiracles holopneustic, variously reduced; radius of fore wing normal, but developing a typical basal kink in higher forms; ground plan male genitalia with basimeres and parameres, but basimeres become fused and immovable in Cyclorrhapha; male abdomen primitively unrotated but variously rotated 180° or 360° in various forms; embryonic abdominal leg rudiments present in ground plan, becoming abdominal prolegs in Blepharoceridae, but embryonic prolegs suppressed in other Diptera; "pseudopods," similar to prolegs, sometimes develop in larvae postembryonically, and may be secondary.

b. Subtercohort *Trichopterida*. (= Amphiesmenoptera Hennig, 1969). New supercohort.

Synapomorphies. Wings densely clothed with setae; anal veins fused terminally (or reduced); wing coupling via jugum (variously modified); aculeate setae forming a holding area for folded fore wings; a pair of setiferous protuberances on prothorax; furcal arms fused with epimeron; a pair of ventral glands or fifth abdominal sternum; sternal apodemes in eighth and ninth abdominal segments; female heterogamy; increased number of chromosomes; spermatozoa lack acrosome; outer ring of axoneme doublets large, containing glycogen; larval silk produced by active larvae; goblet cells in larval mid intestine; lower cervical muscles insert at a single point on tentorium.

Plesiomorphies. (Contrast to Mecopterida). Labral muscles; cardo not fused to stipes; larval postmentum normal; muscles in larval palpi; larval hypopharyngeal and salivarium muscles present; sperm with 9 + 9 + 2 axoneme pattern; pupae decticous; larvae with six instars; six malpighian tubules (four instars and four malpighian tubules in Mecopterida).

i. Order Trichoptera Kirby, 1826. Permian to Recent. Caddisflies.

Synapomorphies. Larvae aquatic, apneustic, respiration epidermal; anal prolegs (future cerci) grasping; larval antennae simple, papillalike; adult mandibulate mouthparts reduced; labium short, joining hypopharynx in adult forming a sucking tube; adult galea reduced, lacinia suppressed.

Plesiomorphies. Wings with generalized venation; male genitalia with basimeres and telomeres; simple cerci in adult; three ocelli (reduced or none in some); wing setae rarely scaly.

ii. Order Lepidoptera Linnaeus, 1758. Upper Triassic to Recent. Moths, butterflies, skippers.

Synapomorphies. Setae of wings modified into scales; methathorax reduced; hind wing largely driven by coupling with fore wings; median ocelli suppressed; telomeres suppressed; phallic protractor muscle in paramere; adult females without cerci; tenth abdominal tergum bilobed; apical labial segment with recessed sensory setae; fore wing holding area of metathorax well developed (suppressed in nonwing-folders); corporotentorium with median posterior process; no mandibular tentorial adductor muscle; absence of antagonistic maxillary palpal muscles; no dorsal muscle on salivarium; labral nerve close to frontal ganglion; ventral nerve cord solid (not double); recurrent nerve inside aorta.

Plesiomorphies. Adult mouthparts functional, biting, but modified into siphoning organ (galea) in higher forms; jugum for wing coupling (lost in most, frenulum acquired in higher forms or substituted for by other setae); abdominal prolegs simple papillae with sclerotized terminal spine (musculated and with crochets in higher forms); larvae terrestrial, originally holopneustic (Micropterygidae), metathoracic spiracle closed in other larvae; no adfrontal sulcus originally, secondary adfrontal sulcus in most.

References/Author Index

Asterisks denote references not seen; *fide* = "having faith in" or "trusting" the source cited. Italic numbers at the end of each reference refer to pages in the text.

Achtelig, M. and N. P. Kristensen. 1973. "A Re-examination of Relationships of the Raphidioptera (Insecta)." *Z. Zool. Syst. Evol. Forsch.* 11:268–274. *246*

Anderson, D. T. 1973. *Embryology and Phylogeny in Annelids and Arthropods.* Pergamon Press, Oxford. 495 pp. *12, 35, 38, 39, 99, 123, 154*

Baccetti, B. 1970. "The Spermatozoa of Arthropoda. IX. The Sperm Cell as an Index of Arthropod Phylogenesis." In: B. Baccetti, Ed., *Comparative Spermatology.* Academic Press, New York. Pp. 169–182. *135, 252, 255*

Baccetti, B. 1979. "Ultrastructure of Sperm and Its Bearing on Arthropod Phylogeny." In: A. P. Gupta, Ed., *Arthropod Phylogeny.* Van Nostrand-Reinhold, New York. Pp. 609–644. *135, 232, 252*

Barlet, J. 1950. "La Question des Pieces Pleurales du Thorax des Machilides (Thysanoures)." *Bull. Ann. Soc. Entomol. Belg.* 86:179–190. *130, 170*

Barlet, J. 1977. "Thorax d'Apterygotes et de Pterygotes Holometaboles." *Bull. Ann. Soc. Entomol. Belg.* 113:229–239. *62, 136, 173, 188*

*Berlese, A. 1909. "Monografia dei Myrientomata." *Redia.* 6:1–182. *Fide:* Essig, 1942. *141, 264, 265, 266*

Bitsch, J. 1963. "Morphologie Cephalique des Machilides (Insecta Thysanura)." *Ann. Sci. Nat. Zool. Biol. Anim.* 12 Ser. 5(3/4):501–706. *171*

Bitsch, J. 1973. "Morphologie Abdominale des Machilides (Insecta Thysanura). I. Squellette et Musculature des Segments Pregenitaux." *Ann. Sci. Nat. Zool. Biol. Anim.* 12 Ser. 15(2):173–200. *170*

Bitsch, J. 1974a. "Morphologie Abdominale des Machilides (Thysanura). II. Squellette et Musculature des Segments Genitaux Femelles." *Int. J. Insect Morphol. Embryol.* 3(1):101–120. *159, 169, 172*

Bitsch, J. 1974b. "Morphologie Abdominale des Machilides (Thysanura). III. Squellette et Musculature des Segments Genitaux Males et des Segments Postgenitaux." *Int. J. Morphol. Embryol.* 3(2):203–224. *191*

Blacklith, R. E. and R. M. Blacklith. 1968. "A Numerical Taxonomy of Orthopteroid Insects." *Aust. J. Zool.* 16:111–131. *208*

293

*Blainville-Gervais. 1844. (Origin of the name Diplopoda). *Fide:* Laurentiaux, D. 1953a. *83, 120*

Börner, C., 1904. "Zur Systematik der Hexapoden." *Zool. Anz.* 27:511–533. *141, 142, 169, 228, 264, 267, 279, 281*

Börner, C. 1910. "Die Phylogenetische Bedeutung der Protura." *Biol. Zentralbl.* 30:636–641. (Origin of the name Ellipura). *141, 264*

Boudreaux, H. B. 1979. "Significance of Intersegmental Tendon System in Arthropod Phylogeny and a Monophyletic Classification of Arthropoda." In: A. P. Gupta, Ed., *Arthropod Phylogeny.* Van Nostrand-Reinhold, New York. Pp. 551–586. *53, 54, 55, 56, 57, 58, 59, 60, 61, 62, 63, 64, 65, 66, 67, 68, 69, 70, 71, 72, 73, 82, 83, 115, 116, 117, 118, 119, 120, 121*

*Brauer, F. 1885. "Systematische-Zoologische Studien." *Sitz. Math. Naturwiss. Cl. Akad. Wiss. (Vienna).* 91:237–413. (Origin of name Pterygota). *Fide:* Wilson and Doner, 1937. *141, 268, 286*

Brinck, P. 1955. "The Reproductive System and Mating in Plecoptera. I." *Opusc. Entomol.* 21(1):57–127. *203*

Britton, E. F. 1970. "Coleoptera (Beetles)." In: D. F. Waterhouse, Ed., *The Insects of Australia.* CSIRO, Melbourne University Press. Carlton, Victoria. Pp. 495–621. *237*

Broili, F. 1933. "Ein Zweites Examplar von Cheloniellon." *Bayer. Akad. Wiss. Munchen, Math. Naturwiss. Kl. Sitzungsber.* Abt. 1933:11–32. *83, 96, 97, 119*

*Brongniart, C. 1885. "Les Insectes Fossiles des Terrains Primaires." *Bull. Soc. Sci. Nat. Rouen.* 1885:50–68. *Fide:* Sharov, 1966a. *166*

*Brongniart, C. 1893. *Recherches pour Servir a L'Histoire des Insectes Fossiles des Temps Primaires.* St. Etienne. 493pp. (Origin of names Megasecoptera and Protodonata). *Fide:* Sharov, 1966a; Jeannel, 1949; Laurentiaux, 1953b. *141, 193, 269*

Brues, C. T. and A. L. Melander. 1932. "Classification of Insects." *Bull. Mus. Comp. Zool. Harv. Coll.* 73:1–672. *142, 276*

*Brulle, A. 1832. "Memoire sur une Nouvelle Disposition de l'Ordre des Neuropteres." *Ann. Soc. Entomol. France.* 1:239–240. (Origin of the name Isoptera). *Fide:* Essig, 1942. *142, 277*

*Brunner, C. de W. 1882. *Prodromus der Europaischen Orthopteren.* Leipzig. 466pp. (Origin of the name Blattodea). *Fide:* Essig, 1942. *142, 277*

*Burmeister, H. C. C. 1829. *De Insectorum Systemati Naturali,* Dissertat Inaug. Grunert, Halle. 48pp. (Origin of name Blattaria). *Fide:* Essig, 1942. *142, 278*

*Burmeister, H. C. C. 1838. "Orthoptera." In: *Handbuch der Entomologie.* Vol. 2. Part 1. 1st half. Reimer, Berlin (Origin of name Mantodea). *Fide:* Essig, 1942. Pp. 397–756. *142, 278*

*Burmeister, H. C. C. 1839. "Neuroptera." In: *Handbuch der Entomologie.* Vol. 2. Part 1. 2nd half. Reimer, Berlin (Origin of names Plecoptera, Corrodentia, Raphidiodea). *Fide:* Essig, 1942. Pp. 757–1050. *141, 142, 143, 200, 273, 280, 286*

*Burmeister, H. C. C. 1843. (Origin of the name Eurypterida). *Fide:* Waterlot, 1953. *83, 117*

Carpenter, F. M., 1943a. "Carboniferous Insects from the Vicinity of Mazon Creek, Illinois." *Sci. Pap. Ill. State Mus.* 3:9–20. *184, 185*

Carpenter, F. M. 1943b. "Studies on Carboniferous Insects from Commentary, France. Part 1. Introduction and Families Protagriidae, Meganeuridae and Campylopteridae. *Bull. Geol. Soc. Am.* 54:527–554. *193*

Carpenter, F. M. 1953. "The Geological History and Evolution of Insects." *Am. Sci.* 41(2): 256–270. *192*

Carpenter, F. M. 1963a. "Studies on Carboniferous Insects from Commentry, France: Part IV. The Genus *Triplosoba.*" *Psyche.* 70: 120–128. *196*

Carpenter, F. M. 1963b. "Studies on Carboniferous Insects from Commentry, France: Part V. The Genus *Diaphanoptera* and the Order Diaphanopterodea." *Psyche.* 70: 240–256. *189, 192*

Carpenter, F. M. 1966. "The Lower Permian Insects of Kansas. Part. 11. The Orders Protorthoptera and Orthoptera." *Psyche.* 73: 46–88. *179, 202*

Carpenter, F. M. 1977. "Geological History and the Evolution of the Insects." *Proc. 15th Int. Congr. Entomol.* Wash., D. C. Pp. 63–70. *166*

Carpenter, F. M. 1978. Personal communication about new species of *Dasyleptus* from North America and Czechoslovakia. *166*

Carpenter, F. M. and J. Kukalova. 1964. "The Structure of the Protelytroptera, with Description of a New Genus from Permian Strata of Moravia." *Psyche.* 71: 179–183. *208*

Carpenter, F. M. and E. S. Richardson. 1968. "Megasecopterous Nymphs in Pennsylvanian Concretions from Illinois." *Psyche.* 75: 295–309. *190, 192*

Cassagnau, P. 1971. "Le Spermatophore des Collemboles Neanuridae." *Rev. Ecol. Biol. Sol.* 8: 609–616. *147*

China, W. E. 1962. "South American Peloridiidae (Hemiptera-Homptera: Coleorrhyncha)." *Trans. R. Entomol. Soc. London.* 114(5): 131–161. *231, 232, 233*

Chopard, L. 1949a. "Supre-ordre des Blattopteroides." In: P. Grasse, Ed., *Traité de Zoologie.* 9: 355–555. Masson, Paris. *213*

Chopard, L. 1949b. "Ordre des Cheleutopteres." In: P. Grasse, Ed., *Traité de Zoologie.* 9: 594–616. Masson, Paris. *216*

Cisne, J. L. 1974. Trilobites and the Origin of Arthropods." *Science.* 186(4158): 13–18. *58, 79, 93*

*Clairville, J. de. 1798. *Entomologie Helvetique, ou un Catalogus des Insectes (Coleopteres) de la Suisse, Ranges d'apres une Nouvelle Methode, avec Descriptions et Figures.* Ovel. Zurich. Vol. 1. 149 pp. (Origin of name Haustellata). *Fide:* Wilson and Doner, 1937. *83, 143, 253, 289*

Clark, R. B., 1964. Dynamics in Metazoan Evolution: *The Origin of the Coelom and Segments.* Clarendon Press, Oxford. 313 pp. *16, 19*

Clay, T. 1970. "The Amblycera (Phthiriaptera: Insecta)." *Bull. Brit. Mus. Nat. Hist. (Entomol.).* 25(3): 75–78. *227, 228*

Comstock, J. H. 1918. *The Wings of Insects.* Comstock Publishing Co., Ithaca. 430 pp. *179, 180*

Comstock, J. H. and A. B. Comstock. 1895. *Manual for the Study of Insects.* Comstock Publishing Co., Ithaca. 701 pp. *226*

Corbet, P. S. 1962. *A Biology of Dragonflies.* Quadrangle Books, Chicago. 247 pp. *194*

Cracraft, J. 1974. "Phylogenetic Models and Classification." *Syst. Zool.* 23: 71–90. *9, 125*

Crampton, J. 1915. "The Thoracic Sclerites and the Systematic Position of *Grylloblatta campodeiformis* Walker, a Remarkable Annectant Orthopteroid Insect." *Entomol. News.* 26: 337–350. (Origin of the name Cheleutoptera). *274*

Crampton, G. C. 1917. "A Phylogenetic Study of the Lateral Head, Neck and Prothoracic Regions in Some Apterygota and Lower Pterygota." *Entomol. News.* 28: 398–412. *272, 273*

Crowson, R. A. 1955. *The Natural Classification of the Families of Coleoptera.* Reprint, 1967. E. W. Classey, Ltd., Hampton. 214 pp. *239*

CSIRO. 1970. *The Insects of Australia.* Melbourne University Press. Carlton, Victoria. 1029 pp. *47, 218*

Cuenot, L. 1949. "Les Onychophores." In: P. Grasse, Ed., *Traité de Zoologie.* 6:3-37. Masson, Paris. *33*

Daly, H. V. and A. Sokoloff. 1965. "Labiopedia, a Sex-linked Mutant in *Tribolium confusum* Duval (Coleoptera Tenebrionidae)." *J. Morphol.* 117:251-270. *223, 235*
Dawydoff, C. 1943. "Observations sur le Developpement des Scolopendrelles (Symphyla) Indochinoises." *Bull. Biol. France Belg.* 77:1-28. *29*
DeBeer, G. 1971. "Homology, an Unsolved Problem." *Oxford Biology Readers No. 11.* Oxford, University Press, London. 16 pp. *11*
Debouttteville, C. D. 1954. "Recherches sur les Crustaces Souterrains, III. Developpment Postembryonaire des Mystacocarides." *Arch. Zool. Exp. Gen.* 91:25-34. *78*
Dechaseaux, C. 1953. "Classe des Crustaces." In: J. Piveteau, Ed., *Traité de Paleontologie.* 3:255-256. Masson, Paris.
*DeGeer, C. 1778. *Memoires pour Servir a l'Histoire des Insectes.* Vol. 7. Stockholm. (Origin of the name Suctoria). *Fide:* Wilson and Doner, 1937; Essig, 1942. *289*
Delany, M. J. 1959. "The Life Histories and Ecology of Two Species of *Petrobius* Leach, *P. brevistylis* and *P. maritimus.*" *Trans. R. Soc. Edinb.* 63:501-533. *171*
Denis, R. 1949. "Sous-classe des Apterygotes. Anatomie, Biologie, Systematique." In: P. Grasse, Ed., *Traité de Zoologie.* 9:111-275. Masson, Paris. *134, 135, 146*
Duporte, E. M. 1962. "Origin of the Gula in Insects." *Can. J. Zool.* 40:381-384. *238*

*Enderlein, V. G. 1903. "Ueber die Morphologie, Gruppierung und Systematische Stellung der Corrodentien." *Zool. Anz.* 26:423-437. (Origin of the name Embiidina). *Fide:* Essig, 1942. *141, 273*
Essig, E. O. 1942. *College Entomology.* MacMillan, New York. 900 pp. *155*

*Fabricius, J. C. 1792. *Entomologie Systemica Emendata et Aucta, Secundus Classes, Ordines, Genera, Species, Adjectis Synonymis, Locus, Observationibus, Descriptibus.* Vol. 2. Hafniae, Paris. (Origin of the name Odonata). *Fide:* Wilson and Doner, 1937. *141, 269*
Fage, L. 1949a. "Classe des Merostomaces." In: P. Grasse, Ed., *Traité de Zoologie.* 6:219-262. Masson, Paris. *75*
Fage, L. 1949b. "Classe des Pycnogonides." In: P. Grasse, Ed., *Traité de Zoologie.* 6:906-941. Masson, Paris. *72*
Fauchald, K. 1974. "Polychaete Phylogeny: A Problem in Protostome Evolution." *Syst. Zool.* 23:493-506. *19, 22*
Ferris, G. F. 1931. "The Louse of Elephants. *Haematomyzus elephantis* Piaget (Mallophaga) Haematomyzidae." *Parasitology.* 23:112-127. *228, 281*
Forbes, W. T. M. 1943. "The Origin of Wings and Venational Types in Insects." *Amer. Midl. Nat.* 29:381-405. *179, 180*
Fraser, F. C. 1938. "A Note on the Fallaciousness of the Theory of Pretracheation in the Venation of Odonata." *Proc. R. Entomol. Soc. London.* (A). 13:60-70. *179*

Gay, F. J. 1970. "Isoptera (Termites)." In: D. F. Waterhouse, Ed., *The Insects of Australia.* pp. 275-293. CSIRO, Melbourne University Press, Carlton, Victoria. *219*
Giles, E. T. 1963. "The Comparative External Morphology and Affinities of the Dermaptera." *Trans. Roy. Entomol. Soc. London.* 115:95-164. *208, 210, 212*
*Goldenberg, F. 1854. *Die fossilen Insecten der Kohlenformation von Saarbrucken.* Fischer, Cassel. 24 pp. (Origin of name Paleodictyoptera). *Fide:* Laurentiaux, 1953b. *141, 193, 268*

Goodchild, A. J. P. 1966. "Evolution of the Alimentary Canal in the Hemiptera." *Biol. Rev.* 41:97–140. *233*

Grasse, P. P., Ed. 1949. *Traité de Zoologie.* Vol. 9. Insectes. Masson, Paris. 1117 pp. *156*

*Grassi, B. 1889. "Les Ancetres des Myriapods et des Insectes." *Attl. Accad. Lincei. Mem.* 5:543–606. (Origin of the name Endotropha). *Fide:* Wilson and Doner, 1937. *149, 264*

Griffiths, G. C. D. 1972. *The Phylogenetic Classification of Diptera Cyclorrhapha, with Special Reference to the Structure of the Male Postabdomen.* Series Entomol. Vol. 8. Junk, The Hague. 340 pp. *9, 258*

*Grube, A. E. 1853. (Origin of the name Onychophora). *Fide:* Cuenot, L. 1949. *82, 115*

Hackman, R. H. and M. Goldberg. 1975. "Peripatus: Its Affinities and its Cuticle." *Science.* 190:582–583. *23*

Hagen, H. A. 1862. *Bibliotheca Entomologica.* Englemann, Leipzig. Vol. 1, A–M, 566 pp. Vol. 2, N–Z, 512 pp.

*Haliday, A. H. 1836. "An Epitome of the British Genera in the Order Thysanoptera, with Indication of a Few of the Species." *Entomol. Mag.* 3:439–451. (Origin of name Thysanoptera). *Fide:* Essig, 1942. *143, 282*

Hamilton, K. G. A. 1971. "The Insect Wing. Part I. Origin and Development of Wings from Notal Lobes." *J. Kansas Entomol. Soc.* 44:421–433. *181, 182, 205, 222, 235, 273, 278*

Hamilton, K. G. A. 1972a. "The Insect Wing. Part II. Vein Homology and Archetypal Insect Wing." *J. Kansas Entomol. Soc.* 45:54–58. *179, 180, 181, 201, 229*

Hamilton, K. G. A. 1972b. "The Insect Wing. Part III. Venation of the Orders." *J. Kan. Entomol. Soc.* 45:145–162. *181, 232*

Hamilton, K. G. A. 1972c. "The Insect Wing. Part IV. Venational Trends and the Phylogeny of the Winged Orders." *J. Kan. Entomol. Soc.* 45:296–308. *181, 211*

*Handlirsch, A. 1903. "Zur Phylogenie der Hexapoden." *Sitz. Nat. Kl. Akad. Wiss.* (Wien). 112:716–738. (Origin of name Machiloidea). *Fide:* Remington, 1954. *169, 266, 267*

Handlirsch, A. 1908. *Die fossilen Insekten und die Phylogenie der rezenten Formen. Ein Handbuch für Paleontologen und Zoologen.* 2 Vols. Engelman, Leipzig. 1430 pp. (Origin of order names Protephemerida, Protohemiptera, Protorthoptera, Protoblataria, and many superorders). *141, 142, 143, 212, 268, 270, 271, 272, 273, 274, 276, 279, 283, 285, 286, 287*

*Hennig, W. 1953. "Kritische Bemerkungen zum Phylogenetischen System der Insekten." *Beitr. Entomol.* (Sonderh). 3:1–85. *Fide:* Hennig, 1969. *3, 141, 264, 265, 266*

Hennig, W. 1966. *Phylogenetic Systematics.* University of Illinois Press, Urbana. 263 pp. *172*

Hennig, W. 1969. *Die Stammesgeschichte der Insekten.* Kramer, Frankfurt am Main. 436 pp. *141, 149, 161, 208, 216, 217, 222, 231, 232, 244, 251, 258, 267, 287, 288, 290*

Hepburn, H. R. 1969. "The Proventriculus of Mecoptera." *J. Ga. Entomol. Soc.* 4:159–167. *255, 256*

Herman, W. S. 1967. "The Ecdysial Glands of Arthropods." *Int. Rev. Cytol.* 22:269–349. *46*

Hessler, R. R. 1964. "The Cephalocarida. Comparative Skeletomusculature." *Mem. Conn. Acad. Arts Sci.* 16:1–97. *54*

*Heymons, R. 1899. "Beitrage zur Morphologie und Entwicklungsgeschichte der Rhyncho-ten." *Nov. Acta Abh. Kais. Leopold Carol. Deut. Akad. Naturf.* 74:349-356. *Fide:* Snodgrass, 1927. *127*

*Heymons, R. 1901. "Die Entwicklungsgeschichte der Scolopender." *Zool. Stuttg.* 13:1-244. (Origin of the names Atelocerata and Chelicerata). *Fide:* Sharov, 1966a. *37, 83, 117, 119*

Hinton, H. E. 1955. "On the Structure, Function and Distribution of the Prolegs of the Panorpoidea, with a Criticism of the Berlese-Imms Theory." *Trans. R. Entomol. Soc. (London).* 106:455-545. *255*

Hinton, H. E. 1958. "The Phylogeny of the Panorpoid Orders." *Ann. Rev. Entomol.* 3:181-206. *252, 253, 257*

*Hirst, S. and S. Maulik. 1926. "On some Arthropod Remains from the Rhynie Chert (Old Red Sandstone). *Geol. Mag.* 63:69-71. *Fide:* Scourfield, 1940. *150*

Hupe, P. 1953. "Classe des Trilobites." In: J. Piveteau, Ed., *Traité de Paleontologie.* 3:44-246. Masson, Paris. *74*

Hutchinson, G. E. 1930. "Restudy of some Burges Shale Fossils." *Proc. U. S. Nat. Mus.* 78(11):1-24. *78*

Ide, F. P. 1937. "The Subimago of Ephoron leukon Will., and a Discussion of the Imago Instar. (Ephem.)." *Can. Entomol.* 69:25-29. *197*

Imms, A. D. 1925. *A General Textbook of Entomology.* Dutton, New York. 720 pp. *264*

Ivanov, P. O. 1933. "Die Embryonale Entwicklung von Limulus molluccanus." *Zool. J. Ont.* 56:163-348. *19*

Janetschek, H. 1957. "Uber die mogliche phyletische Reversion eines Merkmals bei Felsenspringern mit einigen Bemerkungen über die Natur der Styli der Thysanuren (Insekten)." *Broteria, Ser. Cien. Nat.* 26:1-22. *171*

*Jeannel, R. 1945. *Introduction a l'Entomologie. Vol. 1. Anatomie Generale et Classi-fication.* Boubee, Paris. (Origin of name Psocopteroidea). *Fide:* Jeannel, 1949. *279*

Jeannel, R. 1949. "Les Insectes." In: P. P. Grasse, Ed., *Traité de Zoologie.* 9:1-110. Masson, Paris.

*Kellog, V. L. 1896. (Origin of names Amblycera and Ischnocera). *Fide:* Essig, 1942. *280, 281*

Kinzelbach, R. K. 1971. "Morphologische Befunde an Fächerflügern und ihre Phylo-genetische Bedeutung (Insecta: Strepsiptera)." *Zoologica.* 41:1-256. *9, 238, 239, 240, 245, 251*

Kirby, W. F. 1813. "Strepsiptera a New Order of Insects Proposed; and the Characters of the Order, with those of its Genera Laid Down. (Stylops; Xenos.)" *Trans. Linn. Soc. London.* 11:86-123. *143, 284*

*Kirby, W. F. 1826. *An Introduction to Entomology, or Elements of the Natural History of Insects, with Plates.* Vol. 4, 1826. Longmans, London. 634 pp. (Origin of the name Trichoptera). *Fide:* Wilson and Doner, 1937. *143, 290*

Kishinouye, K. 1891. "On the Development of Limulus longispinus." *J. Coll. Sci. Imp. Univ. Japan, Tokyo.* 5:53-100. *71, 85*

Kleinow, W. 1966. "Untersuchungen zum Flügelmechanismus der Dermapteren." *Z. Morphol. Ökol. Tiere.* 56:363-416. *209*

Konigsman, E. 1960. "Zur Phylogenie der Parametabola unter besonderer Berücksichtigung der Phthiriaptera." *Beitr. Entomol.* 10:705-744. *227, 228, 280*

Krishnakumaran, A. and H. A. Schneiderman. 1970. "Control of Molting in Mandibulate and Chelicerate Arthropods by Ecdysones." *Biol. Bull.* 139:520-538. *46*

Krishnan, G. 1953. "On the Cuticle of the Scorpion *Palaemneus swammerdami.*" *Q. J. Micr. Sci.* 94:11-21. *46*

Kristensen, N. P. 1971. "The Systematic Position of the Zeugloptera in the Light of Recent Anatomical Investigations." *Proc. 13th Int. Congr. Entomol., Moscow, 1968.* 1:261. *260*

Kristensen, N. P. 1975. "The Phylogeny of Hexapod 'orders.' A Critical Review of Recent Accounts." *Z. Zool. Syst. Evol. Forsch.* 13:1-44. *9, 199, 216, 217, 231, 232, 233, 247, 248, 251, 260*

Kuijten, P. J. 1974. "On the Occurrence of a Hitherto Unknown Wing-Thorax Coupling Mechanism in Lepidoptera." *Neth. J. Zool.* 24:317-332. *258, 260*

Kukalova, J. 1959. "On the Family Blattinopsidae Bolton, 1925. (Insecta, Protorthoptera)." *Rozpr. Cesk. Akad. Ved.* 69:1-32. *212, 213*

Kukalova, J. 1968. "Permian Mayfly Nymphs." *Psyche.* 75:310-327. *198*

Kukalova, J. 1970. "Revisional Study of the Order Paleodictyoptera in the Upper Carboniferous Shales of Commentry, France. Part III." *Psyche.* 77:1-44. *191*

Kukalova-Peck, J. 1971. "The Structure of *Dunbaria* (Paleodictyoptera)." *Psyche.* 78:306-308. *191*

Kukalova-Peck, J. 1972. "Unusual Structures in the Paleozoic Insect Orders Megasecoptera and Paleodictyoptera, with a Description of a New Family." *Psyche.* 79:243-268. *162, 190, 202*

Kukalova-Peck, J. 1974. "Pteralia of the Paleozoic Insect Orders Paleodictyoptera, Megasecoptera and Diaphanopterodea (Paleoptera)." *Psyche.* 81:416-430. *182, 183, 201*

Kukalova-Peck, J. 1976. "Evolution in Ontogeny of Paleoptera." Unpublished paper read at the *15th Int. Congr. Entomol., Washington, D. C.,* August 1976. *192, 198*

Kukalova-Peck, J. and S. B. Peck. 1976. "Adult and Immature Calvertiellidae (Insecta: Paleodictyoptera) from the Upper Paleozoic of New Mexico and Czechoslovakia." *Psyche.* 83:79-93. *192, 198*

Lafon, M. 1943. "Sur la Structure et la Composition du Tegument de la Limule *(Xiphosura polyphemus* L.)." *Bull. Inst. Oceanogr. Monaco.* 40:No. 850. *46*

Lameere, A. 1917. "Etude sur l'Evolution des Ephemeres. Part 1." *Bull. Soc. Zool. France.* 52:41-59. *180, 184, 197*

Lameere, A. 1923 (1922). "On the Wing-Venation of Insects." *Psyche.* 30:123-132. (Translated from French by A. M. Brues). (Lameere, A. 1922. "Sur la Nervation Alaire des Insectes." *Bull. Acad. R. Belg.* Vol. 5). *179, 180, 184*

*Latreille, P. A. 1796. *Precis des Caracteres Generiques des Insectes, Disposes dans un Ordre Naturel par le Citoyen Latreille.* Brive, Bordeaux. 210 pp. (Origin of name Thysanura). *Fide:* Wilson and Doner, 1937. *141, 267*

*Latreille, P. A. 1802. *Histoire Naturele, Generale et Particuliere des Crustaces et des Insectes. Vol. 3. Familles Naturelles et Genres.* Dufart, Paris. 467 pp. (Origin on names Xiphosurida and Megaloptera). *Fide:* Wilson and Doner, 1937; Hagen, 1862, Essig, 1942. *83, 117, 143, 287*

*Latreille, P. A. 1810. *Considerations Generale sur l'Ordre Naturel des Animaux Composant les Classes des Crustaces, des Arachnides et des Insectes avec un Tableau Methodique de leurs Genres Disposes en Familles.* Schoell, Paris. 444 pp. (Origin of names Arachnida, Pycnogonida, Heteroptera). *Fide:* Wilson and Doner, 1937; Hagen, 1862. *83, 116, 117, 143, 283*

Latreille, P. A. 1817. "Crustaces, Arachnides, Insectes." In: G. Cuvier, *Le Regne Animale.* Vol. 3. Paris. (Origin of the name Plannipennes). *249, 274, 287*

*Latreille, P. A. 1825. *Familles Naturelles du Regne Animale, Exposees Succinctement et dans un Ordre Analytique, avec l'Indication de leurs Genres.* Bailliere, Paris. 570 pp. (Origin of the names Hexapoda, Siphunculata, Siphonaptera). *Fide:* Wilson and Doner, 1937; Essig, 1942. *122, 143, 228, 263, 281, 289*

Laurentiaux, D. 1952. "Decouverte d'un Rostre chez *Stenodictya lobata* Brgt. (Paleodictyoptera, Stenodictyidae) et le Probleme des Protohemipteres." *Bull. Soc. Geol. Fr.* 6e Ser. 2:233–247. *191*

Laurentiaux, D. 1953a. "Classe des Myriapodes." In: J. Piveteau, Ed., *Traité de Paleontologie.* 3:385–396. Masson, Paris.

Laurentiaux, D. 1953b. "Classe des Insectes." In: J. Piveteau, Ed., *Traité de Paleontologie.* 3:397–527. Masson, Paris. *185, 193, 271*

*Leach, W. E. 1814. (Origin of the names Myriapoda and Chilopoda). *Fide:* Laurentiaux, 1953a. *83, 120, 121*

*Leach, W. E. 1815. Entomology. In: *Brewster Edinburgh Encyclopedia,* Vol. 9. Edinburgh. Pp. 57–172. (Origin of names Phasmida, Anoplura, Dermaptera, Homoptera). *Fide:* Wilson and Doner, 1937; Essig, 1942. *142, 143, 227, 228, 274, 275, 280, 282*

*Leach, W. E. 1817. Zoological Miscellany. Vol. 3. *Classification of Insects.* London. 151 pp. (Origin of names Ephemerida, Dictuoptera). *Fide:* Hagen, 1862; Essig, 1942. *141, 200, 271*

Lehman, W. M. 1959. "Neue Entdeckungen an *Palaeoisopus.*" *Paläontol. Z.* 33:96–103. *85*

Lemche, H. 1940. "The Origin of Winged Insects." *Vidensk. Meddel. Dansk. Natur. Foren.* 104:127–168. *141, 189, 268, 270*

Leppla, N. C., T. C. Carlysle and R. H. Guy. 1975. "Reproductive Systems and the Mechanics of Copulation in Plecia nearctica Hardy (Diptera: Bibionidae)." *Int. J. Insect Morphol. Embryol.* 4:299–306. *251*

Linnaeus, C. 1758. *Systema Naturae. Vol. 1, Regnum Animale.* Ed. 10. Holmiae, Upsala. 823 pp. *1, 83, 122, 141, 143, 263, 267, 282, 284, 285, 287, 289, 290*

Lubbock, J. 1866. "On Pauropus, a New Type of Centipede." *Trans. Linn. Soc. London.* 26:181–190. *83, 120*

Lubbock, J. 1873. *Monograph of the Collembola and Thysanura.* Ray Society, London. 276 pp. *83, 120, 141, 265*

Mackerras, I. M. 1970. "Evolution and Classification of the Insects." In: D. F. Waterhouse, Ed., *The Insects of Australia.* CSIRO, Melbourne University Press, Carlton, Victoria. Pp. 152–167. *123*

Maki, T. 1938. "Studies of the Thoracic Musculature of Insects." *Mem. Fac. Sci. Agr. Taihoku Imp. Univ.* 24:1–344. *200*

Manton, S. M. 1950. "The Evolution of Arthropodan Locomotory Mechanisms. Part 1. The Locomotion of *Peripatus.*" *J. Linn. Soc. (Zool.).* 41:529–570. *47, 124*

Manton, S. M. 1952a. "The Evolution of Arthropodan Locomotory Mechanisms. Part 2. General Introduction to the Locomotory Mechanisms of the Arthropoda." *J. Linn. Soc. (Zool.).* 42:93–117. *47, 124*

Manton, S. M. 1952b. "The Evolution of Arthropodan Locomotory Mechanisms. Part 3. The Locomotion of the Chilopoda." *J. Linn. Soc. (Zool.).* 42:118–166. *47, 124*

Manton, S. M. 1954. "The Evolution of Arthropodan Locomotory Mechanisms. Part 4. The Structure, Habits and Evolution of the Diplopoda." *J. Linn. Soc. (Zool.).* 42:299–368. *47, 124*

Manton, S. M. 1956. "The Evolution of Arthropodan Locomotory Mechanisms. Part 5. The Structure, Habits and Evolution of the Pselaphognatha (Diplopoda)." *J. Linn. Soc. (Zool.).* 43:153–189. *47, 50, 55, 62, 108, 124*

Manton, S. M. 1958. "The Evolution of Arthropodan Locomotory Mechanisms. Part 6. Habits and Evolution of the Lysiopetaloidea (Diplopoda), some Principles of Leg Design in Diplopoda and Chilopoda, and Limb Structure in Diplopoda." *J. Linn. Soc. (Zool.).* 43:487–556. *47, 124*

Manton, S. M. 1961. "The Evolution of Arthropodan Locomotory Mechanisms. Part 7. Functional Requirements and Body Design in Cobolognatha (Diplopoda), Together with a Comparative Account of Diplopod Burrowing Techniques, Trunk Musculature and Segmentation." *J. Linn. Soc. (Zool.).* 44:383–461. *47, 124*

Manton, S. M. 1964. "Mandibular Mechanisms and the Evolution of Arthropods." *Phil. Trans. R. Soc. London. B. Biol. Sci.* 247:1–183. *38, 47, 52, 55, 105, 124, 144, 146*

Manton, S. M. 1965. "The Evolution of Arthropodan Locomotory Mechanisms. Part 8. Functional Requirements and Body Design in Chilopoda, Together with a Comparative Account of Their Skeletomuscular Systems and an Appendix on a Comparison Between Burrowing Forces of Annelida and Chilopoda and its Bearing upon the Evolution of the Arthropodan Haemocoel. *J. Linn. Soc. (Zool.).* 45:251–484. *47, 124*

Manton, S. M. 1966. "The Evolution of Arthropodan Locomotory Mechanisms. Part 9. Functional Requirements and Body Design in Symphyla and Pauropoda and the Relationships between Myriapoda and Pterygota." *J. Linn. Soc. (Zool.).* 46:103–141. *47, 124, 128*

Manton, S. M. 1967. "The Polychaete *Spinther* and the Origin of the Arthropoda." *J. Nat. Hist.* 1:1–22. *33, 38, 47, 124*

Manton, S. M. 1972. "The Evolution of Arthropodan Locomotory Mechanisms. Part 10. Locomotory Habits, Morphology and Evolution of the Hexapod Classes." *J. Linn. Soc. (Zool.).* 51:203–400. *36, 47, 48, 63, 123, 124, 127, 128, 129, 152, 154, 170*

Manton, S. M. 1973a. "Arthropod Phylogeny – A Modern Synthesis." *J. Zool., London.* 171:111–130. *24, 47, 124*

Manton, S. M. 1973b. "The Evolution of Arthropodan Locomotory Mechanisms. Part 11. Habits, Morphology and Evolution of the Uniramia (Onychophora, Myriapoda, Hexapoda) and Comparisons with Arachnida, Together with a Functional Review of Uniramian Musculatulature." *J. Linn. Soc. (Zool.).* 53:257–375. *47, 49, 123, 124*

Manton, S. M. 1977. *The Arthropoda. Habits, Functional Morphology, and Evolution.* Clarendon Press, Oxford. 527 pp. *47, 124*

*Martynov, A. B. 1925. "Uber zwei Grundtypen der Flugel bei den Insekten und ihre Evolution." *Z. Morphol. Okol. Tiere.* 4:465–501. (Origin of the names Neoptera, Polyneoptera, Paraneoptera, Oligoneoptera, Dermapteroidea). *Fide:* Wilson and Doner, 1937. *141, 142, 179, 271, 275, 279, 283*

Martynov, A. B. 1927. "Ueber eine neue Ordnung Fossilen Insekten, Miomoptera nov." *Zool. Anz.* 72:99–109. *272*

Martynov, A. B. 1930. (1924). "The Interpretation of the Wing Venation of the Odonata and Agnatha." *Psyche.* 37:245–280. (Translated from the Russian, with introductory note by F. M. Carpenter. Originally published in *Russk. Ent. Obozr.* 18:145–174, 1924). *180*

*Martynov, A. B. 1932. "New Permian Paleoptera with the Discussion of some Problems of their Evolution." *Inst. Paleontol. Acad. Sci. USSR., Trav.,* 1:1–44. (Origin of the name Meganisoptera). *Fide:* Carpenter, 1943b. *141, 193, 269*

Martynov, A. B. 1938a. "Etudes sur l'Histoire Geologique et de Phylogenetique des Ordres des Insectes (Polyneoptera)." *Trav. Inst. Paleontol. Acad. Sci. URSS* (Leningrad).

7: 1–150. (Origin of names Odonatoptera, Caloneurodea, Paraplecoptera). (In Russian). *141, 269, 272, 274*

*Martynov, A. B. 1938b. "On a New Permian Order of Orthopteroid Insects, Glosselytrodea." *Bull. Acad. Sci. USSR, Ser. Biol.* 1938:157–206. *Fide:* Laurentiaux, 1953b. *142, 274*

Matsuda, R. 1970. *Morphology and Evolution of the Insect Thorax.* Mem. Entomol. Soc. Can. No. 76. D. P. Pielou, Ed., Entomology Society of Canada, Ottawa. 431 pp. *127, 147, 172, 186, 189, 190, 194, 202, 205, 208, 226, 245, 247, 254*

Matsuda, R. 1976. *Morphology and Evolution of the Insect Abdomen, with Special Reference to Developmental Patterns and Their Bearings upon Systematics.* Int. Ser. Pure Appl. Biol., Zool. Div. Vol. 56. Pergamon, Oxford and New York. 534 pp. *210, 213, 223, 224, 226, 236, 249, 255*

Mayr, E. 1969. *Principles of Systematic Zoology.* McGraw-Hill, New York. 428 pp. *1*

Meglitsch, P. A. 1967. *Invertebrate Zoology.* 2nd ed., 1972. Oxford University Press, New York, London, Toronto. 834 pp. *38*

Mickoleit, G. 1973. "Uber den Ovipositor der Neuropteroidea und Coleoptera und seine Phylogenetische Bedeutung (Insecta, Holometabola)." *Z. Morphol. Tiere.* 74:37–64. *225, 238, 245, 246*

Millot, J. 1949a. "Ordre des Amblypyges." In: P. Grasse, Ed., *Traité de Zoologie,* 6:563–588. Masson, Paris. *63, 72*

Millot, J. 1949b. "Ordre des Araneides (Araneae)." In: P. Grasse, Ed., *Traité de Zoologie,* 6:589–743. Masson, Paris. *54*

Nelson, G. J. 1969. "Gill Arches and the Phylogeny of Fishes, with Note on the Classification of Vertebrates." *Bull. Am. Mus. Hist.* 141:475–552. *9*

Neville, A. C. 1975. *Biology of the Arthropod Cuticle.* Springer-Verlag, New York, Heidelberg, Berlin. 448 pp. *46, 90, 221*

*Nitzsch, C. L. 1818. "Die Familien und Gattungen der Thierinsekten (Insecta Epizoica) als ein Prodromus der Naturgeschichte derselben." *Germars Mag. Entomol.* 3:261–316. (Origin of the name Mallophaga). *Fide:* Wilson and Doner, 1937. *142, 280*

*Olivier, A. G. 1789. *Encyclopedie Methodique, Dictionnaire des Insectes.* Vols. 2–6. Pankouke, Paris. (Origin of the name Orthoptera). *Fide:* Wilson and Doner, 1937; Essig, 1942. *142, 274*

*Packard, A. S. 1886. "A new Arrangement of the Orders of Insects." *Am. Nat.* 20:808. (Origin of the names Plectoptera, Mecoptera). *Fide:* Wilson and Doner, 1937. *143, 200, 271, 288*

Paclt, J. 1956a. *Biologie der primar flugellosen Insekten.* Gustave Fischer, Jena. 258 pp. *147*

Paclt, J. 1956b. "Nochmals über das System der niederen Insekten." *Zool. Anz.* 156:272–276. *176*

*Pennant, T. 1777. *British Zoology.* B. White, London. 4 Vols. 1460 pp. (Origin of the name Crustacea). *Fide:* Dechaseaux, 1953. *83, 119*

Penny, N. D. 1975. "Evolution of the Extant Mecoptera." *J. Kans. Entomol. Soc.* 48: 331–350. *9*

Pocock, R. J. 1893. "On the Classification of the Tracheate Arthropoda." *Zool. Anz.* 16:271–275. *111*

Ravoux, P. 1962. "Etude sur la Segmentation des Symphyles." *Ann. Sci. Nat. Zool.* 12: 141-472. *113, 135*

*Redtenbacher, J. 1886. "Vergleichende Studien über das Flugelader der Insekten." *Ann. K. K. Naturh. Hofmuseum.* 1: 153-232. *Fide:* Comstock. 1918. *179*

Remington, C. L. 1954. "The Suprageneric Classification of the Order Thysanura (Insecta)." *Ann. Entomol. Soc. Am.* 47:277-286. *169*

Remy, P. A. 1950. "Les Millitauropus, Types d'un Nouveau Groupe de Pauropodes." *C. R. Acad. Sci. Paris.* 230:472-474. *66, 105, 107*

Richards, P. A. and A. G. Richards. 1969. "Acanthae: A New Type of Cuticular Process in the Proventriculus of Mecoptera and Siphonaptera." *Zool. Jahrb., Ant.* 86: 158-176. *255*

Riek, E. F. 1970a. (Illustration of prototype insect wing). In: D. F. Waterhouse, Ed., *The Insects of Australia,* Chapter 1, CSIRO, Melbourne University Press, Carlton, Victoria. P. 18. *179, 180*

Riek, E. F. 1970b. "Fossil History." In: D. F. Waterhouse, Ed., *The Insects of Australia,* CSIRO, Melbourne University Press, Carlton, Victoria. Pp. 168-186. *204*

Riek, E. F. 1970c. "Mecoptera." In: D. F. Waterhouse, Ed., *The Insects of Australia,* CSIRO, Melbourne University Press, Carlton, Victoria. Pp. 636-655. *253*

Riek, E. F. 1970d. "Lower Cretaceous Fleas." *Nature.* 227:746-747. *257*

Roger, J. 1953. "Sous-classe des Malacostraces." In: J. Piveteau, Ed., *Traité de Paleontologie,* 3:309-378. Masson, Paris. *76*

Rohdendorf, B. B. 1961. "Description of the First Winged Insect from the Devonian Beds of the Timan." *Entomol. Rev.* 40:260-262. *178*

Rohdendorf, B. B. 1970. "A Second Find of Remains of Winged Devonian Insects." *Entomol. Rev.* 49:508-509. *178*

Rohdendorf, B. B. 1972. "The Devonian Eopteridae are not Insects but Crustacean Eumalacostraca." *Ent. Obozr. USSR* 51:96-97, (In Russian). *178*

Rohdendorf, B. B., Y. E. Bekker-Magdisova, O. M. Martynova and A. G. Sharov. 1961. "Paleozoiskie Nasjekomye Kuznetzkogo Basseina." *Trudy Paleontol. Inst. Akad. Nauk. USSR.* 85:1-705. (Paleozoic Insects of the Kuznetsk Basin). *130*

Rolfe, W. D. I. 1967. "Arthropleurida." In: R. C. Moore, Ed., *Treatise on Invertebrate Paleontology,* Part R, Vol. 2, Geological Society of America and University of Kansas Press, New York. pp. R607-R620. *103*

Rolfe, W. D. I. and J. K. Ingham. 1967. "Limb Structure, Affinity and Diet of the Carboniferous 'Centipede' *Arthropleura.*" *Scottish J. Geol.* 3:118-124. *77, 103, 128*

Ross, H. H. 1955. "The Evolution of the Insect Orders." *Entomol. News.* 66:197-208. *169*

Ross, H. H. 1965. *A Textbook of Entomology.* 3rd ed. Wiley, New York. 539 pp. *151, 163, 179, 180, 184*

Ross, H. H., 1967. "The Evolution and Past Dispersal of the Trichoptera." *Ann. Rev. Entomol.* 12:169-203. *259*

Rothschild, M. 1965. "Fleas." *Sci. Amer.* 213(6):44-53. *251*

Rothschild, M. 1975. "Recent Advances in our Knowledge of the Order Siphonaptera." *Ann. Rev. Entomol.* 20:241-259. *254, 255, 256*

Rousset, A. 1973. "Squellette et Musculature des Regions Genitales et Postgenitales de la Femelle de *Thermobia domestica* (Packard). Comparison avec la Region Genitale de *Nicoletia* sp. (Insecta: Apterygota: Lepismatida)." *J. Insect Morphol. Embryol.* 2:55-80. *173*

Rudall, K. M. 1955. "The Distribution of Collagen and Chitin." *Symp. Soc. Exper. Biol.* 9:49-70. *23*

Rudall, K. M. 1963. "The Chitin/Protein Complex of Insect Cuticle." *Adv. Ins. Physiol.* 1:257-313. *22*

Ryder, J. A. 1880. "Scolpendrella as the Type of a New Order of Articulata." *Amer. Nat.* 14:375-376. *83, 121*

Sanders, H. L. 1963. The Cephalocarida, Functional Morphology, Larval Development, Comparative External Anatomy. *Mem. Conn. Acad. Arts Sci.* 15:1-180. *99*

Schaller, F. 1971. "Indirect Sperm Transfer by Soil Arthropods." *Ann. Rev. Entomol.* 16:407-446. *147*

Schwanwitsch, B. N. 1956. "Alary Musculature as a Basis of the System of Pterygote Insects." *Proc. 10th Int. Congr. Entomol.* Vol. 1, 1956:605-610. *189, 268, 270*

Scourfield, D. J. 1940. "The Oldest Known Fossil Insect (*Rhyniella praecursor* Hirst and Maulik). Further Details from Additional Specimens." *Proc. Linn. Soc. London.* 152:113-131. *151*

Sharif, M. 1935. "On the Presence of Wing Buds in the Pupa of Aphaniptera." *Parasitology* 27:461-464. *254*

Sharov, A. G. 1957. "Peculiar Paleozoic Wingless Insects of the New Order Monura (Insecta, Apterygota.)" *Dokl. Akad. Nauk USSR* 115:796-798. *141, 166, 265*

Sharov, A. G. 1966a. *Basic Arthropodan Stock with Special Reference to Insects.* Pergammon Press, Oxford. 271 pp. *27, 33, 37, 78, 86, 95, 123, 127, 130, 159, 166, 179, 180, 235*

Sharov, A. G. 1966b. "The Position of the Orders Glosselytrodea and Caloneurodea in the System of the Insecta." *Paleontol. Zhn.* 3:84-93. *206*

Sharp, D. 1899 (1898). "Some Points in the Classification of the Insecta Hexapoda." *Proc. 4th Int. Congr. Zool.* 4:246-249. *143, 283*

Shipley, A. E. 1904. "The Orders of Insects." *Zool. Anz.* 28:259-262. *200, 204, 226, 271, 273, 280*

*Siebold, C. T. W. von, and H. Stannius. 1848. *Lehrbuch der vergleichenden Anatomie der Wirbellosen Tiere.* Veit, Berlin. 679 pp. (Origin of the name Arthropoda). *Fide:* Wilson and Doner, 1937. *42, 82, 116*

*Silvestri, F. 1907. "Descrizione di un Nuovo Genera d'Insetti Apterigoti Reppresentante di un Nuovo Ordine." *Boll. Lab. Zool. Gen. Agr. Portici.* 1:296-311. (Origin of name Protura). *Fide:* Börner, 1910. *141, 265*

*Silvestri, F. 1913. "Descrizione di un Nuovo Ordine di Insetti." *Boll. Lab. Zool. Gen. Agr. Portici.* 7:192-209. (Origin of the name Zoraptera). *Fide:* Essig, 1942. *142, 276*

Smith, E. L. 1969. "Evolutionary Morphology of External Insect Genitalia. 1. Origin and Relationships to other Appendages." *Ann. Entomol. Soc. Am.* 62:1051-1079. *161, 199, 203, 213, 222*

Smith, E. L. 1970a. "Evolutionary Morphology of External Insect Genitalia. 2. Hymenoptera." *Ann. Entomol. Soc. Am.* 63:1-27. *222*

Smith, E. L. 1970b. "Biology of Some California Bristletails and Silverfish (Apterygota: Microcoryphia, Thysanura)." *Pan-Pacific Entomol.* 46:212-225. *170*

Snodgrass, R. E. 1909. "The Thorax of Insects and the Articulation of the Wings." *Proc. U. S. Nat. Mus.* 36:511-595. *179*

Snodgrass, R. E. 1927. "Morphology and Mechanism of the Insect Thorax." *Smiths. Misc. Coll.* 80(1):1-108. *127*

Snodgrass, R. E. 1935. *Principles of Insect Morphology.* McGraw-Hill, New York. 667 pp. *53, 62, 130, 153, 165, 201*

Snodgrass, R. E. 1938. "Evolution of the Annelida, Onychophora and Arthropoda." *Smiths. Misc. Coll.* 97(6):1-159. *23, 32, 83, 115, 118*

Snodgrass, R. E. 1952. *A Textbook of Arthropod Anatomy.* Cornell University Press, Ithaca. 363 pp. *51, 87, 89, 91, 92, 94, 100, 101, 107, 109, 111, 112, 114, 129, 145, 175, 179, 180*

Snodgrass, R. E. 1957. "A Revised Interpretation of the External Reproductive Organs of Male Insects." *Smiths. Misc. Coll.* 135(6):1-60. *163, 164, 202, 213, 222, 223*

Snodgrass, R. E. 1958. "Evolution of Arthropod Mechanisms." *Smiths. Misc. Coll.* 138(2):1-77. *32, 34, 76, 91, 100*

Snodgrass, R. E. 1960. "Facts and Theories Concerning the Insect Head." *Smiths. Misc. Coll.* 142(1):1-61. *104*

Stoll, N. R., R. P. Dollfus, J. Forest, N. D. Riley, C. W. Sabrosky, C. W. Wright, and R. V. Mellville, Eds. 1964. *International Code of Zoological Nomenclature.* International Trust for Zoological Nomenclature, London. 176 pp. *139*

Størmer, L. 1939. "Studies on Trilobite Morphology, I. The Thoracic Appendages and their Significance." *Norsk Geol. Tidsskr.* 10:143-273. *127*

Størmer, L. 1944. "On the Relationships and Phylogeny of Fossil and Recent Arachnomorpha." *Skrift. Norske Videns. Akad. Oslo Mat. Nat. Kl.* 1944(5):1-158. *52, 83, 94, 118*

Størmer, L. 1949. "Classe des Trilobites." In: P. Grasse, Ed., *Traité de Zoologie.* 6:160-197. Masson, Paris.

Størmer, L. 1959. "Trilobitomorpha." In: R. C. Moore, Ed., *Treatise on Invertebrate Paleontology.* Part O. Arthropoda. Geological Society of America, New York, and University of Kansas Press, Lawrence. Pp. 22-37. *83, 118*

Taylor, R. L. and A. G. Richards. 1963. "The Subimaginal Cuticle of the Mayfly *Callibaetis* sp. (Ephemeroptera)." *Ann. Entomol. Soc. Am.* 56:418-426. *197*

Tiegs, O. W. 1940. "The Embryology and Affinities of the Symphyla Based on a Study of *Hanseniella agilis.*" *Q. J. Micr. Sci.* 82:1-225. *28, 72, 104, 111*

Tiegs, O. W. 1947. "The Development and Affinities of the Pauropoda, Based on a Study of *Pauropus silvaticus.*" *Q. J. Micr. Sci.* 88:165-336. *28*

Tiegs, O. W. and S. M. Manton. 1958. "The Evolution of the Arthropoda." *Biol. Rev.* 33:255-337. *75*

Tillyard, R. J. 1917. "Mesozoic Insects of Queensland. No. 1. Plannipennia, Trichoptera, and the New Order Protomecoptera." *Proc. Linn. Soc. New South Wales.* 42:175-200. *142, 288*

Tillyard, R. J. 1924. "Upper Permian Coleoptera and a New Order from the Belmont Beds, New South Wales." *Proc. Linn. Soc. New South Wales.* 49:429-435. (Origin of name Protocoleoptera). *275*

Tillyard, R. J. 1926. "Kansas Permian Insects. Part 10. The New Order Protoperlaria; A Study of the Typical Genus Lemmatophora Sellards." *Am. J. Sci.* 16:185-220. *272*

Tillyard, R. J. 1931. "Kansas Permian Insects. Part 13. The New Order Protelytroptera, with a Discussion of its Relationships. *Am. J. Sci.* 21:232-266. *275*

Tillyard, R. J. 1932. "Kansas Permian Insects. Part 15. The Order Plectoptera." *Am. J. Sci.* 23:97-134, 237-272. (Origin of the name Permoplectoptera). *271*

Trimble, J. J. and S. A. Thompson. 1974. "Fine Structure of the Sperm of the Lovebug *Plecia nearctica* Hardy (Diptera: Bibionidae)." *Int. J. Insect Morphol. Embryol.* 3:425-432. *252, 255*

Tschernova, O. A. 1970. "On the Classification of Fossil and Recent Ephemeroptera." *Biol. Rev. (Washington).* 49:71-81. *196*

Tuxen, S. L. 1959. "The Phylogenetic Significance of Entognathy in Entognathous Apterygotes." *Smiths. Misc. Coll.* 137:379-416. *124, 132, 144, 154*

Tuxen, S. L. 1970. "Protura." In: S. L. Tuxen, Ed., *Taxonomist's Glossary of the Genitalia of Insects.* 2nd ed. Munksgaard, Copenhagen. Pp. 21-24. *134*

*Verhoeff, K. W. 1903. "Uber die Nerven des Metachephalsegmentes und die Insectordnus Oothecaria." *Zool. Anz.* 26:20-31. (Origin of the name Oothecaria). *Fide:* Wilson and Doner, 1937. *277*

*Verhoeff, K. W. 1904. *Nov. Acta Abhl. Kais. Leopold Carol Deut. Akad. Naturf. Halle.* 84:109. (Origin of the name Microcoryphia). *Fide:* Remington, 1954. *141, 169, 266*

*Walch, J. E. E. 1771. (Origin of the name Trilobita). *Fide:* Stφrmer, 1949. *83, 118*

Walcott, C. D. 1911a. "Cambrian Geology and Paleontology. 11. No. 4. Middle Cambrian Merostomata." *Smiths. Misc. Coll.* 57:17-56. *94*

Walcott, C. D. 1911b. "Middle Cambrian Annelids." *Smiths. Misc. Coll.* 57:109-144. *29, 34*

Walker, C. M. 1922. "The Terminal Structures of Orthopteroid Insects: A Phylogenetic Study." *Ann. Entomol. Soc. Am.* 15:1-76. *213*

*Waterlot, G. 1934. "Etude de la Faune Continentale du Terrain Houillier Sarro-lorrain." *Etud. Git. Min. France.* 2:1-320. (Origin of the name Arthropleurida). *Fide:* Rolfe and Ingham, 1967. *83, 119*

Waterlot, G. 1953. "Classe des Merostomes." In: J. Piveteau, Ed., *Traité de Paleontologie.* 3:529-554. Masson, Paris.

Weber, H. 1933. *Lehrbuch der Entomologie.* Fischer, Jena. 726 pp. *142, 279, 282*

Weber, H. 1939. "Zur Eiblage und Entwicklung der Elefantlaus *Haematomyzus elephantis* Piaget." *Biol. Zbl.* 59:98-109. *142, 280*

Weber, H. 1954. *Grundriss der Insektenkunde.* G. Fischer, Stuttgart. 428 pp. *116*

*Werner. 1906. (Origin of the name Blattiformia). *Fide:* Chopard, 1949a; Laurentiaux, 1953b. *277*

Westwood, J. O. 1839. *An Introduction to the Modern Classification of Insects, Founded on the Natural Habits and Corresponding Organisation of the Different Families.* Vol. 1. Longmans, London. 462 pp. (Origin of name Cursoria). *213, 277*

Whitten, J. M. 1962. "Homology and Development of Insect Wing Tracheae." *Ann. Entomol. Soc. Am.* 55:288-295. *179*

Whittington, H. B. 1975. "The Enigmatic Animal *Opabinia regalis,* Middle Cambrian, Burgess Shale, British Columbia." *Phil. Trans. R. Soc. B.* 271:1-43. *78, 98*

Wilson, H. F., and M. H. Doner. 1937. *The Historical Development of Insect Classification.* J. S. Swift Co. Inc., St. Louis. 133 pp.

Wygodzinsky, P. 1961. "On a Surviving Representative of the Lepidothricidae (Thysanura)." *Ann. Entomol. Soc. Am.* 54:621-627. *177*

General Index

Abdomen, 134
Acanthae, 255
Acarina, 21, 77
Acercaria, 279
Acercarida, 142, 165, 215, 216, 222, 223,
    224, 234, 236, 279
  apomorphies of, 225, 279
  convergencies with Zoraptera, 215
  plesiomorphies of, 279
  spermatozoa of, 232
Acerentomidae, 157
Acerentomon, 156
  doderoi, 155
Achorutes armatus, 151
Acrididae, 206
Acron, 11, 12, 20, 26, 28, 36, 79
Actaletes, 152
Aculeae, 260
Adephaga, 225, 238, 284
Aedeagus, 223
Agulla adnixa, 255
Algae, 17
Allopauropus brevisetis, 107
Amblycera, 227, 280
Amblypigida, 124
Amblypygi, 72, 111
Amnion, 5, 138
Amphiesmenoptera, 258, 290
Amphipoda, 102
Anapleurite, 128, 186
Anaspidacea, 69
Anaspides, 100
  tasmaniae, 69, 101
Anax junius, 183, 185
Anepimeron, 186

Anepisternum, 186
Anisoptera, 194
Anisozygoptera, 194
Annelida, 16-22, 24
  apomorphies of, 40
  plesiomorphies of, 40
Anoplura, 38, 135, 137, 142, 158, 202,
    225, 227, 228, 229, 263, 280, 281
  apomorphies of, 228, 281
  plesiomorphies of, 281
Antenna, 28, 29
  of Ectognathata, 158
Antennomere, 124, 136
Antennule, 11, 29
Antliophora, 25, 288
Apheloria coriacea, 109
Apodeme, occipital, 228
  tentorial, 133
Apolysis, 46
Apomorphy, 4
Appendage, gnathal, 23
  locomotor, 4
  metameric, 22
Aptera, 5, 174, 267
Apterata, 141, 173, 174, 267
Apterobittacus, 253
Apteropanorpa, 253
Apterygota, 5, 6, 9, 178
Arachnida, 5, 24, 54, 83, 89, 92, 102, 120,
    252
  apomorphies of, 90, 117
Archedictyon, 192
Archenteron, 17, 29
Archentomata, 141, 166, 167, 265
Archeognatha, 266